GE◉METRIC
MECHANICS

Part III: Broken Symmetry and Composition of Maps

GE◉METRIC
MECHANICS

Part III: Broken Symmetry and Composition of Maps

$$L: TM \to \mathbb{R} \xleftarrow{\text{Legendre Transform}} H: T^*M \to \mathbb{R}$$

Diagram: $L: TM \to \mathbb{R}$ (via HP, TM/G) → EL eqns ⟺ Ham eqns; $\{\cdot,\cdot\}_{can}$; **Reduction via Momentum Map** $J(q,p) \to \mathfrak{g}^* \simeq T^*M/G$; Noether's Theorem — $\ell: \mathfrak{g} \to \mathbb{R} \xleftarrow{\text{Reduced LT}} h: \mathfrak{g}^* \to \mathbb{R}$ — Noether's Theorem; TM/G; $J(q,p)$; EP HP; Euler–Poincaré eqns ⟺ Lie-Poisson eqns; $\{\cdot,\cdot\}_{LP}$

DARRYL D HOLM

Imperial College London, UK

World Scientific

NEW JERSEY · LONDON · SINGAPORE · BEIJING · SHANGHAI · HONG KONG · TAIPEI · CHENNAI · TOKYO

Published by

World Scientific Publishing Europe Ltd.

57 Shelton Street, Covent Garden, London WC2H 9HE

Head office: 5 Toh Tuck Link, Singapore 596224

USA office: 27 Warren Street, Suite 401-402, Hackensack, NJ 07601

Library of Congress Control Number: 2008298655

British Library Cataloguing-in-Publication Data
A catalogue record for this book is available from the British Library.

GEOMETRIC MECHANICS
Part III: Broken Symmetry and Composition of Maps

ISBN 978-1-80061-659-2 (hardcover)
ISBN 978-1-80061-678-3 (paperback)
ISBN 978-1-80061-660-8 (ebook for institutions)
ISBN 978-1-80061-661-5 (ebook for individuals)

For any available supplementary material, please visit
https://www.worldscientific.com/worldscibooks/10.1142/Q0489#t=suppl

Desk Editors: Aanand Jayaraman/Shi Ying Koe

Typeset by Stallion Press
Email: enquiries@stallionpress.com

Guiding Principles

We are dealing with Lie group invariant variational principles. [No1918]

Emmy Noether

Geometrical truths are not essentially different from physical ones in any aspect and they are established in the same way. [Hi1935]

David Hilbert

It is only slightly overstating the case to say that Physics is the study of symmetry. [An1972]

Philip Anderson

The most difficult task is to think of workable examples that will reveal something new. [Fl2015]

Hermann Flaschka

Preface

To paraphrase Isaac Newton, a motion with symmetry retains its symmetry unless acted on by an external force. That is, broken symmetry reveals the presence of an external force. The response to the force may often be described as a composition of maps. For example, the stepwise precession of the swing plane of a Foucault's pendulum reveals that the pendulum experiences the Coriolis force of the rotation of the Earth in whose frame of motion the pendulum swings. As another example, waves arising on the sea surface reveal the wind forcing which created them. Surface waves propagating on the moving currents in the Earth's rotating frame can be described by the composition of two smooth transformations acting on the flat sea surface at rest. The first transformation maps the sea at rest into the Earth's rotating frame. The second transformation composed with the first one maps the sea resting in the Earth's rotating frame into the flow of the currents whose paths are bent by the Coriolis force of the Earth's rotation. The wave dynamics then takes place in the frame of motion defined by the composition of the first two maps. This broad dynamical range of broken symmetry and composition of maps is the purview of geometric mechanics.

Thus, besides its mathematical depth and beauty, geometric mechanics is a framework for investigating how symmetry breaking and composition of maps deal with the multi-scale, multi-physics sequences of challenges arising in modern investigations of dynamics in nature.

As reviewed in Ref. [Ma2013], one may say that geometric mechanics arose in 1901 when Poincaré [Po1901] introduced a Lie group invariant variational principle which extended Euler's approach to the rigid body (resp., heavy top) from $SO(3)$ (resp., $SE(3)$) to arbitrary Lie groups. V. I. Arnold [Ar1966] extended these results to include Euler's ideal fluid equations. Subsequently, many symmetry-breaking variants of Euler's ideal fluid equations have been included in the realm of geometric mechanics. These include the equations of ocean and atmosphere dynamics, for example, as well as the equations of ideal plasma physics, which form the basis of astrophysics and also guide the design of magnetic confinement devices, such as tokamaks.

This textbook for learning geometric mechanics is written in the format of lecture notes, which are meant to convey mathematical and physical insight through clear definitions and workable examples. The lecture notes have evolved over decades of teaching and research, as well as writing several other textbooks surveying geometric mechanics in other formats [Ho2011b, Ho2011c, HoScSt2009]. The lecture format adopted here is intended to convey the immediacy of the taught course and to be useful as a basis for other courses. The lecture notes comprise:

AP – Applications of Pure maths, e.g., Noether's theorem: Lie group symmetry of Hamilton's variational principle implies conservation laws for its equations of motion.

PA – Purifications of Applied maths, e.g., Euler fluid dynamics describes geodesic flow on the manifold of smooth invertible maps acting on the domain of flow.

Both AP and PA appear here, though the difference is not mentioned. It is left to the reader to decide whether it was AP or PA in each of the lectures.

In following the theme of reduction by symmetry from the Euler–Poincaré–Noether viewpoint of Lie group invariant variational principles in geometric mechanics, the lecture notes select only the material that readers will find absolutely necessary for dealing with the almost two hundred exercises with explicit solutions in this textbook at the level of a beginning postgraduate student.

Note: the exercises and their explicit solutions are *not optional reading* in this textbook. They are essential to understanding the lectures. The exercises are intended to reveal something new in the steps towards learning geometric mechanics. The exercises and their solutions are indented and marked with ★ and ▲, respectively. Moreover, the careful reader will find that many of the exercises are answered in passing somewhere later in the text in a more developed context.

Many excellent and more encyclopaedic texts have been published on the foundations of the subject of geometric mechanics and its links to symplectic and Poisson geometry. See, for example, Abraham and Marsden [AbMa1978], Arnold [Ar1989], Guillemin and Sternberg [GuSt1984], José and Saletan [JoSa1998], Libermann and Marle [LiMa1987], Marsden and Ratiu [MaRa1994], McDuff and Salamon [McSa1995] and many more.

The classic references for courses at this level are Marsden [Ma1992], Marsden and Ratiu [MaRa1994], Lee [Le2003], Bloch [Bl2003], and Ratiu, Tudoran, Sbano, Sousa Dias and Terra [RaTuSoTe2005]. Other very useful references are Arnold and Khesin [ArKh1998] and Olver [Ol1993]. The reader may see the strong influences of all these references in these lecture notes, but expressed perhaps at a lower level of mathematical sophistication than most of the originals. See also other thoughtful geometric mechanics textbooks, such as Ref. [Ma1997].

For brevity, the scope of the lectures is quite limited. In particular, the necessary elements of variational calculus on smooth manifolds and the basics of Lie group theory are only briefly described. And the analysis of the solution behaviour of the partial differential equations derived here is absent. Perhaps the study of these more fundamental and advanced aspects of geometric mechanics may be facilitated after learning their basic elements in the context of the workable examples explained here.

Geometric mechanics deals with dynamical systems defined by Lie group invariant variational principles, such as geodesic motion on a Lie group G whose metric is invariant under the action of the Lie group G. An example is the variational formulation of Euler's rigid body equations in three dimensions, whose solutions are then seen to be geodesics on the rotation group $SO(3)$. Another example is the variational formulation of Euler's fluid equations in three dimensions, whose solutions are then seen to be geodesics on the manifold of smooth invertible maps (diffeomorphisms) with respect to the metric given by the fluid's kinetic energy.

In more detail, the lecture notes here consider Lie group invariant Lagrangians in Hamilton's principle defined on the tangent space TG of a Lie group, G. Invariance of such a Lagrangian under the action of G leads to the symmetry-reduced Euler–Lagrange equations called the Euler–Poincaré equations. In this case, the G-invariant Lagrangian is defined on the Lie algebra $\mathfrak{g}$ of the group, and the variables in its Euler–Poincaré equations are defined on a dual Lie algebra, $\mathfrak{g}^*$, where "dual" is defined by the pairing obtained in the operation of taking variational derivative. On the Hamiltonian side, the Euler–Poincaré equations are expressed in terms of Lie–Poisson brackets among the accompanying momentum maps, which encode both the conservation laws and the geometry of their solution space.

Thus, these lecture notes use the Lie–Noether symmetries of Hamilton's variational principle to derive momentum maps leading to symmetry-reduced Euler–Poincaré equations whose dynamical variables evolve by coadjoint motion. This coadjoint motion endows all these symmetry-reduced equations with the same type of solution properties. The symmetry-reduced Legendre transformation provides the Hamiltonian formulation of these equations in terms of Lie–Poisson brackets. The Lie–Poisson brackets are defined on the dual of the Lie algebra of the Lie–Noether symmetry group that is used in reduction by symmetry of the Lagrangian in the starting Hamilton principle.[1]

The standard Euler–Poincaré examples are treated in the lecture notes. These examples include, for example, particle dynamics, Foucault's pendulum, rigid body, heavy top and geodesic motion

---

[1]For an interesting discussion of Noether's theorems from the Hamiltonian side, see Ref. [Ba2020].

on Lie groups. Additional topics deal with the Lie symmetry reduction of Fermat's principle in geometric optics. The topic of nonlinear water waves deals with the completely integrable system known as the Camassa–Holm equation, whose soliton solutions are called *peakons* because their velocity profile develops sharp C^1 peaks. The lectures also include the semidirect-product Euler–Poincaré reduction theorem for ideal fluid dynamics. The Euler–Poincaré theorem leads to the Kelvin–Noether circulation theorem and the semidirect-product Lie–Poisson Hamiltonian formulation for incompressible and compressible motions of ideal fluid dynamics and plasma physics [HoMaRa1998a].

The lecture series for this textbook is arranged in three parts, as follows:

Part I: Basic Elements of the geometric mechanics approach to finite-dimensional dynamical systems.

Part II: Geometric Mechanics on Manifolds.

Part III: Euler–Poincaré Framework of Continuum Partial Differential Equations.

Part I of these lecture notes is designed to introduce undergraduate mathematics and physics students to the applications of geometric mechanics in finite-dimensional dynamical systems of ordinary differential equations.

Part II discusses the minimal amount of the theory of manifolds and Lie groups needed to prepare senior undergraduates and graduate students for the modern applications of geometric mechanics introduced in the third part.

Part III discusses the geometric mechanics of the partial differential equations governing the dynamics of ideal continuum mechanics for fluids and plasmas at the modern research level.

An aspect of modern applications emphasised here is the use of the composition of evolutionary maps for multi-physics, multi-timescale interactions. These applications include waves interacting with flows in the Euler–Poincaré framework in geophysical fluid dynamics for ocean and atmosphere dynamics. They also include waves in magnetohydrodynamics for applications in plasma physics, such as magnetic confinement fusion, and also in astrophysical processes, such as Alfvén waves and gravity waves propagating on the solar tachocline [HoHuSt2023]. The topics covered in each lecture can be gleaned from its table of contents listed at the onset of each lecture.

See Appendix A for a detailed topical outline of the course of lectures and some suggestions for postgraduate research projects in geometric mechanics.

About the Author

Darryl Holm is a professor of applied mathematics at Imperial College London (2005–present) and a laboratory fellow (lifetime position) at the Los Alamos National Laboratory (1972–2005). His research primarily focuses on the applications of geometric mechanics, which involves using geometric and algebraic structures to formulate and solve physical problems.

Holm's work spans several key areas, summarised as follows. **Geometric mechanics** is a broad area of research that applies geometric methods to study the dynamics of physical systems. Holm's contributions in this field include developing mathematical frameworks that use geometric structures to describe and analyse dynamical systems, particularly those governed by Hamiltonian and Lagrangian mechanics. These frameworks are essential for understanding complex systems that exhibit both order and chaos. In **nonlinear dynamics**, Holm has made significant contributions to the understanding of nonlinear phenomena in physics and engineering. His research explores how systems evolve over time under nonlinear influences, which can lead to behaviours such as turbulence and soliton formation. These studies are crucial for fields ranging from meteorology to engineering, where predicting system behaviour is complex but essential. A substantial portion of Holm's research is devoted to the study of **fluid dynamics**, with a particular focus on **geophysical flows**. This involves understanding the behaviour of fluid flows on a global scale, such as ocean currents and atmospheric circulation, and how they are influenced by factors like rotation and stratification. His work often involves developing new mathematical models to simulate these complex systems more accurately. Holm's research also delves into **stochastic processes**, which are used to model systems that evolve in a probabilistic manner. He investigates how randomness affects the evolution of physical systems, such as in fluid dynamics, where turbulence can introduce stochastic behaviour into otherwise deterministic systems. Holm applies **variational principles** to derive equations that describe the dynamics of physical systems. This approach is fundamental in mechanics and physics, where the behaviour of a system can be understood as minimising or extremising some quantity, such as energy or action. His work in this area includes the development of new methods for dealing with systems that are influenced by both deterministic and stochastic factors.

In addition to these topics, Holm's research extends into areas such as **shape analysis**, where he studies the geometric properties of objects and their transformations, and nonlinear waves, which include phenomena like solitons and shock waves. His interdisciplinary approach combines theoretical developments with practical applications, making his work influential across multiple scientific domains.

For more detailed information about Darryl Holm's research, you can visit his profile at Imperial College London (`https://profiles.imperial.ac.uk/d.holm`), or to review his comprehensive research description (`https://profiles.imperial.ac.uk/d.holm/grants`) see `https://orcid.org/0000-0001-6362-9912`.

Acknowledgements

I am enormously grateful to many friends and colleagues whose encouragement, advice and support have helped sustain my interest in this field. I am particularly grateful to my students and postdocs for their participation in my journey of learning and developing a sequential approach to teaching them geometric mechanics. On this journey, I have learned a lot and I have had a lot of fun. Thanks to them!

I also happily acknowledge the love, encouragement and support that my life partner, Justine Jones, has so generously given me to help me finish this effort.

Contents

Part III Euler–Poincaré Framework of Continuum Partial Differential Equations 357

Part I

Basic Elements

$$L : TM \to \mathbb{R} \xleftarrow{\text{Legendre Transform}} H : T^*M \to \mathbb{R}$$

EL eqns $\Longleftrightarrow$ Ham eqns

Noether's Theorem —————— $\ell : \mathfrak{g} \to \mathbb{R} \xleftarrow{\text{Reduced LT}} h : \mathfrak{g}^* \to \mathbb{R}$ —— Noether's Theorem

Euler–Poincaré eqns $\Longleftrightarrow$ Lie-Poisson eqns

HP TM/G $\{\cdot,\cdot\}_{can}$ **Reduction via**

Momentum Map

$J(q,p) \to \mathfrak{g}^* \simeq T^*M/G$

TM/G $J(q,p)$

EP HP $\{\cdot,\cdot\}_{LP}$

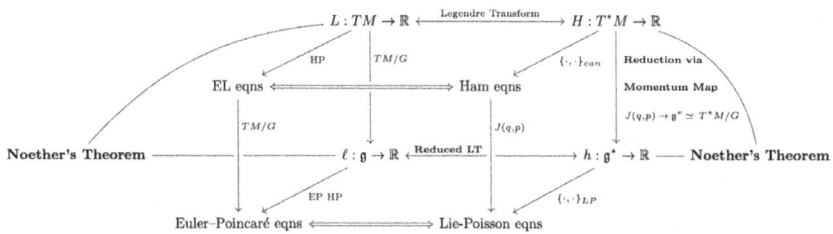

Envisioning geometric mechanics as a cube of six commuting diagrams.

1

INTRODUCTION

Contents

1.1 Geometric mechanics framework (GMF)

Classical mechanics may be visualised as a commuting diagram on the top face of the GMF cube.

Definition 1.1.1 *As shown in Figures 1.1 and 1.2, geometric mechanics is a **framework of relationships** in dynamical systems following from Lie group invariant variational principles.*[1]

Remark 1.1.1

1. Geometric mechanics can be used to formulate and analyse systems with constraints, controls, chaos or stochasticity.

2. Geometric mechanics has applications in many fields, e.g., robotics, fluid dynamics, plasma physics, celestial mechanics

---

[1]A Lie group is a group of transformations that depends smoothly on a set of parameters. A variational principle extremalises an integral. Well-known examples of Lie group invariant variational principles include Fermat's principle for geometric optics, Euler's equations for rigid body motion, Euler's equations for ideal fluid dynamics and Einstein's equations for general relativity.

$$L : TM \to \mathbb{R} \xleftrightarrow{\quad\text{Legendre transform}\quad} H : T^*M \to \mathbb{R}$$

Lagrange variational principle   Hamilton variational principle

Euler–Lagrange eqns $\Longleftrightarrow$ Hamilton's eqns

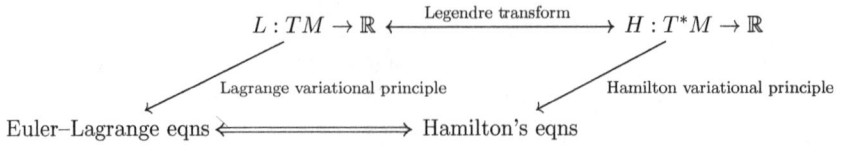

Figure 1.1. Envisioning classical mechanics as a commuting diagram.

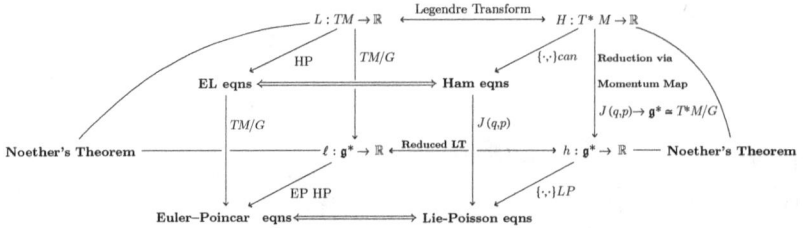

$L : TM \to \mathbb{R} \xleftarrow{\text{Legendre Transform}} H : T^*M \to \mathbb{R}$

HP TM/G $\{\cdot,\cdot\}can$ Reduction via

EL eqns $\Longleftrightarrow$ Ham eqns   Momentum Map

TM/G $J(q,p)$ $J(q,p) \to \mathfrak{g}^* \cong T^*M/G$

Noether's Theorem $\ell : \mathfrak{g}^* \to \mathbb{R} \xleftarrow{\text{Reduced LT}} h : \mathfrak{g}^* \to \mathbb{R}$ —— Noether's Theorem

EP HP $\{\cdot,\cdot\}LP$

Euler–Poincaré eqns $\Longleftrightarrow$ Lie–Poisson eqns

Figure 1.2. Envisioning geometric mechanics as a cube of six commuting diagrams.

and complex fluids — including liquid crystals and superfluids — as well as quantum molecular chemistry. □

Remark 1.1.2 The reader will do well to remember this page and occasionally revisit the GMF sketched here to identify their current position within the framework as they progress in their study of these notes. □

2

COUNTERPOINTS BETWEEN MATHEMATICS AND PHYSICS

Contents

What is this lecture about? This lecture comprises a series of different problem formulations and workable solutions, which together are meant to welcome the brave new reader into the richly varied world of geometric mechanics. This lecture provides a fittingly eclectic introduction to the wide range of applications of geometric mechanics covered throughout the rest of the book. It is meant to be *explored* to provide context for *future* understanding. The remaining lectures are laid out progressively, as is typical for textbooks. As they progress, readers might consider returning to this lecture occasionally to measure for themselves how far their understanding has extended into the realm of geometric mechanics.

Definition 2.0.1

Mathematics is the study of emergent logical relationships.

Physics is the study of experimentally observable phenomena.

Because these definitions of these two endeavours appear to be fundamentally different, one might conclude that the two fields might not be related to each other at all! Hence, it is all the more remarkable that mathematical models should be so effective in *describing* experimental observations such as the rotational dynamics of a rigid body, the flow of water in the ocean or the movement of air in the atmosphere [Wi1990].

In these notes, we will follow the guiding principles enunciated by Noether, Hilbert, Anderson and Flaschka, quoted before the Preface to take advantage of the *counterpoints* between geometric mechanics and physics by using each field of study to interpret the vocabulary of the other, as though geometric mechanics and physics were complementary languages [No1918, Hi1935, An1972]. We will be dealing with Lie group invariant variational principles. We will blend the viewpoints of mathematics and physics, and in the process we will find workable examples that will reveal new things.

2.1   Poisson manifold: Phase space of a rigid body

A *Poisson manifold* is a manifold equipped with a Poisson bracket structure, which maps two functions on the manifold into a third one. Poisson manifolds provide an arena for formulating classical mechanics, especially Hamiltonian mechanics.

Poisson brackets describe symmetries, conservation laws and dynamics.

In the example of rigid body rotation, the state space of configurations of the rigid body can be recognised as being isomorphic to the Lie group $SO(3)$.

A Poisson bracket structure governs the dynamics of rigid bodies, whose Poisson manifold is isomorphic to the dual Lie algebra $\mathfrak{so}(3)^*$ of the Lie group $SO(3)$.[1]

2.1.1   Geometric mechanics concept: Poisson manifold and Poisson bracket

Definition 2.1.1 *A **manifold** is a space which is locally indistinguishable from $\mathbb{R}^n$ and on which the rules of calculus apply.*

*A **Poisson manifold**, denoted as $(P, \{\cdot, \cdot\})$, is a manifold endowed with an operation*

$$\{\cdot, \cdot\} : C^\infty(P) \times C^\infty(P) \to C^\infty(P)$$

---

[1]As we shall see, the dual Lie algebra of a Lie group is a Poisson manifold.

*called the **Poisson bracket**, which satisfies the following:*

1. $\{F + G, H\} = \{F, H\} + \{G, H\}$ *(bilinear).*

2. $\{F, H\} = -\{H, F\}$ *(skew symmetric).*

3. $\{F, \{G, H\}\} + \{H, \{F, G\}\} + \{G, \{H, F\}\} = 0$ *(Jacobi identity).*

4. $\{FG, H\} = \{F, H\}G + F\{G, H\}$ *(Leibniz rule).*

Definition 2.1.2 ***Hamiltonian dynamics*** *is defined on a Poisson manifold* $(P, \{\cdot, \cdot\})$ *as the relation*

$$\frac{dF}{dt} = \{F, H\}$$

for a certain Hamiltonian H and an arbitrary function $F \in C^\infty(P)$*, which may vary with time.*

2.1.2   Corresponding physical concept

Poisson manifold: A Poisson manifold in classical mechanics comprises the position variables q in the configuration manifold $q \in Q$ along with their canonically conjugate momentum variables p in the cotangent bundle $(q, p) \in T^*Q$. The variables $(q, p) \in T^*Q$ satisfy the Poisson bracket relation $\{q, p\} = Id$, whose dynamics preserves the symplectic form $dq \wedge dp$. The Poisson manifold for angular momentum dynamics need not be symplectic though. Instead, the Poisson manifold can be identified with the dual $\mathfrak{so}(3)^*$ of the Lie algebra $\mathfrak{so}(3)$, which is the tangent space at the identity of the rotation group $SO(3)$. This observation in physics goes back to Pauli [Pa1953] in the 1950s for quantum mechanics and to Sudarshan [SuMu1974] in the 1970s for classical mechanics although had Lie introduced it in [Li1890].

Poisson brackets: In rigid body rotation, Poisson brackets arise as relations among different physical quantities (such as angular momentum and energy) within the system. Poisson brackets provide a method for determining the dynamics of functions of physical quantities, in particular, for calculating and analysing conserved quantities in rigid body dynamics.

Exercise. Suppose the Poisson bracket for rigid body dynamics of body angular momentum $\mathbf{\Pi} \in \mathbb{R}^3$ may be written in $\mathbb{R}^3$ vector form for real functions defined on $\mathbb{R}^3$, denoted C, F, H and volume element $d^3\Pi$, as

$$\frac{dF}{dt} d^3\Pi = \{F, H\} d^3\Pi := \frac{\partial C}{\partial \mathbf{\Pi}} \cdot \frac{\partial F}{\partial \mathbf{\Pi}} \times \frac{\partial H}{\partial \mathbf{\Pi}} d^3\Pi$$

$$= J\left(\frac{dC, dF, dH}{d\Pi_1, d\Pi_2, d\Pi_3}\right) d^3\Pi = dC \wedge dF \wedge dH,$$

where $d^3\Pi := d\Pi_1 \wedge d\Pi_2 \wedge d\Pi_3$ for $\mathbf{\Pi} \in \mathbb{R}^3$ is the volume element. The wedge ($\wedge$) notation for differential forms is explained later in Chapters 18 and 19. For now, we can think in terms of this Poisson bracket as being associated with the Jacobian determinant for a change of variables $(d\Pi_1, d\Pi_2, d\Pi_3) \to (dC, dF, dH)$.

1. What is the value of the Poisson bracket $\{C, H\}$?

2. How would this Poisson bracket look if it were restricted to a level set of the function $C(\Pi_1, \Pi_2, \Pi_3)$?

3. Would this dynamics preserve the $\mathbb{R}^3$ volume element $d^3\Pi = d\Pi_1 \wedge d\Pi_2 \wedge d\Pi_3$?   ★

Answer.

1. The Poisson bracket $\{C, H\}$ defined above vanishes for every function H.

2. When restricted to a level set of the function C, parameterised by level set coordinates (x, y), this Poisson bracket would become $\{F, H\} = J(F, H) = F_x H_y - F_y H_x$.

3. The dynamics of this Poisson bracket is given by

$$V := \frac{d\mathbf{\Pi}}{dt} = -\frac{\partial C}{\partial \mathbf{\Pi}} \times \frac{\partial H}{\partial \mathbf{\Pi}}.$$

Since $\mathrm{div}_\Pi V = 0$, this flow preserves volume in $\mathbb{R}^3$ with coordinates (Π_1, Π_2, Π_3). ▲

2.2 Hamilton's principle

2.2.1 Geometric mechanics concept

The Euler–Lagrange equations for the Lagrangian $L(q, v_q) : TQ \to \mathbb{R}$ for an arbitrary configuration manifold Q with solution curve $q(t) \in Q$ follow from **Hamilton's principle** as

$$0 = \delta S = \delta \int_0^T L(q, v_q) + \left\langle p, \frac{dq}{dt} - v_q \right\rangle dt$$

$$= \int_0^T \left\langle \frac{\partial L}{\partial q} - \frac{dp}{dt}, \delta q \right\rangle + \left\langle \frac{\partial L}{\partial v_q} - p, \delta v_q \right\rangle \tag{2.2.1}$$

$$+ \left\langle \delta p, \frac{dq}{dt} - v_q \right\rangle dt + \underbrace{\langle p, \delta q \rangle \Big|_0^T}_{\text{Noether term}}.$$

Here, the variation of the solution path $\delta q(t)$, for example, is defined by the tangent to a local deformation $q(t, \epsilon) = \phi_\epsilon q(t)$ of the path parameterised by $\epsilon \in \mathbb{R}$, with $q(t, \epsilon)|_{\epsilon=0} = \phi_0 q(t) = q(t)$ at fixed time t. That is, one defines the variation of the solution path $q(t)$ as

$$\delta q(t) := \frac{dq(t, \epsilon)}{d\epsilon} \Big|_{\epsilon=0} = \frac{d\phi_\epsilon}{d\epsilon} \Big|_{\epsilon=0} q(t), \tag{2.2.2}$$

and similarly for the other variations in Hamilton's principle (2.2.1).

Also, the quantity

$$\frac{\partial L}{\partial v_q} \tag{2.2.3}$$

arising in the functional variation of the action integral S with respect to v_q in equation (2.2.1) is called the *fibre derivative* of the Lagrangian L.

Upon assuming that δq vanishes at the endpoints in time, collecting terms in (2.2.1) yields the *Euler–Lagrange equations*:

$$\frac{d}{dt}\frac{\partial L}{\partial v_q}\bigg|_{v_q=\frac{dq}{dt}} - \frac{\partial L}{\partial q} = 0. \tag{2.2.4}$$

Theorem 2.2.1 *Noether's theorem [No1918] states the following.*

Suppose the Lagrangian $L(q, v_q)$ in Hamilton's principle (2.2.1) is invariant under the tangent-lifted action $G : TQ \to TQ$ of a one-parameter Lie group G, and δq is given by the infinitesimal transformation $\delta q \in T_e G(q)$ of the Lie symmetry group G acting on $q \in Q$ linearised around its identity transformation, e.

That is, suppose the Lagrangian $L(q, v_q)$ is invariant under a particular δq given by

$$\delta q := \frac{d\phi_\epsilon}{d\epsilon}\bigg|_{\epsilon=0} q(t) =: -\pounds_\xi q, \tag{2.2.5}$$

where $-\pounds_\xi q$ defines (minus) the Lie derivative of $q \in Q$ with respect to the Lie algebra element $\xi \in \mathfrak{g} \simeq T_e G$.[2] Then, Hamilton's principle $\delta S = 0$ with $S := \int_0^T L(q, \dot q)dt$ implies conservation of the endpoint term $\langle p, \delta q \rangle$ in (2.2.1), provided the Euler–Lagrange equations in (2.2.4) hold.[3]

Proof. Noether's theorem holds by inspection since the other terms in (2.2.1) vanish when the Euler–Lagrange equations in (2.2.4) hold. ∎

Remark 2.2.1 The Noether endpoint term in Hamilton's principle (2.2.1) may be rewritten equivalently as

$$\langle p, \delta q \rangle_{T^*Q \times TQ} := \left\langle p, \frac{dq(t, \epsilon)}{d\epsilon}\bigg|_{\epsilon=0} \right\rangle_{T^*Q \times TQ} =: \langle p, -\pounds_\xi q \rangle_{T^*Q \times TQ}.$$
$$\tag{2.2.6}$$

---

[2]The reason for the minus sign in this definition of Lie derivative will become clearer later. Basically, the sign distinguishes between the left and right actions of Lie groups on manifolds, but we are not there yet. For an example, see the paragraph about signs in the actions of body and space angular velocities on rotating rigid bodies discussed in Section 2.5. For those who are already familiar with manifold theory, the sign also distinguishes between tangents to pull-backs of smooth mappings of manifolds and tangents to their corresponding push-forwards.

[3]One of the goals of this text is to explain what the mathematical terms in the statement of Noether's theorem mean in the context of geometric mechanics. For an interesting discussion of Noether's theorems from the Hamiltonian side, see Ref. [Ba2020].

The tangent-lifted action of G on TQ is defined by (minus) the Lie derivative $-\mathcal{L}_\xi q$ of $q \in Q$ with respect to the Lie algebra element $\xi \in \mathfrak{g}$ in (2.2.5). In combination with the dual pairing $\langle \cdot , \cdot \rangle_{T^*Q \times TQ}$, one induces the cotangent-lifted action of G on T^*Q, denoted by $T^*Q \to \mathfrak{g}^*$, as follows:

$$\langle \delta p , q \rangle_{T^*Q \times TQ} := \left\langle \left. \frac{dp(t,\epsilon)}{d\epsilon} \right|_{\epsilon=0} , q \right\rangle_{T^*Q \times TQ} = \langle -\mathcal{L}_\xi^T p , q \rangle_{T^*Q \times TQ}.$$

(2.2.7)

□

Upon introducing a natural, real non-degenerate symmetric pairing $\langle \cdot , \cdot \rangle_{\mathfrak{g}^* \times \mathfrak{g}}$, the Noether endpoint term in (2.2.6) induces a *cotangent lift momentum map*, denoted

$$J(q,p) : T^*Q \to \mathfrak{g}^*.$$

(2.2.8)

This is accomplished by defining the diamond ($\diamond$) operation as

$$\langle p , -\mathcal{L}_\xi q \rangle_{T^*Q \times TQ} =: \langle p \diamond q , \xi \rangle_{\mathfrak{g}^* \times \mathfrak{g}} =: \langle J(q,p) , \xi \rangle_{\mathfrak{g}^* \times \mathfrak{g}} =: J^\xi(q,p).$$

(2.2.9)

Example 2.2.1 The *Noether term* for infinitesimal rotations of the Lie group $SO(3)$ acting on vectors $\mathbf{q} \in \mathbb{R}^3$ induces the following tangent and cotangent lifts under the Euclidean pairing (dot product) on $\mathbb{R}^3$:

$$\langle p , \delta q \rangle_{T^*\mathbb{R}^3 \times T\mathbb{R}^3} := \langle p , -\mathcal{L}_\xi q \rangle_{T^*\mathbb{R}^3 \times T\mathbb{R}^3} = \mathbf{p} \cdot (-\boldsymbol{\xi} \times \mathbf{q})$$

$$= -\boldsymbol{\xi} \cdot (\mathbf{q} \times \mathbf{p}),$$

$$\langle \delta p , q \rangle_{T^*\mathbb{R}^3 \times T\mathbb{R}^3} = \langle -\mathcal{L}_\xi^T p , q \rangle_{T^*\mathbb{R}^3 \times T\mathbb{R}^3} = \boldsymbol{\xi} \times \mathbf{p} \cdot \mathbf{q}$$

(2.2.10)

$$= (-\boldsymbol{\xi} \times \mathbf{p}) \cdot \mathbf{q}.$$

Exercise. Show that the transformations in (2.2.10) also follow by computing the canonical equations for the Hamiltonian $J^\xi(\mathbf{q}, \mathbf{p}) = -\boldsymbol{\xi} \cdot (\mathbf{q} \times \mathbf{p})$,

$$\delta \mathbf{q} = \frac{\partial J^\xi}{\partial \mathbf{p}} = -\boldsymbol{\xi} \times \mathbf{q} \quad \text{and} \quad \delta \mathbf{p} = -\frac{\partial J^\xi}{\partial \mathbf{q}} = -\boldsymbol{\xi} \times \mathbf{p}.$$

Explain why $\delta \mathbf{q}$ and $\delta \mathbf{p}$ should have the same form in this case. ★

Example 2.2.2 (Geodesics) As mentioned earlier, a manifold is a space where the rules of calculus apply. A Riemannian manifold is a space on which geometric notions, such as distance, metric, angles, length, volume and curvature, are also defined.[4] Optimal paths in a Riemannian configuration manifold are called geodesics.

When the configuration space (Q, g) is a *Riemannian manifold* with *metric* g, the natural Lagrangian on (TQ, g) is given by the *kinetic energy* $K(v) := \frac{1}{2}g(q)(v_q, v_q)$, for $q \in Q$ and $v_q \in T_qQ$. In finite dimensions, in a local chart, this is written as $K(q, v_q) = \frac{1}{2}g_{ij}(q)v_q^i v_q^j$.

Exercise. Show that for the kinetic energy $\frac{1}{2}g(v_q, v_q)$ with $v_q = \dot{q}$ and Riemannian metric g, the Euler–Lagrange equations (2.2.4) become (for finite-dimensional Q in a local chart) the *geodesic equations*

$$\ddot{q}^i + \Gamma^i_{jk}(q)\dot{q}^j\dot{q}^k = 0, \quad i = 1, \ldots, n, \qquad (2.2.11)$$

where one sums repeated indices over their range and the three-index quantities $\Gamma^i_{jk}(q) = \Gamma^i_{kj}(q)$ are symmetric in their lower indices and defined in terms of the metric by

$$\Gamma^i_{jk}(q) = \frac{1}{2}g^{il}\left(\frac{\partial g_{jl}}{\partial q^k} + \frac{\partial g_{kl}}{\partial q^j} - \frac{\partial g_{jk}}{\partial q^l}\right), \quad \text{with } g_{hi}g^{il} = \delta^l_h. \qquad (2.2.12)$$

These are the *Christoffel symbols* of the Levi–Civita connection on (Q, g). ★

---

[4]See, e.g., Wikipedia https://en.wikipedia.org/wiki/Riemannian_manifold.

2.2.2 Corresponding physical concept: Motion on a curved surface

Geodesics can be thought of as trajectories of free particles in a manifold. Indeed, the geodesic equation (2.2.11) implies that the acceleration vector of the curve has no components in the direction of the surface (and therefore it is perpendicular to the tangent plane of the surface at each point of the curve). Consequently, the motion is determined by the bending of the surface. Geodesic motion in general is the natural path of motion in the absence of external forces, following Hamilton's principle of stationary action. This is also the idea of general relativity, according to which particles move on geodesics and the bending of space-time is associated with gravity.

Exercise. Consider a free particle of mass m moving on the Lobachevsky half-plane, $\mathbb{H}^2$, with coordinates $q = (x, y)$ and $y \geq 0$. Its Lagrangian is the kinetic energy $K(q, v_q) = \frac{1}{2} g_{ij}(q) v_q^i v_q^j$, corresponding to the Lobachevsky metric $g_{ij} = \delta_{ij}/y^2$. Namely,

$$L = \frac{m}{2} \left(\frac{\dot{x}^2 + \dot{y}^2}{y^2} \right). \tag{2.2.13}$$

1. Write the fibre derivative defined in (2.2.3) of the Lagrangian (2.2.13).
2. Compute its Euler–Lagrange equations.
3. Evaluate the Christoffel symbols. ★

Answer.

1. Fibre derivatives:

$$\frac{\partial L}{\partial \dot{x}} = \frac{m\dot{x}}{y^2} =: p_x \quad \text{and} \quad \frac{\partial L}{\partial \dot{y}} = \frac{m\dot{y}}{y^2} =: p_y.$$

2. The Euler–Lagrange equations $\frac{d}{dt}\frac{\partial L}{\partial \dot{x}} = \frac{\partial L}{\partial x}$ and $\frac{d}{dt}\frac{\partial L}{\partial \dot{y}} = \frac{\partial L}{\partial y}$ yield, respectively,

$$\frac{d}{dt}\left(\frac{\dot{x}}{y^2}\right) = 0 \quad \text{and} \quad \frac{d}{dt}\left(\frac{\dot{y}}{y^2}\right) = -\frac{\dot{x}^2 + \dot{y}^2}{y^3}.$$
(2.2.14)

3. Expanding these equations yield the Christoffel symbols for the geodesic motion:

$$\ddot{x} - \frac{2}{y}\dot{x}\dot{y} = 0, \quad \ddot{y} + \frac{1}{y}\dot{x}^2 - \frac{1}{y}\dot{y}^2 = 0$$

$$\Longleftrightarrow \quad \Gamma^1_{12} = -\frac{2}{y} = \Gamma^1_{21}, \quad \Gamma^2_{11} = \frac{1}{y},$$

$$\Gamma^2_{22} = -\frac{1}{y},$$

for $(p_1, p_2) = (p_x, p_y)$ and $(q_1, q_2) = (q_x, q_y)$ in $(x_1, x_2) = (x, y)$ planar components.   ▲

As we will see later, it is easy to compute that the geodesics of the two-dimensional Riemannian manifold $\mathbb{H}^2$ are circles and straight lines perpendicular to the x-axis. See Exercise 16.2.4.

The isometric transformations of the manifold $\mathbb{H}^2$ are linear fractional (Moebius) transformations of the complex plane with coordinates $z = x + iy$:

$$z \rightarrow \frac{az + b}{cz + d} \quad \text{with } ad - bc = 1,$$
(2.2.15)

with real coefficients a, b, c and d, This is the subgroup of the Moebius transformations that maps the upper half-plane $\mathbb{H}^2$ (given by $z = x + iy, y > 0$) into itself.

These isometric transformations of $\mathbb{H}^2$ are significant in physics. For example, they correspond to Lorentz transformations of space-time associated with Huygens waves [Sa2023].

2.3   Lie group S^1: Elroy and his beanie — Two planar coupled rigid bodies

Continuing with the overview in this lecture of counterpoints between mathematics and physics, "Elroy and his beanie" is the title of a beautiful example in Ref. [MaMoRa1990], which reveals that a mathematical property of geometric mechanics called *holonomy* also conveys the true nature of the geometric phase measured in the Aharonov–Bohm effect in quantum physics. This result recalls Hermann Flaschka's remark quoted before the Preface:

> The most difficult task is to think of workable examples that will reveal something new [Fl2015].

2.3.1   Geometric mechanics concept: Holonomy of a connection 1-form

Consider two *planar* rigid bodies joined together by an axis linking their centres of masses.

Let I_1 and I_2 be their principal moments of inertia, and let θ_1 and θ_2 be the angles they make with a fixed direction in an inertial frame.[5]

Conservation of the total angular momentum for the planar rotations of this system states that

$$I_1\dot{\theta}_1 + I_2\dot{\theta}_2 = \mu, \tag{2.3.1}$$

where μ is a constant of the motion. The *shape space* of a system is the space which describes the configuration of the system. In this case, the shape space is the circle S^1 specified by the relative angle $\psi = \theta_2 - \theta_1$. The formula (2.3.1) for the conservation of angular

---

[5]The principal moments of inertia I_1 and I_2 of an elliptical body are the second spatial moments of its area weighted by its density. For constant density, I_1 and I_2 are essentially the semi-major and semi-minor axes, respectively.

momentum then reads

$$I_1\dot{\theta}_1 + I_2(\dot{\psi} + \dot{\theta}_1) = \mu \quad \text{or} \quad A_{\text{mech}} := d\theta_1 + \frac{I_2}{I_1 + I_2}d\psi = \frac{\mu}{I_1 + I_2}dt.$$

$$(2.3.2)$$

Suppose that the total angular momentum μ is zero and Body #2 goes through one full revolution so that the relative angle ψ increases from 0 to 2π. Then, Body #1 rotates by

$$\theta_1 = -\frac{I_2}{I_1 + I_2}\int_0^{2\pi} d\psi = -\left(\frac{I_2}{I_1 + I_2}\right)2\pi. \qquad (2.3.3)$$

Since θ_1 is measured relative to a fixed frame, this is the amount that the entire system rotates relative to the fixed frame each time Body #2 completes one revolution.

Here is a geometric interpretation of this calculation. The 1-form A_{mech} in equation (2.3.2) can be regarded as a flat connection for the trivial principal S^1 bundle $\mathcal{B} : S^1 \times S^1 \to S^1$ given by the projection $\mathcal{B}(\theta_1, \psi) = \theta_1$. Formula (2.3.3) is the *holonomy* of the connection A_{mech}, when the shape variable (relative angle) makes a periodic traversal of the base circle $0 \leq \psi \leq 2\pi$.

2.3.2   Corresponding physical concept: Geometric phase/ Berry–Hannay angle

The connection A_{mech} in equation (2.3.2) is the same connection that appears in identifying *holonomy* in the Aharonov–Bohm effect in physics.

Holonomy can also explain why a falling cat can always land on its feet (paws). It can reorient itself, even though it is falling with zero angular momentum. However, the holonomy of a falling cat is defined in the more complicated group bundle $SO(3) \times SO(3) \to SO(3)$. The *shape variable* for this case is again the relative angle, but this time in $SO(3)$ [Mo1993]. In the following section, we will find that a planar model of the Foucault spherical pendulum possesses a similar holonomy to that for Elroy & his beanie.

Exercise.

1. Discuss the holonomy of bundle $\mathcal{B}_2$: $S^1 \times S^1 \times S^1 \to S^1 \times S^1$, in which three planar massive ellipses are physically pinned together at their centres of mass, and there exist two shape parameters corresponding to differences in sequential azimuthal angles.

2. Write the dynamical equations and determine the solution behaviour when the azimuthal shape variables of this system have a linear restoring force whose natural frequency can resonate with the others.   ★

2.4 S^1 Foucault pendulum: Planar oscillation/ precession in a rotating frame

2.4.1   Geometric mechanics concept: Composition of maps $S^1_{\text{oscillation}} \circ S^1_{\text{rotation}}$

Hamilton's action principle for the Foucault pendulum is given in planar polar coordinates $(r, \theta, \dot{r}, \dot{\theta}) \in T\mathbb{R}^2$ by

$$0 = \delta S = \delta \int_0^T L \, dt = \delta \int_0^T \frac{m}{2} \left(\dot{r}^2 + r^2 (\dot{\theta} + \Omega)^2 - \omega^2 r^2 \right) dt,$$

$$(2.4.1)$$

with constant parameters: rotation rate Ω about the vertical axis (e.g., Earth's rotation) and oscillation frequency $\omega = \sqrt{g/l_0}$, with gravitational acceleration g and pendulum length l_0.

Taking the variations $\delta r := \partial_\epsilon r(t, \epsilon)|_{\epsilon=0}$ and $\delta\theta := \partial_\epsilon \theta(t, \epsilon)|_{\epsilon=0}$ of the action integral S yields

$$
\begin{aligned}
0 = \delta S = m \int_0^T & \langle \dot{r}, \delta\dot{r} \rangle + \left\langle r\left((\dot{\theta} + \Omega)^2 - \omega^2\right), \delta r \right\rangle \\
& + \left\langle r^2(\dot{\theta} + \Omega), \delta\dot{\theta} \right\rangle \, dt \\
= m \int_0^T & \left\langle -\ddot{r} + r\left((\dot{\theta} + \Omega)^2 - \omega^2\right), \delta r \right\rangle \\
& - \left\langle \frac{d}{dt}(r^2(\dot{\theta} + \Omega)), \delta\theta \right\rangle \, dt \\
& + \underbrace{\langle m\dot{r}, \delta r \rangle \Big|_0^T + \left\langle mr^2(\dot{\theta} + \Omega), \delta\theta \right\rangle \Big|_0^T}.
\end{aligned}
$$
(2.4.2)

Noether endpoint terms

The Noether endpoint terms define linear momentum and angular momentum, given by

$$
p_r := \frac{\partial L}{\partial \dot{r}} = m\dot{r} \quad \text{and} \quad p_\theta := \frac{\partial L}{\partial \dot{\theta}} = mr^2(\dot{\theta} + \Omega),
$$

which satisfy

$$
\frac{dp_r}{dt} = mr((\dot{\theta} + \Omega)^2 - \omega^2) = \frac{p_\theta^2}{mr^3} - m\omega^2 r \quad \text{and} \quad \frac{dp_\theta}{dt} = 0.
$$
(2.4.3)

The Legendre transform to the Hamiltonian $H(r, \theta, p_r, p_\theta)$ is given by

$$
\begin{aligned}
H(r, \theta, p_r, p_\theta) := & \langle p_r, \dot{r} \rangle + \left\langle p_\theta, \dot{\theta} \right\rangle - L(r, \dot{r}, \theta.\dot{\theta}) \\
= & \frac{p_r^2}{2m} + \frac{p_\theta^2}{2mr^2} - \Omega p_\theta + \frac{m}{2}\omega^2 r^2.
\end{aligned}
$$
(2.4.4)

Hamilton's canonical equations for the Foucault pendulum are given by

$$
\dot{r} = \frac{\partial H}{\partial p_r} = p_r/m, \quad \dot{\theta} = \frac{\partial H}{\partial p_\theta} = \frac{p_\theta}{mr^2} - \Omega,
$$

$$
\dot{p}_r = -\frac{\partial H}{\partial r} = \frac{p_\theta^2}{mr^3} - m\omega^2 r, \quad \dot{p}_\theta = -\frac{\partial H}{\partial \theta} = 0.
$$
(2.4.5)

Exercise. Show that the solutions of (2.4.5) represent divergence-free flows in $\mathbb{R}^3$-preserving $d^3x = dp_\theta \wedge dr \wedge dp_r$ along the intersections of the level sets of H and p_θ.   ★

Answer. If we define $(x_1, x_2, x_3) := (r, p_r, p_\theta)$ and $C(\boldsymbol{x}) := x_3 = p_\theta$, one may write equations (2.4.5) as

$$\frac{dF}{dt} = \nabla C \cdot \nabla F \times \nabla H = \hat{e}_3 \cdot \nabla F \times \nabla H$$

$$= F_{,1}H_{,2} - F_{,2}H_{,1} =: \{F, H\}_{x_3},$$

(2.4.6)

which represents canonical Hamiltonian dynamics of the radial degree of freedom on a level set of $x_3 = p_\theta$. After solving for the radial solution, $r(t)$, one integrates the θ equation to reconstruct the azimuthal angle θ as a function of time. In the classical dynamics literature, the integral solution for $\theta(t)$ is called a *quadrature*.   ▲

In using conservation of p_θ to reduce the $\theta(t)$ solution to a quadrature, one finds the following periodic solution for $r(t)$ with period $T = \omega^{-1}$:

$$r(t) = r(0)\sqrt{\cos^2(\omega t) + (\Omega^2/\omega^2)\sin^2(\omega t)}.$$

(2.4.7)

However, the quadrature solution for $\theta(t)$, given by integration as

$$\int_0^T \dot{\theta}(t)dt = \theta(T) - \theta(0) = \int_0^T \frac{p_\theta}{mr^2(t)}\,dt - \Omega T,$$

(2.4.8)

is *not* periodic. Instead, at the end of each closed orbit $r(t)$, the angle $\theta(t)$ has precessed by the cumulative angle $\Delta\theta = \Omega T$.

The rotation of the moving frame with angular frequency $\Omega \neq 0$ causes a perturbation in the evolution of the angular orbit $\theta(t)$, and as a result, when the radial orbit $r(t)$ closes at time $T = \omega^{-1}$, the angular orbit does not close. The discrepancy $\Delta\theta = \Omega T$ in the closure of the angular orbit after one period T of the radial orbit arises because adding the term $p_\theta\Omega$ to the Hamiltonian $H(r,\theta,p_r,p_\theta)$ places the Foucault pendulum into a rotating frame with frequency Ω. Thus, one says that the motion is a *composition of maps*, $S_{\mathrm{osc}}^1 \circ S_{\mathrm{rot}}^1$ for oscillation and rotation.

2.4.2 Corresponding physical concept: Geometric phase/ Berry–Hannay angle

The angular discrepancy $\Delta\theta = \Omega T$ in the angular orbit after one period T of the radial orbit is called the Berry–Hannay angle, corresponding to the angular displacement of the oscillation plane over one radial period. The physics concept underlying the Berry–Hannay deviation angle is known mathematically as *holonomy*. Besides Foucault's measurement of the latitude on Earth, the physical phenomenon his pendulum demonstrates is that *the vibrational degree of freedom in a dynamical system always excites the system's rotational degree of freedom.*

Exercise. Use Hamilton's principle to derive the equations of motion and discuss the solution behaviour of a Foucault pendulum in planar polar coordinates when coupled to a planar azimuthal degree of freedom $\Theta(t)$, comprising a disk of fixed radius R and mass M with a linear angular restoring force given by

$$M\ddot{\Theta}(t) + \kappa\Theta(t) = 0,$$

for constant parameter $\varpi^2 = \kappa/M$. In particular, what conditions are required for the planar Foucault pendulum in an oscillatory rotating frame to precess by regular steps in its azimuthal angle?   ★

Answer. The Lagrangian for this situation is independent of the azimuthal angle of the planar Foucault pendulum motion:

$$L(\Theta, \dot{\Theta}, r, \dot{r}, \dot{\theta}) = \frac{1}{2} MR^2 (\dot{\Theta}^2 - \varpi^2 \Theta^2)$$

$$+ \frac{m}{2}\left(\dot{r}^2 + r^2(\dot{\theta} + \dot{\Theta})^2 - \frac{1}{2}\omega^2 r^2\right).$$

This Lagrangian implies two canonical momentum variables for the radial planar pendulum:

$$p_\theta := \frac{\partial L}{\partial \dot{\theta}} = mr^2(\dot{\theta} + \dot{\Theta}) \quad \text{and} \quad p_r := \frac{\partial L}{\partial \dot{r}} = m\dot{r}.$$

A Legendre transformation in these canonical pendulum variables produces the following Hamilton principle, cf. equation (2.4.4):

$$0 = \delta S = \delta \int_0^T \frac{1}{2} MR^2 (\dot{\Theta}^2 - \varpi^2 \Theta^2)$$

$$- p_\theta \dot{\Theta} + \frac{p_\theta^2}{2mr^2} + \frac{p_r^2}{2m} + \frac{m}{2}\omega^2 r^2 \, dt.$$

The rest of the solution follows as above, but with the constant rotation rate Ω replaced by $\dot{\Theta}(t)$, so that finally the solution for $\theta(t)$ is given by integration as

$$\int_0^T \frac{d}{dt}(\theta(t) + \Theta(t)) \, dt = \int_0^T \frac{p_\theta}{mr^2(t)} \, dt. \qquad (2.4.9)$$

Thus, at the end of each closed periodic orbit of $r(t)$, the angle $\theta(t)$ is precessed by the cumulative angle $\Delta\theta = \Theta(T) - \Theta(0)$, where $\Theta(t)$ is periodic with period $1/\varpi$.

Thus, the Foucault pendulum in an oscillatory rotating frame does not precess regularly, except in the resonant case of $n\omega + m\varpi = 0$, in which m and n are integers. ▲

2.5   Lie group $SO(3)$: Rotational states of a rigid body

The Lie group $SO(3)$ represents all possible rotations of a rigid body, providing a global description of its rotational states.

2.5.1   Geometric mechanics concept

Definition 2.5.1 (Group) *A group G is a set of elements possessing:*

1. *a binary product $G \times G \to G$ such that:*

   a. *the product of g and h is written gh for left group action (hg for right group action);*

   b. *the product is associative, i.e., $(gh)k = g(hk)$ for left action and $k(hg) = (kh)g$ for right.*

2. *a unique identity element, denoted by e, such that $ge = eg = g$ $\forall g \in G$.*

3. *an inverse operation $G \to G$ such that $g^{-1}g = gg^{-1} = e$.*

Definition 2.5.2 (Lie group) *A **Lie group** is a smooth manifold which is also a group, so that the binary product and inversion are smooth functions.*

Example 2.5.1

1. The manifold of invertible square $n \times n$ real matrices is a Lie group denoted by GL$(n, \mathbb{R})$.

2. The manifold of invertible square $n \times n$ matrices with unit determinant is a Lie group denoted by SL$(n, \mathbb{R})$ and called the *special linear group.*

3. The manifold of rotation matrices in n dimensions is a Lie group denoted by SO$(n, \mathbb{R})$ and called the *special orthogonal group.*

2.5.2   Corresponding physical concept: Body and space angular velocities

Recall that a *Lie group* is a group of transformations that depend smoothly on a set of parameters.

The Lie group $SO(3)$ faithfully represents the domain of orientation states of a rigid body. Therefore, the Lie group $SO(3)$ is the natural configuration manifold for rigid-body dynamics. Consider a body whose centre of mass is fixed at a point $P \in \mathbb{R}^3$. The most general allowed motion is a rotation about P. To describe this motion, we specify positions in a fixed space frame with orthonormal basis $\tilde{\mathbf{e}}_a$, with $a = 1, 2, 3$. One then transforms the motion to a moving frame $\mathbf{e}_a(t) = O_{ab}(t)^{-1}\tilde{\mathbf{e}}_b$ in the body so that $\mathbf{e}_a(t)$ moves with the body as it rotates.

Both sets of axes are orthogonal, so we have

$$\tilde{\mathbf{e}}_a \cdot \tilde{\mathbf{e}}_b = \delta_{ab}, \quad \text{and} \quad \mathbf{e}_a(t) \cdot \mathbf{e}_b(t) = \delta_{ab}.$$

Any point $\mathbf{r}(t)$ in the moving body can be expanded in either the space frame or the body frame:

$$\mathbf{r}(t) = \tilde{r}_b(t)\tilde{\mathbf{e}}_b \quad \text{in the space frame and}$$

$$\mathbf{r}(t) = r_a\mathbf{e}_a(t) \quad \text{in the body frame,}$$

where $\tilde{r}_b(t) = r_a O_{ab}(t)$ with constant r_a.

In the body frame, the moving and fixed bases are related by the left action $\mathbf{e}_a(t) = O_{ab}^{-1}(t)\tilde{\mathbf{e}}_b$. Consequently, viewed from the body, we have

$$\frac{d\mathbf{e}_a(t)}{dt} = \frac{d}{dt}O_{ab}^{-1}(t)\tilde{\mathbf{e}}_b = -\left(O^{-1}\dot{O}\right)_{ab}O_{bc}^{-1}\tilde{\mathbf{e}}_c = -\left(O^{-1}\dot{O}\right)_{ab}\mathbf{e}_b(t).$$
$$\tag{2.5.1}$$

In vector form, this equation is written as

$$\frac{d\mathbf{e}(t)}{dt} = -\,\mathbf{\Omega} \times \mathbf{e}(t), \tag{2.5.2}$$

where $\mathbf{\Omega}$ is the *left-invariant* body angular velocity vector of a point in space as observed from the rotating body.

One transforms from motion in the body frame to motion in the spatial frame via the *adjoint action* Ad, which defines how the Lie group $SO(3)$ acts on its Lie algebra $\mathfrak{so}(3)$, as follows. Namely, the adjoint action $\mathrm{Ad}_{O(t)}$ of $SO(3)$ on the apparent rotation of the fixed spatial basis $\tilde{\mathbf{e}}$ as viewed from the body basis $\mathbf{e}_a(t) = O_{ab}^{-1}(t)\tilde{\mathbf{e}}_b \in \mathbb{R}^3$ yields the following expression for the rotation of the body, as viewed from the spatial basis

$$\left(\mathrm{Ad}_{O(t)}\, \mathbf{e}(t) \right)_b := \left(O(t)\mathbf{e}(t)O^{-1}(t) \right)_b = \tilde{\mathbf{e}}_a O_{ab}^{-1}(t) =: \mathbf{E}_b(t).$$

Taking the time derivative yields the *right-invariant* angular velocity in the spatial frame given by

$$\frac{d}{dt}\mathbf{E}_b(t) = \frac{d}{dt}\left(\tilde{\mathbf{e}}_a O_{ab}^{-1}(t) \right) = -\left(\mathbf{E}_a(t)(\dot{O}O^{-1})_{ab} \right). \tag{2.5.3}$$

In vector form, this equation is written as, cf. equation (2.5.2),

$$\frac{d\mathbf{E}(t)}{dt} = \boldsymbol{\omega} \times \mathbf{E}(t), \tag{2.5.4}$$

where $\boldsymbol{\omega}$ is the *right-invariant* spatial angular velocity of the rotating body as observed from fixed space.

Remark 2.5.1 (Hat map) Note that orthogonality $O^T O = Id$ implies that both angular velocities $O^{-1}\dot{O}$ and $\dot{O}O^{-1}$ are skew-symmetric 3×3 matrices, and the $\mathbb{R}^3$ vector cross-product is equivalent to skew matrix multiplication. In particular, we write the *hat map* $(\hat{\cdot}) : \mathbb{R}^3 \mapsto \mathfrak{so}(3)$ as

$$\widehat{\Omega}_{ij} = -\,\epsilon_{ijk}\Omega^k \quad \text{and} \quad \widehat{\Omega} = \Omega\times, \quad \text{or} \quad (\Omega\times)_{ij} = \widehat{\Omega}_{ij}.$$

Here, $\epsilon_{123} = 1$ and $\epsilon_{213} = -1$, with cyclic permutations, and vanishes otherwise. □

Exercise. (Skew symmetry of the hat map) Show that the angular velocities $\widehat{\Omega} = O^{-1}\dot{O}$ and $\widehat{\omega} = \dot{O}O^{-1}$ are both skew symmetric. ★

Answer. Since $O^T O = O^{-1} O = Id$, we have $\dot{O}^T = -O^T \dot{O} O^T$. Consequently,

$$(\widehat{\Omega})^T = (O^T \dot{O})^T = \dot{O}^T O = -O^T \dot{O} = -\widehat{\Omega},$$

$$(\widehat{\omega})^T = (\dot{O} O^T)^T = O \dot{O}^T = -\dot{O} O^T = -\widehat{\omega},$$

so both $\widehat{\Omega} = O^{-1} \dot{O}$ and $\widehat{\omega} = \dot{O} O^{-1}$ are skew symmetric, although $\widehat{\Omega}$ is left invariant and $\widehat{\omega}$ is right invariant under the respective actions of $SO(3)$. ▲

Exercise. (Properties of the hat map) The hat map arises in the infinitesimal rotations given by

$$\widehat{\Omega}_{jk} = (O^{-1} dO/ds)_{jk}|_{s=0} = -\Omega_i \epsilon_{ijk}.$$

The hat map is an isomorphism between $\mathbb{R}^3$ and $\mathfrak{so}(3)$:

$$(\mathbb{R}^3, \times) \mapsto (\mathfrak{so}(3), [\,\cdot\,,\,\cdot\,]).$$

That is, the hat map identifies the composition of two vectors in $\mathbb{R}^3$ using the cross product with the commutator $[\widehat{P}, \widehat{Q}] = \widehat{P}\widehat{Q} - \widehat{Q}\widehat{P}$ of two skew-symmetric 3×3 matrices, $\widehat{P}$ and $\widehat{Q}$. Specifically, we write for any two vectors $\mathbf{Q}, \boldsymbol{\Omega} \in \mathbb{R}^3$,

$$-(\mathbf{Q} \times \boldsymbol{\Omega})_k = \epsilon_{klm} \Omega^l Q^m = \widehat{\Omega}_{km} Q^m.$$

Here, the upper (resp., lower) indices denote vector (resp., matrix) components. That is,

$$\boldsymbol{\Omega} \times \mathbf{Q} = \widehat{\Omega} \, \mathbf{Q} \quad \text{for all } \boldsymbol{\Omega}, \mathbf{Q} \in \mathbb{R}^3.$$

Verify the following formulas for $\mathbf{P}$ and $\mathbf{Q}, \boldsymbol{\Omega} \in \mathbb{R}^3$:

$$\left\langle \widehat{P}, \widehat{Q} \right\rangle = -\frac{1}{2} \operatorname{trace}\left(\widehat{P}\widehat{Q}\right) = \mathbf{P} \cdot \mathbf{Q},$$

$$\left[\widehat{P}, \widehat{Q}\right]_{ik} = -\epsilon_{ikj}(\mathbf{P} \times \mathbf{Q})^j, \quad \left[\widehat{P}, \widehat{Q}\right] = (\mathbf{P} \times \mathbf{Q})\widehat{},$$

$$\left[\hat{P}, \hat{Q}\right]\Omega = (\mathbf{P} \times \mathbf{Q}) \times \Omega,$$

$$\left[\hat{Q}, \hat{\Omega}\right]\mathbf{P} + \left[\hat{\Omega}, \hat{P}\right]\mathbf{Q} + \left[\hat{P}, \hat{Q}\right]\Omega = 0. \tag{2.5.5}$$

Hint: A useful identity here is $\epsilon_{ijk}\epsilon_{klm} = \delta_{il}\delta_{jm} - \delta_{im}\delta_{jl}$.

★

2.6   Lie algebra $\mathfrak{so}(3) \simeq \mathbb{R}^3$: Infinitesimal generators of rotation

Lie algebra $\mathfrak{so}(3)$ contains angular velocity vectors, representing infinitesimal rotations in the body's motion.

2.6.1   Geometric mechanics concept

If G is a Lie group, then $T_e G$ (the tangent space of G at the identity) is a vector space which possesses a remarkable structure called a *Lie algebra*.

Definition 2.6.1 (Lie algebra) *A **Lie algebra**, denoted $\mathfrak{g} = T_e G$, is a vector space endowed with a product known as the **commutator** (or **Lie bracket**), which is a skew bilinear map*

$$[\cdot, \cdot] : \mathfrak{g} \times \mathfrak{g} \to \mathfrak{g},$$

that satisfies the Jacobi identity: $[\xi, [\eta, \zeta]] + [\eta, [\zeta, \xi]] + [\zeta, [\xi, \eta]] = 0$ *for* $(\xi, \eta, \zeta) \in \mathfrak{g}$.

   Examples of matrix Lie algebras under matrix commutator are as follows:

1. *The Lie algebra* $\mathfrak{gl}(n, \mathbb{R}) := T_e GL(n, \mathbb{R})$ *can be identified with the vector space of real **square** $n \times n$ matrices.*

2. *The Lie algebra $\mathfrak{sl}(n, \mathbb{R}) := T_e SL(n, \mathbb{R})$ can be identified with the vector space of real **traceless** square matrices.*

3. *The Lie algebra $\mathfrak{so}(3) = T_e SO(3)$ can be identified with the vector space of real **skew-symmetric** matrices.*

2.6.2   Corresponding physical concept

Lie algebra elements generate group trajectories by exponentiation. For $\xi \in \mathfrak{g}$, one has

$$g_t = \exp t[\xi, \cdot] = Id + t[\xi, \cdot] + \frac{1}{2!}[\xi, [\xi, \cdot]] + \cdots .$$

Thus, the Lie algebra is the linearisation of the Lie group near the identity, and the Lie group transformation generated by a given Lie algebra element can be represented in exponential form.

2.7   Left action: Composition of maps

2.7.1   Geometric mechanics concept

Let M be a manifold, and let G be a Lie group. A *(left) action* of the Lie group G on M is a smooth mapping $\phi : G \times M \to M$ such that:

(1) $\phi(e, x) = x \quad \forall x \in M$;

(2) $\phi(g, \phi(h, x)) = \phi(gh, x) \quad \forall g, h \in G, \quad \forall x \in M$;

(3) for every $g \in G$, the map

$$\phi_g : M \to M, \qquad \phi_g(x) = \phi(g, x)$$

is a diffeomorphism (i.e. smooth and invertible).

Concatenation notation: We write gx for $\phi(g, x)$ and consider the group element g acting on the point x. Then, (2) becomes $g(hx) = (gh)x$.

2.7.2   Corresponding physical concept

In the context of rigid-body rotation, *left action* is observed look-
ing outwards from the rigid body's own local reference frame. For
example, left action can represent the rotation of stars in the night
sky relative to fixed coordinates on Earth.

2.8   Right action: Composition of maps

2.8.1   Geometric mechanics concept

Right Lie group action is similar to left action but with multiplication from the opposite direction. A *right Lie group action* of G on M
satisfies properties (1) and (3), while (2) is replaced with

$$\phi(g, \phi(h, x)) = \phi(hg, x) \quad \forall g, h \in G, \quad \forall x \in M.$$

Concatenation notation: $\phi(g, x)$ is denoted by xg, and (2) becomes
$(xh)g = x(hg)$.

Example 2.8.1 (Left vs. right) The left action $(g, x) \mapsto gx$ becomes
right action via $(g, x) \mapsto g^{-1}x$. That is, $g^{-1}x$ is a right action and
xg^{-1} is a left action.

2.8.2   Corresponding physical concept

Right action corresponds to observing the rotation of a rigid body
from a fixed external reference frame. For example, consider the
rotation of a gyroscope, where each rotation is observed relative
to the gyroscope's previous state in the fixed coordinate system.
It can also be regarded as a change of frame in which the second dynamics takes place in the frame of reference of the first
dynamics. (An example is waves carried on currents in the ocean
[HoHuSt2023].)

2.9 Lie group action on itself and on its Lie algebra: Group action on a rigid body

Lie group action, e.g., of $SO(3)$ on a rigid body, governs the body's orientation in space.

2.9.1 Geometric mechanics concept for matrix Lie groups

Definition 2.9.1 (Conjugation action) *Let $g \in G$, with G being a matrix Lie group. Then, the **conjugation action** of G on itself is given by the operation*

$$AD_g h : h \mapsto ghg^{-1} \qquad \forall h \in G.$$

Note that AD_g changes a right action into a left action.

   Take an arbitrary curve $h(t) \in G$ such that $h(0) = e$. Then, upon denoting

$$\xi = \dot{h}(0) \in T_e G, \quad \text{the derivative of } AD_g h \text{ defines}$$

$$\frac{d}{dt}\bigg|_{t=0} AD_g h(t) =: Ad_g\, \xi = g\xi g^{-1} \in T_e G.$$

Definition 2.9.2 (Adjoint and coadjoint actions of G on $\mathfrak{g}$ and $\mathfrak{g}^*$)
*The **adjoint action** of the matrix Lie group G on its matrix Lie algebra $\mathfrak{g}$ is a map:*

$$\mathrm{Ad} : G \times \mathfrak{g} \to \mathfrak{g}, \quad \mathrm{Ad}_g\, \xi = g\xi g^{-1}.$$

The dual map with symmetric, non-degenerate, real-valued pairing $\langle \cdot, \cdot \rangle : \mathfrak{g}^* \times \mathfrak{g} \to \mathbb{R}$ enables one to introduce Ad_g^*:

$$\langle \mathrm{Ad}_g^*\, \mu, \xi \rangle = \langle \mu, \mathrm{Ad}_g\, \xi \rangle,$$

*which is called the **coadjoint action** of G on $\mathfrak{g}^*$, the dual to the Lie algebra $\mathfrak{g}$ with respect to the pairing $\langle \cdot, \cdot \rangle$. Take $g(t) \in G$ such that $g(0) = e$, and denote*

$$\eta = \dot{g}(0) \in T_e G.$$

Then, one defines

$$ad_\eta \, \xi := \left. \frac{d}{dt} \right|_{t=0} \mathrm{Ad}_{g(t)} \, \xi \quad \forall \xi \in \mathfrak{g}.$$

Definition 2.9.3 (Adjoint and coadjoint action of $\mathfrak{g}$ on $\mathfrak{g}$ and $\mathfrak{g}^*$)
*The **adjoint action** of the matrix Lie algebra on itself is given as a map:*

$$\mathrm{ad} : \mathfrak{g} \times \mathfrak{g} \to \mathfrak{g}$$

$$\mathrm{ad}_\eta \, \xi = [\eta, \xi] := \eta\xi - \xi\eta.$$

The dual map

$$\langle \mathrm{ad}_\eta^* \, \mu, \, \xi \rangle = \langle \mu, \, \mathrm{ad}_\eta \, \xi \rangle$$

*is the **coadjoint action** of $\mathfrak{g}$ on $\mathfrak{g}^*$.*

Exercise. Write these adjoint and coadjoint operations for the Lie group of upper-triangular 3×3 matrices. Upper-triangular 3×3 matrices represent the action of the Heisenberg group. See, e.g., Ref. [Ho2011c] for more discussion of the adjoint and coadjoint actions of upper-triangular 3×3 matrices.   ★

2.9.2 Corresponding physical concept

In rigid-body rotation, coadjoint action describes the dynamics of angular momentum under rotation. It reflects transformations between two different reference frames, called the body and space frames. The angular velocities in the body and space frames are denoted

Body: $\widehat{\Omega} = O^{-1}\dot{O} \in \mathfrak{so}(3)$ and   Space: $\widehat{\omega} = \dot{O}O^{-1} \in \mathfrak{so}(3),$

$$(2.9.1)$$

for $O(t) \in SO(3)$.

Exercise. Show that the skew-symmetric space and body angular velocities $\widehat{\omega} = \dot{O}O^{-1}$ and $\widehat{\Omega} = O^{-1}\dot{O}$, respectively, are related by $\widehat{\omega} = \mathrm{Ad}_{O(t)}\widehat{\Omega}$.   ★

2.10   Coadjoint motion: Angular momentum state space

2.10.1   Geometric mechanics concept

In geometric mechanics, *coadjoint motion* satisfies

$$\mathrm{Ad}^*_{g^{-1}(t)}\frac{\partial \ell}{\partial \xi}(t) = const.$$

for a left-invariant Lagrangian function defined on the Lie algebra $\ell : \mathfrak{g} \to \mathbb{R},\ \xi(t) = g^{-1}\dot{g}(t) \in \mathfrak{g}$.

For a fixed $\zeta \in \mathfrak{g}$, one may calculate the time derivative of the following pairing:

$$\frac{d}{dt}\left\langle \mathrm{Ad}^*_{g^{-1}}\frac{\partial \ell}{\partial \xi}, \zeta \right\rangle$$

$$= \frac{d}{dt}\left\langle \frac{\partial \ell}{\partial \xi}(t), \mathrm{Ad}_{g^{-1}(t)}\zeta \right\rangle$$

$$= \left\langle \frac{d}{dt}\frac{\partial \ell}{\partial \xi}, \mathrm{Ad}_{g^{-1}(t)}\zeta \right\rangle + \left\langle \frac{\partial \ell}{\partial \xi}, \frac{d}{dt}(g^{-1}(t)\zeta g(t)) \right\rangle$$

$$= \left\langle \frac{d}{dt}\frac{\partial \ell}{\partial \xi}, \mathrm{Ad}_{g^{-1}(t)}\zeta \right\rangle + \left\langle \frac{\partial \ell}{\partial \xi}, -g^{-1}\dot{g}g^{-1}\zeta g + g^{-1}\zeta \dot{g} \right\rangle$$

$$= \left\langle \frac{d}{dt}\frac{\partial \ell}{\partial \xi}, \mathrm{Ad}_{g^{-1}(t)}\zeta \right\rangle + \left\langle \frac{\partial \ell}{\partial \xi}, -\xi \mathrm{Ad}_{g^{-1}}\zeta + (g^{-1}\zeta g)(g^{-1}\dot{g}) \right\rangle$$

$$= \left\langle \frac{d}{dt}\frac{\partial \ell}{\partial \xi}, \mathrm{Ad}_{g^{-1}(t)}\zeta \right\rangle - \left\langle \frac{\partial \ell}{\partial \xi}, \mathrm{ad}_\xi(\mathrm{Ad}_{g^{-1}}\zeta) \right\rangle$$

$$= \left\langle \frac{d}{dt}\frac{\partial \ell}{\partial \xi} - \mathrm{ad}_{\xi}^{*}\frac{\partial \ell}{\partial \xi}, \mathrm{Ad}_{g^{-1}}\zeta \right\rangle$$

$$= \left\langle \mathrm{Ad}_{g^{-1}}^{*} \left(\frac{d}{dt}\frac{\partial \ell}{\partial \xi} - \mathrm{ad}_{\xi}^{*}\frac{\partial \ell}{\partial \xi} \right), \zeta \right\rangle = 0. \qquad (2.10.1)$$

Since the fixed Lie algebra element ζ is arbitrary, we have

$$\frac{d}{dt}\left(\mathrm{Ad}_{g^{-1}(t)}^{*}\frac{\partial \ell}{\partial \xi} \right) = \mathrm{Ad}_{g^{-1}(t)}^{*}\left(\frac{d}{dt}\frac{\partial \ell}{\partial \xi} - \mathrm{ad}_{\xi}^{*}\frac{\partial \ell}{\partial \xi} \right) = 0.$$

Consequently, the solutions of the *Euler–Poincaré equation*, expressed as

$$\left(\frac{d}{dt}\frac{\partial \ell}{\partial \xi} - \mathrm{ad}_{\xi}^{*}\frac{\partial \ell}{\partial \xi} \right) = 0, \qquad (2.10.2)$$

satisfy $\mathrm{Ad}_{g^{-1}(t)}^{*}\partial \ell/\partial \xi(t) = const$ and thus describe coadjoint motion. The solutions of the Euler–Poincaré equation in (2.10.2) comprise a set of *coadjoint orbits*.

Exercise. Verify that

$$\frac{d}{dt}\Big|_{t=0}\mathrm{Ad}_{g_t}^{-1}\zeta = -\,\mathrm{ad}_{\xi}\,\zeta$$

for any *fixed* $\zeta \in \mathfrak{g}$ and right action $\xi = \dot{g}_t g_t^{-1}\big|_{t=0}$.
Verify that

$$\frac{d}{dt}\mathrm{Ad}_{g_t}^{-1}\zeta = -\,\mathrm{ad}_{\xi_t}\,\mathrm{Ad}_{g_t}^{-1}\zeta$$

for any fixed $\zeta \in \mathfrak{g}$ and $\xi_t = \dot{g}_t g_t^{-1}$.
Hint: Follow the calculations in (2.10.1).   ★

Answer. One begins with the second part because the first part follows by evaluating the result of the second part at $t = 0$:

$$\frac{d}{dt}\mathrm{Ad}_{g_t^{-1}}\zeta = \frac{d}{dt}(g_t^{-1}\zeta g_t) = -(g_t^{-1}\dot{g}_t)g_t^{-1}\zeta g_t + g_t^{-1}\zeta \dot{g}_t$$

$$= -\xi_t(\mathrm{Ad}_{g_t^{-1}}\zeta) + (g_t^{-1}\zeta g_t)(g_t^{-1}\dot{g}_t)$$

$$= -\xi_t(\mathrm{Ad}_{g_t^{-1}}\zeta) + (\mathrm{Ad}_{g_t^{-1}}\zeta)\xi_t$$

By (2.10.1) $= -\mathrm{ad}_{\xi_t}(\mathrm{Ad}_{g_t^{-1}}\zeta).$

Then, evaluating $\mathrm{Ad}_{g_t^{-1}}\big|_{t=0} = Id$ yields the result for the first part.   ▲

As we shall discuss in detail later, the Euler–Poincaré equation (2.10.2) follows from a constrained Hamilton's variational principle, namely, the *Hamilton–Pontryagin* variational principle, which yields in this case

$$0 = \delta S = \delta \int_0^T \ell(\xi) + \langle \mu, g^{-1}\dot{g} - \xi \rangle \, dt$$

$$= \int_0^T \left\langle \frac{\partial \ell}{\partial \xi} - \mu, \delta\xi \right\rangle + \left\langle \mu, \frac{d\eta}{dt} + \mathrm{ad}_{g^{-1}\dot{g}}\eta \right\rangle$$

$$+ \langle \delta\mu, g^{-1}\dot{g} - \xi \rangle \, dt \qquad (2.10.3)$$

$$= \int_0^T \left\langle \frac{\partial \ell}{\partial \xi} - \mu, \delta\xi \right\rangle - \left\langle \frac{d\mu}{dt} - \mathrm{ad}^*_{g^{-1}\dot{g}}\mu, \eta \right\rangle$$

$$+ \langle \delta\mu, g^{-1}\dot{g} - \xi \rangle \, dt + \langle \mu, \eta \rangle \big|_0^T,$$

where $\xi = g^{-1}\dot{g}$ and the quantity $\eta := g^{-1}\delta g$ vanishes at the endpoints in time, and we have used the relation

$$\delta\xi = \frac{d\eta}{dt} + \mathrm{ad}_\xi \eta. \qquad (2.10.4)$$

Exercise. Derive the Euler–Poincaré relation in (2.10.4).

★

Answer. Define "prime" notation $\eta := g^{-1}\delta g = g^{-1}g'$ and compute

$$\xi' = (g^{-1}\dot{g})' = -(g^{-1}g')(g^{-1}\dot{g}) + g^{-1}\dot{g}' = -\eta\xi + g^{-1}\dot{g}',$$

$$\dot{\eta} = (g^{-1}g')^{\cdot} = -(g^{-1}\dot{g})(g^{-1}g') + g^{-1}g'^{\cdot} = -\xi\eta + g^{-1}g'^{\cdot}.$$

Subtracting the second equation from the first, resulting in the cancellation of equal second-order cross derivatives $\dot{g}' = g'^{\cdot}$, and collecting the terms yields the Euler–Poincaré equation for coadjoint motion in (2.10.2).   ▲

Exercise. Prove the following identities with $\xi_t = g_t^{-1}\dot{g}_t$ for *fixed* $\eta \in \mathfrak{g}$ and $\mu \in \mathfrak{g}^*$:

$$\left\langle \frac{d}{dt}\left(\mathrm{Ad}_{g_t}\eta\right), \mu \right\rangle = \left\langle \mathrm{Ad}_{g_t}\left(\mathrm{ad}_{g_t^{-1}\dot{g}_t}\eta\right), \mu \right\rangle$$

$$= \left\langle \mathrm{Ad}_{g_t}\left(\mathrm{ad}_{\xi_t}\eta\right), \mu \right\rangle$$

$$\left\langle \frac{d}{dt}\left(\mathrm{Ad}_{g_t^{-1}}\eta\right), \mu \right\rangle = \left\langle -\mathrm{ad}_{g_t^{-1}\dot{g}_t}\left(\mathrm{Ad}_{g_t^{-1}}\eta\right), \mu \right\rangle$$

$$= \left\langle -\mathrm{ad}_{\xi_t}\left(\mathrm{Ad}_{g_t^{-1}}\eta\right), \mu \right\rangle.$$

$$(2.10.5)$$

★

Answer. By direct calculation,

$$\left\langle \frac{d}{dt}\left(\mathrm{Ad}_{g_t}\eta\right), \mu \right\rangle$$

$$= \left\langle \frac{d}{dt}\left(g_t\eta g_t^{-1}\right), \mu \right\rangle$$

$$= \left\langle \dot{g}_t\eta g_t^{-1} - g_t\eta g_t^{-1}\dot{g}_t g_t^{-1}, \mu \right\rangle$$

$$= \left\langle \mathrm{Ad}_{g_t}\left((g_t^{-1}\dot{g}_t)\eta\right) - \mathrm{Ad}_{g_t}\left(\eta(g_t^{-1}\dot{g}_t)\right),\, \mu \right\rangle$$

$$= \left\langle \mathrm{Ad}_{g_t}\left(\mathrm{ad}_{g_t^{-1}\dot{g}_t}\,\eta\right),\, \mu \right\rangle$$

$$= \left\langle \mathrm{Ad}_{g_t}\left(\mathrm{ad}_{\xi_t}\,\eta\right),\, \mu \right\rangle$$

$$\mathrm{Dual:} = \left\langle \eta,\, \mathrm{ad}_{\xi_t}^*\left(\mathrm{Ad}_{g_t}^*\,\mu\right)\right\rangle. \qquad (2.10.6)$$

A similar calculation yields

$$\left\langle \frac{d}{dt}(\mathrm{Ad}_{g_t^{-1}}\,\eta),\, \mu \right\rangle = \left\langle -\,\mathrm{ad}_{\xi_t}(\mathrm{Ad}_{g_t^{-1}}\,\eta),\, \mu \right\rangle. \qquad \blacktriangle$$

2.10.2   Corresponding physical concept

Euler–Poincaré Equations and Rigid Body Dynamics — Euler–Poincaré equations for rigid-body rotation describe the dynamics considering the body's symmetry and frame of motion, leading in particular to the classical Euler rigid body equations in the body frame. The Euler–Poincaré equations will appear prominently in the remainder of this text.

2.11   Reviewing the diamond operator and cotangent lift momentum map

The diamond operator and cotangent lift momentum map were introduced in Section 2.2 in the context of Noether's theorem for the Lie symmetry of the Lagrangian in Hamilton's principle. Here, we say a bit more about their properties on the Hamiltonian side.

2.11.1   Geometric mechanics concept

The diamond operator, denoted as $(\diamond)$, appears in Noether's end-point term $\langle p,\, \delta q\rangle\big|_0^T$ in Hamilton's principle (2.2.1) when the variation δq is an infinitesimal left action of a Lie symmetry of the configuration space (Q, q),

$$\delta q := \tfrac{d}{d\epsilon}\big|_{\epsilon=0}q(t, \epsilon) =: -\,\pounds_\xi q, \qquad (2.11.1)$$

where the notation $\mathcal{L}_\xi$ for the Lie derivative is defined as the tangent at the identity $\epsilon = 0$ of the ϵ-variational group action. One may compare relation (2.11.1) for a left Lie action with equations (2.5.1) and (2.5.3), which contrast the left-invariant body representation and right-invariant spatial representation of angular velocity.

Theorem 2.11.1 (Noether's theorem) *Let the Lagrangian $L : TQ \to \mathbb{R}$ in Hamilton's principle (2.2.1) be invariant under a Lie group left action $g_\epsilon q(t) = q(t, \epsilon) : G \times Q \to Q$, where the curve $g_\epsilon \in G$ is the flow of G parameterised by ϵ, which becomes the identity transformation at $\epsilon = 0$. The corresponding infinitesimal Lie G-symmetry $\delta q = -\mathcal{L}_\xi q$, in which the left Lie algebra action of $\xi \in \mathfrak{g} \simeq T_e G$ leaves the Lagrangian L invariant and implies the conservation of the endpoint term,*

$$\left\langle \frac{\partial L}{\partial \dot{q}} , \, \delta q \right\rangle_{TQ} =: \langle p, \, \delta q \rangle_{TQ} = \langle p, \, -\mathcal{L}_\xi q \rangle_{TQ} \tag{2.11.2}$$

$$=: \langle p \diamond q, \, \xi \rangle_{\mathfrak{g}} =: \langle J(q,p), \, \xi \rangle_{\mathfrak{g}} =: J^\xi(q,p),$$

where $\langle \cdot, \, \cdot \rangle_{TQ}$ is a symmetric, non-degenerate, real-valued pairing $T^*Q \times TQ \to \mathbb{R}$. The middle equation defines the diamond operator $(\diamond)$, which is of central importance. The quantity

$$J(q,p) = p \diamond q = -q \diamond p \tag{2.11.3}$$

is the cotangent-lift momentum map $J : T^*(Q) \to \mathfrak{g}^*$ by the left Lie algebra action.

Remark 2.11.1 From its definition, the properties of the diamond operator $(\diamond)$ are inherited from those of the Lie derivative. In particular, the quantity

$$J^\xi(q,p) = \langle p \diamond q, \, \xi \rangle_{\mathfrak{g}}$$

is the Hamiltonian defined by the ξ-component of the momentum map $J(q,p) = p \diamond q \in \mathfrak{g}^*$.

Under the *canonical Poisson bracket*, one has

$$\delta q = \{q, J^\xi(q,p)\}_{can} = -\mathcal{L}_\xi q \quad \text{and} \quad \delta p = \{p, J^\xi(q,p)\}_{can} = -\mathcal{L}_\xi^T p.$$

Here,

$$\langle p,\, \delta q\rangle := \langle p,\, -\pounds_\xi q\rangle_{TG} = \langle -\pounds_\xi^T p,\, q\rangle_{TG} = \langle \delta p,\, q\rangle_{TG}$$

and δp is said to be the *cotangent lift* of δq. Hence the name *cotangent-lift momentum map* for $J(q,p)$. □

Exercise. Calculate the Poisson bracket $\{J^\xi(p,q), H(p,q)\}$ for $J^\xi(p,q) = \langle p \diamond q,\, \xi\rangle$ and Hamiltonian $H(p,q)$. Explain how this result is related to Noether's theorem.[6] ★

Answer.

$$-\frac{d}{dt}J^\xi(p,q) = \{H(p,q), J^\xi(p,q)\}_{\text{can}} = \frac{\partial H}{\partial q}\delta q + \frac{\partial H}{\partial p}\delta p$$

$$= -\frac{\partial H}{\partial q}\pounds_\xi q - \frac{\partial H}{\partial p}\pounds_\xi^T p = -\pounds_\xi H(p,q).$$

Hamiltonian Noether theorem. The left invariance of the Hamiltonian $H(p,q)$ under the Lie transformation of phase space generated by the infinitesimal canonical transformation $-\pounds_\xi$ so that $\pounds_\xi H(p,q) = 0$ implies that the phase-space function $J^\xi(p,q)$ which generates that Lie transformation by

$$\{(\,\cdot\,),\, J^\xi(p,q)\}_{\text{can}} = -\pounds_\xi(\,\cdot\,)$$

will be conserved by the dynamics generated by the Hamiltonian $H(p,q)$ via the Poisson bracket $\{J^\xi(p,q), H(p,q)\}$. Namely, the phase-space function $J^\xi(p,q)$ will be invariant under the Poisson bracket operation for Hamiltonian dynamics:

$$\frac{d}{dt} = \{\cdot, H(p,q)\}.$$

▲

Example 2.11.1 (Momentum map for $SO(3)$ acting on $\mathbb{R}^3$) For $Q = \mathbb{R}^3$ and $\mathfrak{g} = \mathfrak{so}(3)$, one finds $\xi_Q(q) = -\pounds_\xi q = -\xi \times q$ by the hat map and

$$\left\langle p \diamond q, \xi \right\rangle = \left\langle p, -\pounds_\xi q \right\rangle_{TQ}$$

$$= -p \cdot (\xi \times q) = -(q \times p) \cdot \xi = \left\langle J(p,q), \xi \right\rangle = J^\xi(p,q),$$

which is the Hamiltonian for an infinitesimal rotation around ξ in $\mathbb{R}^3$. In the case where $\mathfrak{g} = \mathfrak{so}(3)$, the pairing $\langle \cdot, \cdot \rangle$ may be taken as dot products of vectors in $\mathbb{R}^3$, the momentum map $J(p,q) = p \diamond q = p \times q \in \mathbb{R}^3$ is the phase-space expression for angular momentum and the $\diamond$ operation is $\times$, which is the cross product of vectors in $\mathbb{R}^3$. The map $J(p,q) = p \diamond q = -q \times p \in \mathbb{R}^3$ is an example of a *cotangent-lift* momentum map.

2.11.2   Corresponding physical concept

In the case of particle motion for a rotationally left-invariant Lagrangian, one finds

$$\delta\mathbf{q} = \frac{\partial J^\xi}{\partial \mathbf{p}} = -\boldsymbol{\xi} \times \mathbf{q}, \quad \delta\mathbf{p} = -\frac{\partial J^\xi}{\partial \mathbf{q}} = -\boldsymbol{\xi} \times \mathbf{p}, \quad \text{and} \quad p \diamond q = \mathbf{p} \times \mathbf{q}.$$

Thus, the diamond operator for the left Lie algebra action reduces to the vector cross product in $\mathbb{R}^3$ for the case of rotation, and (minus) the particle angular momentum $\mathbf{J} = -\mathbf{q} \times \mathbf{p}$ emerges as an element in the dual space of the Lie algebra $\mathfrak{so}(3)^* \simeq \mathfrak{so}(3) \simeq \mathbb{R}^3$ for the left Lie algebra action of $\mathfrak{so}(3)$ on $\mathbb{R}^3$.

The cross product is very important in understanding the role of the angular momentum in particle dynamics for a rotationally invariant Lagrangian as in Newtonian gravitation. Likewise, we will see that the diamond operation and the cotangent-lift momentum map $J(q,p) = p \diamond q$ are very important in understanding Euler–Poincaré motion for other Lie group invariant Lagrangians.

Exercise. Calculate the Poisson bracket $\{J^\xi(p,q), J^\zeta(p,q)\}$ for $J^\xi(p,q) = \langle p \diamond q, \xi \rangle$ and $J^\zeta(p,q) = \langle p \diamond q, \zeta \rangle = \langle p, -\pounds_\zeta q \rangle$ for infinitesimal left action. ★

Answer.

$$\{J^\xi(q,p), J^\zeta(q,p)\}_{\mathrm{can}} = \{J^\xi(q,p), \langle p \diamond q, \zeta \rangle_{\mathfrak{g}}\}_{\mathrm{can}}$$

By product rule $= \langle \{J^\xi, p\}_{\mathrm{can}} \diamond q + p \diamond \{J^\xi, q\}_{\mathrm{can}}, \zeta \rangle_{\mathfrak{g}}$

$$= \langle (-\pounds_\xi^T p) \diamond q + p \diamond (\pounds_\xi q), \zeta \rangle_{\mathfrak{g}}$$

$$= \langle \pounds_\xi(p \diamond q), \zeta \rangle_{\mathfrak{g}} = \langle \mathrm{ad}_\xi^*(p \diamond q), \zeta \rangle_{\mathfrak{g}}$$

$$= \langle p \diamond q, \mathrm{ad}_\xi \zeta \rangle_{\mathfrak{g}} = \langle p, -\pounds_{[\xi,\zeta]} q \rangle_{TQ}$$

$$= -J^{[\xi,\zeta]}(q,p) \quad \text{(anti-homomorphism)}.$$

▲

Exercise.

1. Derive for a *right-invariant* Lagrangian function the corresponding results of Hamilton's principle in Section 2.10, which are derived there for a left-invariant Lagrangian function.

2. Transform the rigid-body Lagrangian for left-invariant angular velocity in the body frame into the *same* Lagrangian written in terms of the right-invariant angular velocity in the spatial frame.

3. What differences emerge in Noether's theorem for the *same* rigid-body Lagrangian written in either the left-invariant or right-invariant angular velocities? ★

Exercise. Calculate the Poisson bracket $\{J^\xi(p, q), J^\zeta(p, q)\}$ for $J^\xi(p, q) = \langle p \diamond q, \xi \rangle$ and $J^\zeta(p, q) = \langle p \diamond q, \zeta \rangle = \langle p, -\pounds_\zeta q \rangle$ for infinitesimal *right* action. ★

3

PARTICLE MECHANICS OF NEWTON, LAGRANGE AND HAMILTON

Contents

What is this lecture about? This lecture is about the equivalences among the formulations of dynamics in the approaches of Newton, Lagrange and Hamilton.

3.1 Newton's equations for particle motion in Euclidean space

Newton's equations in a fixed inertial frame,

$$m_i \ddot{\mathbf{q}}_i = \mathbf{F}_i, \quad i = 1, \ldots, N, \quad (\text{no sum on } i), \qquad (3.1.1)$$

describe the *accelerations* $\ddot{\mathbf{q}}_i$ of N particles with

$$\text{Masses}\quad m_i, \quad i = 1, \ldots, N,$$

$$\text{Euclidean positions}\quad \mathbf{q} := (\mathbf{q}_1, \ldots, \mathbf{q}_N) \in \mathbb{R}^{3N},$$

in response to *prescribed forces,*

$$\mathbf{F} = (\mathbf{F}_1, \ldots, \mathbf{F}_N),$$

acting on these particles. Suppose the forces arise from a *potential.* That is, let

$$\mathbf{F}_i(\mathbf{q}) = -\frac{\partial V(\{\mathbf{q}\})}{\partial \mathbf{q}_i}, \quad V \colon \mathbb{R}^{3N} \to \mathbb{R}, \tag{3.1.2}$$

where $\partial V/\partial \mathbf{q}_i$ denotes the gradient of the potential with respect to the variable $\mathbf{q}_i$. Then, Newton's equations (3.1.1) become

$$m_i \ddot{\mathbf{q}}_i = -\frac{\partial V}{\partial \mathbf{q}_i}, \quad i = 1, \ldots, N. \tag{3.1.3}$$

Such a Newtonian system in potential form is called a *simple mechanical system.*

Remark 3.1.1 Newton (1620) introduced the gravitational potential for celestial mechanics, now called the *Newtonian potential:*

$$V(\{\mathbf{q}\}) = \sum_{i,j=1}^{N} \frac{-Gm_i m_j}{|\mathbf{q}_i - \mathbf{q}_j|}. \tag{3.1.4}$$

$\square$

3.2   Equivalence of Newton, Lagrange and Hamilton dynamics

Theorem 3.2.1 (Lagrangian and Hamiltonian formulations) *Newton's equations in potential form,*

$$m_i \ddot{\mathbf{q}}_i = -\frac{\partial V}{\partial \mathbf{q}_i}, \quad i = 1, \ldots, N, \tag{3.2.1}$$

for particle motion in Euclidean space $\mathbb{R}^{3N}$ are equivalent to the following four statements:

1. *The Euler–Lagrange equations,*

$$\frac{d}{dt}\left(\frac{\partial L}{\partial \dot{\mathbf{q}}_i}\right) - \frac{\partial L}{\partial \mathbf{q}_i} = 0, \quad i = 1, \ldots, N, \qquad (3.2.2)$$

hold for the Lagrangian $L\colon \mathbb{R}^{6N} = \{(\mathbf{q}, \dot{\mathbf{q}}) \mid \mathbf{q}, \dot{\mathbf{q}} \in \mathbb{R}^{3N}\} \to \mathbb{R}$, *defined by*

$$L(\mathbf{q}, \dot{\mathbf{q}}) := \sum_{i=1}^{N} \frac{m_i}{2} \|\dot{\mathbf{q}}_i\|^2 - V(\mathbf{q}), \qquad (3.2.3)$$

with $\|\dot{\mathbf{q}}_i\|^2 = \dot{\mathbf{q}}_i \cdot \dot{\mathbf{q}}_i = \dot{q}_i^j \dot{q}_i^k \delta_{jk}$ *(no sum on* i*). Lagrangians of the separated form in (3.2.3) are said to govern simple mechanical systems.*

2. *Hamilton's principle of stationary action,* $\delta S = 0$, *holds for the action functional (dropping* i *indices)*

$$S[\mathbf{q}(\cdot)] := \int_a^b L(\mathbf{q}(t), \dot{\mathbf{q}}(t)) + \left\langle \mathbf{p}, \frac{d\mathbf{q}}{dt} - \dot{\mathbf{q}} \right\rangle dt. \qquad (3.2.4)$$

3. *Hamilton's equations of motion,*

$$\dot{\mathbf{q}} = \frac{\partial H}{\partial \mathbf{p}}, \quad \dot{\mathbf{p}} = -\frac{\partial H}{\partial \mathbf{q}}, \qquad (3.2.5)$$

hold for the Hamiltonian resulting from the Legendre transform

$$H(\mathbf{q}, \mathbf{p}) := \mathbf{p} \cdot \dot{\mathbf{q}}(\mathbf{q}, \mathbf{p}) - L(\mathbf{q}, \dot{\mathbf{q}}(\mathbf{q}, \mathbf{p})), \qquad (3.2.6)$$

where $\dot{\mathbf{q}}(\mathbf{q}, \mathbf{p})$ *solves for* $\dot{\mathbf{q}}$ *from the definition* $\mathbf{p} := \partial L(\mathbf{q}, \dot{\mathbf{q}})/\partial \dot{\mathbf{q}}$.

In the case of Newton's equations in potential form (3.2.1), the Lagrangian in equation (3.2.3) yields $\mathbf{p}_i = m_i \dot{\mathbf{q}}_i$, *and the resulting Hamiltonian is (restoring subscript is)*

$$H = \underbrace{\sum_{i=1}^{N} \frac{1}{2m_i} \|\mathbf{p}_i\|^2}_{\text{Kinetic energy}} + \underbrace{V(\mathbf{q})}_{\text{Potential}} .$$

4. *Hamilton's equations in their Poisson bracket formulation,*

$$\dot{F} = \{F, H\} \quad \text{for all } F \in \mathcal{F}(P), \tag{3.2.7}$$

hold, with Poisson bracket defined by

$$\{F, G\} := \sum_{i=1}^{N} \left(\frac{\partial F}{\partial \mathbf{q}_i} \cdot \frac{\partial G}{\partial \mathbf{p}_i} - \frac{\partial F}{\partial \mathbf{p}_i} \cdot \frac{\partial G}{\partial \mathbf{q}_i} \right) \quad \text{for all } F, G \in \mathcal{F}(P). \tag{3.2.8}$$

We will prove this theorem by proving a chain of linked equivalence relations: (3.2.1) ⇔ (i) ⇔ (ii) ⇔ (iii) ⇔ (iv) as propositions. (The symbol ⇔ means "equivalent to".)

Step I: Proof that Newton's equations (3.2.1) ⇔ (i). Check by direct verification. ∎

Step II: Proof that (i) ⇔ (ii). The *Euler–Lagrange equations (3.2.2) are equivalent to Hamilton's principle of stationary action.*

To simplify notation, we momentarily suppress the particle index i.

We need to prove that the solutions of (3.2.2) are critical points $\delta S = 0$ of the *action functional*

$$S[\mathbf{q}(\cdot)] := \int_a^b L(\mathbf{q}(t), \dot{\mathbf{q}}(t)) \, dt, \tag{3.2.9}$$

(where $\dot{\mathbf{q}} = d\mathbf{q}(t)/dt$) with respect to variations on $C^\infty([a, b], \mathbb{R}^{3N})$, the space of smooth trajectories $\mathbf{q}: [a, b] \to \mathbb{R}^{3N}$ with fixed endpoints $\mathbf{q}_a, \mathbf{q}_b$.

In $C^\infty([a, b], \mathbb{R}^{3N})$, consider a *deformation* $\mathbf{q}(t, s)$, $s \in (-\epsilon, \epsilon)$, $\epsilon > 0$, with fixed endpoints $\mathbf{q}_a, \mathbf{q}_b$, of a curve $\mathbf{q}_0(t)$, that is, $\mathbf{q}(t, 0) = \mathbf{q}_0(t)$ for all $t \in [a, b]$ and $\mathbf{q}(a, s) = \mathbf{q}_0(a) = \mathbf{q}_a$, $\mathbf{q}(b, s) = \mathbf{q}_0(b) = \mathbf{q}_b$ for all $s \in (-\epsilon, \epsilon)$.

Define a *variation* of the curve $\mathbf{q}_0(\cdot)$ in $C^\infty([a, b], \mathbb{R}^{3N})$ by

$$\delta \mathbf{q}(\cdot) := \frac{d}{ds}\bigg|_{s=0} \mathbf{q}(\cdot, s) \in T_{\mathbf{q}_0(\cdot)} C^\infty([a, b], \mathbb{R}^{3N}),$$

and define the *first variation* of S at $\mathbf{q}_0(t)$ to be the derivative

$$\delta S := \mathbf{D}S[\mathbf{q}_0(\cdot)](\delta\mathbf{q}(\cdot)) := \frac{d}{ds}\bigg|_{s=0} S[\mathbf{q}(\cdot, s)]. \tag{3.2.10}$$

Note that $\delta\mathbf{q}(a) = \delta\mathbf{q}(b) = \mathbf{0}$. With these notations, *Hamilton's principle of stationary action* states that the curve $\mathbf{q}_0(t)$ satisfies the Euler–Lagrange equations (3.2.2) if and only if $\mathbf{q}_0(\cdot)$ is a critical point of the action functional, that is, $\mathbf{D}S[\mathbf{q}_0(\cdot)] = 0$. Indeed, using the equality of mixed partials, integrating by parts and taking into account that $\delta\mathbf{q}(a) = \delta\mathbf{q}(b) = 0$, one finds

$$\delta S := \mathbf{D}S[\mathbf{q}_0(\cdot)](\delta\mathbf{q}(\cdot))$$

$$= \frac{d}{ds}\bigg|_{s=0} S[\mathbf{q}(\cdot, s)] = \frac{d}{ds}\bigg|_{s=0} \int_a^b L(\mathbf{q}(t, s), \dot{\mathbf{q}}(t, s))\, dt$$

$$= \sum_{i=1}^N \int_a^b \left[\frac{\partial L}{\partial \mathbf{q}_i} \cdot \delta\mathbf{q}_i(t, s) + \left(\frac{\partial L}{\partial \dot{\mathbf{q}}_i} \cdot \delta\dot{\mathbf{q}}_i \right)\bigg|_{\dot{\mathbf{q}}_i = \frac{d\mathbf{q}_i}{dt}} \right] dt$$

$$= -\sum_{i=1}^N \int_a^b \left[\frac{d}{dt}\left(\frac{\partial L}{\partial \dot{\mathbf{q}}_i}\bigg|_{\dot{\mathbf{q}}_i = \frac{d\mathbf{q}_i}{dt}} \right) - \frac{\partial L}{\partial \mathbf{q}_i} \right] \cdot \delta\mathbf{q}_i\, dt$$

$$+ \underbrace{\left(\frac{\partial L}{\partial \dot{\mathbf{q}}_i}\bigg|_{\dot{\mathbf{q}}_i = \frac{d\mathbf{q}_i}{dt}} \cdot \delta\mathbf{q}_i \right)\bigg|_{t=a}^{t=b}}_{\text{Noether quantity}} = 0 \tag{3.2.11}$$

for all smooth $\delta\mathbf{q}_i(t)$ satisfying $\delta\mathbf{q}_i(a) = \delta\mathbf{q}_i(b) = 0$. This proves the equivalence of (i) and (ii) upon restoring particle index i in the last two lines. Note that the Noether quantity here is written relative to fixed coordinates in the inertial frame of Newton's force law.   ∎

Definition 3.2.1 *The conjugate momenta for the Lagrangian in (3.2.3) are defined as*

$$\mathbf{p}_i := \frac{\partial L}{\partial \dot{\mathbf{q}}_i}\bigg|_{\dot{\mathbf{q}}_i = \frac{d\mathbf{q}_i}{dt}} = m_i \frac{d\mathbf{q}_i}{dt} \in \mathbb{R}^3, \quad i = 1, \dots, N, \quad \text{(no sum on i).} \tag{3.2.12}$$

Definition 3.2.2 *The **Hamiltonian** is defined via the change of variables* $(\mathbf{q}, \dot{\mathbf{q}}) \mapsto (\mathbf{q}, \mathbf{p})$, *called the **Legendre transform**, where one assumes that one may recover* $\dot{\mathbf{q}}$ *as a function of* $(\mathbf{q}, \mathbf{p})$,

$$H(\mathbf{q}, \mathbf{p}) := \mathbf{p} \cdot \dot{\mathbf{q}}(\mathbf{q}, \mathbf{p}) - L(\mathbf{q}, \dot{\mathbf{q}}(\mathbf{q}, \mathbf{p}))$$

$$= \sum_{i=1}^{N} \frac{m_i}{2} \|\dot{\mathbf{q}}_i\|^2 + V(\mathbf{q})$$

$$= \underbrace{\sum_{i=1}^{N} \frac{1}{2m_i} \|\mathbf{p}_i\|^2}_{\text{Kinetic energy}} + \underbrace{V(\mathbf{q})}_{\text{Potential}} . \qquad (3.2.13)$$

Remark 3.2.1 The value of the Hamiltonian coincides with the *total energy* of the system. This value will be shown to remain constant under the evolution of the Euler–Lagrange equations (3.2.2). □

Remark 3.2.2 The Hamiltonian H may be obtained from the Legendre transformation as a function of the variables $(\mathbf{q}, \mathbf{p})$, provided one may solve for $\dot{\mathbf{q}}(\mathbf{q}, \mathbf{p})$. Solving for $\dot{\mathbf{q}}(\mathbf{q}, \mathbf{p})$ requires the Lagrangian to be *regular* . In particular, it requires

$$\det \frac{\partial^2 L}{\partial \dot{\mathbf{q}}_i \partial \dot{\mathbf{q}}_i} \neq 0 \quad \text{(no sum on } i\text{)}. \qquad □$$

Step III: Proof that (2) $\Leftrightarrow$ **(3).** (Hamilton's principle of stationary action is equivalent to Hamilton's canonical equations.) Lagrangian (3.2.3) is regular, and the derivatives of the Hamiltonian may be shown to satisfy

$$\frac{\partial H}{\partial \mathbf{p}_i} = \frac{1}{m_i} \mathbf{p}_i = \dot{\mathbf{q}}_i = \frac{d\mathbf{q}_i}{dt} \quad \text{and} \quad \frac{\partial H}{\partial \mathbf{q}_i} = \frac{\partial V}{\partial \mathbf{q}_i} = -\frac{\partial L}{\partial \mathbf{q}_i}.$$

Consequently, the Euler–Lagrange equations (3.2.2) imply

$$\dot{\mathbf{p}}_i = \frac{d\mathbf{p}_i}{dt} = \frac{d}{dt} \left(\frac{\partial L}{\partial \dot{\mathbf{q}}_i} \right) = \frac{\partial L}{\partial \mathbf{q}_i} = -\frac{\partial H}{\partial \mathbf{q}_i}.$$

These calculations show that the Euler–Lagrange equations (3.2.2) are equivalent to Hamilton's canonical equations:

$$\dot{\mathbf{q}}_i = \frac{\partial H}{\partial \mathbf{p}_i}, \quad \dot{\mathbf{p}}_i = -\frac{\partial H}{\partial \mathbf{q}_i}, \tag{3.2.14}$$

where $\partial H/\partial \mathbf{q}_i, \partial H/\partial \mathbf{p}_i \in \mathbb{R}^3$ are the gradients of H with respect to $\mathbf{q}_i, \mathbf{p}_i \in \mathbb{R}^3$, respectively. This proves the equivalence of (ii) and (iii).
∎

Remark 3.2.3 The *Noether quantity* in (3.2.11) may be expressed in terms of the canonically conjugate momentum **p** as (dropping the subscript i)

$$\left(\frac{\partial L}{\partial \dot{\mathbf{q}}}\Big|_{\dot{\mathbf{q}}=\frac{d\mathbf{q}}{dt}} \cdot \delta\mathbf{q}\right)\Big|_{t=a}^{t=b} = \left(\mathbf{p}\cdot\delta\mathbf{q}\right)\Big|_{t=a}^{t=b}. \qquad \square$$

Theorem 3.2.2 *Noether's theorem and tangent-lift momentum maps. In coordinate notation* $(\dot{q}, q) \in TQ$, *consider a Lagrangian* $L : TQ \to \mathbb{R}$ *which is right invariant under the tangent lift of the Lie group G. The tangent lift of the right action of G on the fixed configuration manifold Q to its tangent space TQ induces an equivariant momentum map $J : T^*Q \to \mathfrak{g}^*$ given by the endpoint term in Noether's theorem as*

$$\langle J(v_q), \xi\rangle_{\mathfrak{g}^*\times\mathfrak{g}} = \langle \mathbb{F}L(v_q), \xi_Q(q)\rangle_{T^*Q\times TQ},$$

where $\xi \in \mathfrak{g}$, $v_q \in T_qQ$, $\mathbb{F}L(v_q) := \partial L/\partial v_q \in T^*Q$ *is the fibre derivative of the Lagrangian L and the variation $\delta q = \xi_Q(q) \in T_qQ$ is taken to be the right G-action on the configuration manifold Q. This infinitesimal action is explicitly given by the positive Lie derivative,*

$$\xi_Q(q) = \frac{d}{ds}\Big|_{s=0} g_s q = \pounds_\xi q \quad \text{for } g_s \in G \quad \text{and}$$

$$\xi \in T_e G \quad \text{for subscript } e \text{ at } s = 0.$$

Consequently, one may explicitly write

$$\langle J(v_q), \xi\rangle_{\mathfrak{g}^*\times\mathfrak{g}} = \left\langle \frac{\partial L}{\partial v_q}, \pounds_\xi q\right\rangle =: \left\langle q \diamond \frac{\partial L}{\partial v_q}, \xi\right\rangle_{\mathfrak{g}^*\times\mathfrak{g}},$$

in which the last equality defines the diamond operation $\diamond : T_q^*Q \times T_qQ \to \mathfrak{g}^*$. We may now write the tangent-lift momentum map $J : T^*Q \to \mathfrak{g}^*$ for the right action of G on the fixed configuration manifold Q in terms of $(\diamond)$ as

$$J(v_q) = q \diamond \frac{\partial L}{\partial v_q}.$$

Hence, the infinitesimal action leads directly to the corresponding momentum map.

This theorem may be demonstrated in the familiar example of the angular momentum map which arises when the particle Lagrangian $L(q, \dot{q})$ for $(q, \dot{q}) \in T\mathbb{R}^3$ is invariant under rotations by the Lie group $SO(3)$.

Example 3.2.1 When the Lagrangian $L(q, \dot{q})$ for $q \in \mathbb{R}^3$ is right invariant under infinitesimal rotations about a certain direction $\boldsymbol{\xi} \in \mathbb{R}^3$, say, then the variation may be written as $\delta\mathbf{q} = \boldsymbol{\xi} \times \mathbf{q}$. This variation does not vanish at the endpoints in time. Instead, as Noether described, it is constant in time and yields the following conservation law in this case upon substituting $\mathbb{F}L(v_q) := p_q$:

$$\mathbf{p} \cdot \delta\mathbf{q} = \mathbf{p} \cdot \boldsymbol{\xi} \times \mathbf{q} = \mathbf{q} \times \mathbf{p} \cdot \boldsymbol{\xi} =: \mathbf{J}(\mathbf{q}, \mathbf{p}) \cdot \boldsymbol{\xi} =: J^\xi(\mathbf{q}, \mathbf{p}).$$

This formula implies conservation of the $\boldsymbol{\xi}$-component of the particle angular momentum

$$\mathbf{J}(\mathbf{q}, \mathbf{p}) := \mathbf{q} \times \mathbf{p}.$$

Thus, according to Noether's theorem, invariance of the Lagrangian under infinitesimal rotations around the fixed vector $\boldsymbol{\xi} \in \mathbb{R}^3$ implies conservation of angular momentum in the direction of the rotation axis of symmetry, $\boldsymbol{\xi}$. Quantities arising from the endpoint term for Noether symmetry this way are called *cotangent-lift momentum maps*. The reason for that name can be explained by treating $\mathbf{J}(\mathbf{q}, \mathbf{p}) \cdot \boldsymbol{\xi}$ as a Hamiltonian and calculating its canonical Poisson brackets with $\mathbf{q}$ and $\mathbf{p}$. Namely,

$$\delta\mathbf{q} = \{\mathbf{q}, J^\xi\} = \frac{\partial J^\xi}{\partial \mathbf{p}} = \boldsymbol{\xi} \times \mathbf{q},$$

$$\delta\mathbf{p} = \{\mathbf{p}, J^\xi\} = -\frac{\partial J^\xi}{\partial \mathbf{q}} = \boldsymbol{\xi} \times \mathbf{p}.$$

(3.2.15)

These equations imply that the infinitesimal right action $\delta\mathbf{q}$ to rotate the position variable $\mathbf{q}(t)$ in the fixed inertial frame of the configuration manifold Q has been "lifted" to the corresponding action $\delta\mathbf{p}$ on the canonically conjugate momentum $\mathbf{p}(t)$ in the phase space $T^*\mathbb{R}^3$ via the Poisson bracket operations with the momentum-map Hamiltonian. Moreover, this Hamiltonian is precisely Noether's endpoint term, $\mathbf{p} \cdot \delta\mathbf{q}$. Thus, the Noether endpoint quantity has produced the cotangent-lift momentum map for the right action of the Lie group $SO(3)$ on $Q = \mathbb{R}^3$.

Remark 3.2.4 The Euler–Lagrange equations for a simple mechanical system are second order, and they determine *curves in configuration space* $\mathbf{q}_i \in C^\infty([a, b], \mathbb{R}^{3N})$. In contrast, Hamilton's equations are first order, and they determine *curves in phase space* $(\mathbf{q}_i, \mathbf{p}_i) \in C^\infty([a, b], \mathbb{R}^{6N})$, a space whose dimension is twice the dimension of the configuration space. $\square$

Step IV: Proof that (iii) $\Leftrightarrow$ (iv). (Hamilton's canonical equations may be written using a Poisson bracket.)

By the chain rule and Hamilton's canonical equations in (3.2.14), any $F \in \mathcal{F}(P)$ satisfies

$$\frac{dF}{dt} = \sum_{i=1}^{N} \left(\frac{\partial F}{\partial \mathbf{q}_i} \cdot \dot{\mathbf{q}}_i + \frac{\partial F}{\partial \mathbf{p}_i} \cdot \dot{\mathbf{p}}_i \right)$$

$$= \sum_{i=1}^{N} \left(\frac{\partial F}{\partial \mathbf{q}_i} \cdot \frac{\partial H}{\partial \mathbf{p}_i} - \frac{\partial F}{\partial \mathbf{p}_i} \cdot \frac{\partial H}{\partial \mathbf{q}_i} \right) = \{F, H\}.$$

This completes the proof of the theorem by proving the equivalence of (iii) and (iv). $\blacksquare$

Remark 3.2.5 (Energy conservation) Since the Poisson bracket is skew symmetric, $\{H, F\} = -\{F, H\}$, one finds that $\dot{H} = \{H, H\} = 0$. Consequently, the value of the Hamiltonian is preserved by the evolution. Thus, the Hamiltonian is said to be a *constant of the motion*. $\square$

Exercise. Show that the canonical Poisson bracket is bilinear, skew symmetric, satisfies the Jacobi identity and acts as a derivation on products of functions in phase space.   ★

Exercise. Given two constants of motion, what does the Jacobi identity imply about additional constants of motion?   ★

Exercise. Compute the Poisson brackets among

$$J_i = \epsilon_{ijk} q_j p_k = (\widehat{q} p)_i = (\mathbf{q} \times \mathbf{p})_i$$

in Euclidean space. What Lie algebra do these Poisson brackets do you recall?   ★

Exercise. Verify that Hamilton's equations determined by the function

$$\langle J(\mathbf{p}, \mathbf{q}), \xi \rangle = (\mathbf{q} \times \mathbf{p}) \cdot \xi$$

define infinitesimal rotations about the ξ-axis. Thus, in the case of rotation, the momentum map $J(\mathbf{p}, \mathbf{q})$ is the *angular momentum*.   ★

Exercise. (Rotations and angular momentum) Find the Hamiltonian vector field for

$$J^\xi = \boldsymbol{\xi} \cdot (\mathbf{q} \times \mathbf{p})$$

for $\boldsymbol{\xi} \in \mathbb{R}^3$ and $(\mathbf{q}, \mathbf{p}) \in T^*\mathbb{R}^3 \simeq \mathbb{R}^3 \times \mathbb{R}^3$.

Interpret this vector field geometrically. ★

Answer. (Notations and angular momentum) The Hamiltonian vector field X_{J^ξ} for $J^\xi = \boldsymbol{\xi} \cdot \mathbf{J} = \boldsymbol{\xi} \cdot (\mathbf{q} \times \mathbf{p})$ is obtained from the canonical Poisson bracket by setting

$$X_{J^\xi} = \{\cdot, J^\xi\} = \{q_k, J^\xi\}\frac{\partial}{\partial q_k} + \{p_k, J^\xi\}\frac{\partial}{\partial p_k}.$$

The Poisson brackets among the components of $\mathbf{J}$ may be computed in two stages.

First, use canonical Poisson brackets $\{q_j, p_k\} = \delta_{jk}$ with the definition $J_l = \epsilon_{lmn}q_m p_n$ for $l, m, n = 1, 2, 3$ to find

$$\widehat{q}_{kl} := \{q_k, J_l\} = \epsilon_{lmn}q_m\{q_k, p_n\} = \epsilon_{lmn}q_m\delta_{kn}$$

$$= \epsilon_{lmk}q_m = \epsilon_{klm}q_m.$$

The skew symmetry of $\widehat{q}_{kl} = -\widehat{q}_{lk}$ arises from the skew symmetry of the Poisson bracket.

A similar calculation of the Poisson bracket for the components of the canonically conjugate momentum $\{\mathbf{p}, \mathbf{J}\}$ gives

$$\widehat{p}_{kl} := \{p_k, J_l\} = \epsilon_{klm}p_m.$$

Next, calculate

$$\{q_k, \boldsymbol{\xi} \cdot \mathbf{J}\} = \{q_k, \xi_l J_l\} = \xi_l\epsilon_{lmn}q_m\{q_k, p_n\}$$

$$= \epsilon_{klm}\xi_l q_m = (\boldsymbol{\xi} \times \mathbf{q})_k.$$

Likewise for the conjugate momentum vector **p**, so that

$$\{\mathbf{q}, \boldsymbol{\xi} \cdot \mathbf{J}\} = \boldsymbol{\xi} \times \mathbf{q} \quad \text{and} \quad \{\mathbf{p}, \boldsymbol{\xi} \cdot \mathbf{J}\} = \boldsymbol{\xi} \times \mathbf{p}.$$

The Hamiltonian vector field associated with $J^{\xi} = \boldsymbol{\xi} \cdot \mathbf{J}$ may now be expressed as

$$X_{J^{\xi}} = \{\cdot, J^{\xi}\} = \{\mathbf{q}, J^{\xi}\} \cdot \frac{\partial}{\partial \mathbf{q}} + \{\mathbf{p}, J^{\xi}\} \cdot \frac{\partial}{\partial \mathbf{p}}$$

$$= (\boldsymbol{\xi} \times \mathbf{q}) \cdot \frac{\partial}{\partial \mathbf{q}} + (\boldsymbol{\xi} \times \mathbf{p}) \cdot \frac{\partial}{\partial \mathbf{p}}. \qquad \blacktriangle$$

3.3   Geometrical interpretation

The flow of the vector field $X_{J^{\xi}}$ is a rotation with angular velocity $\boldsymbol{\xi}$. As might be expected,

$$\{|\mathbf{q}|^2, J^{\xi}\} = \{|\mathbf{p}|^2, J^{\xi}\} = 0 = \{\mathbf{q} \cdot \mathbf{p}, J^{\xi}\}$$

because rotations leave invariant the lengths and relative orientations of vectors in $\mathbb{R}^3$. If $\boldsymbol{\xi}$ is a unit vector, the quantity $J^{\xi} = \boldsymbol{\xi} \cdot \mathbf{J}$ is the angular momentum in the direction of $\boldsymbol{\xi}$. Rotationally invariant Hamiltonians Poisson-commute with J^{ξ}. Therefore, rotationally invariant Hamiltonians conserve all three components of the angular momentum, $\mathbf{J} := \mathbf{q} \times \mathbf{p}$.

Remark 3.3.1 (Hat map) The table of Poisson brackets $\widehat{q}_{kl} = \{q_k, J_l\}$ produces the linear invertible map

$$\mathbf{q} := (q_1, q_2, q_3) \in \mathbb{R}^3 \quad \mapsto \quad \{\mathbf{q}, \mathbf{J}\} = \widehat{\mathbf{q}} := \begin{bmatrix} 0 & -q_3 & q_2 \\ q_3 & 0 & -q_1 \\ -q_2 & q_1 & 0 \end{bmatrix}.$$

This *hat map* $\widehat{q}_{ij} = -\epsilon_{ijk} q_k$ is a remarkable isomorphism because the skew-symmetric 3×3 matrices comprise the matrix Lie algebra $\mathfrak{so}(3)$ for the matrix Lie group $SO(3)$ of rotations by right action in three dimensions. $\qquad \square$

Exercise. (Angular momentum Poisson brackets) Compute the Poisson brackets among

$$J_l = \epsilon_{lmn} q_m p_n \quad \text{for} \quad l, m, n = 1, 2, 3,$$

using the canonical Poisson brackets $\{q_k, p_m\} = \delta_{km}$.

Do the Poisson brackets $\{J_l, J_m\}$ close among themselves? If so, what does this imply for the restriction of the dynamics to functions of (J_1, J_2, J_3)?

Compute the Hamiltonian vector field for $J^\xi = \boldsymbol{\xi} \cdot \mathbf{J}$ in terms of $\mathbf{J}$. ★

Answer. (Angular momentum Poisson brackets) The components of the angular momentum $J_l = \epsilon_{lmn} q_m p_n$ do Poisson-commute among themselves. In particular,

$$\widehat{J}_{kl} := \{J_k, J_l\} = \epsilon_{klm} J_m.$$

Proof. The Poisson brackets are computed as

$$\{\mathbf{q} \times \mathbf{p}, \boldsymbol{\xi} \cdot \mathbf{J}\} = \{\mathbf{q}, \boldsymbol{\xi} \cdot \mathbf{J}\} \times \mathbf{p} + \mathbf{q} \times \{\mathbf{p}, \boldsymbol{\xi} \cdot \mathbf{J}\}$$
$$= (\boldsymbol{\xi} \times \mathbf{q}) \times \mathbf{p} + \mathbf{q} \times (\boldsymbol{\xi} \times \mathbf{p})$$
$$= \boldsymbol{\xi} \times (\mathbf{q} \times \mathbf{p}).$$

Consequently,

$$\{\mathbf{J}, \boldsymbol{\xi} \cdot \mathbf{J}\} = \boldsymbol{\xi} \times \mathbf{J},$$

and the result follows for $\widehat{J}_{kl}$. In addition, the Hamiltonian vector field X_{J^ξ} becomes

$$X_{J^\xi} = \{\cdot, J^\xi\} = \{\mathbf{J}, J^\xi\} \cdot \frac{\partial}{\partial \mathbf{J}} = -(\boldsymbol{\xi} \times \mathbf{J}) \cdot \frac{\partial}{\partial \mathbf{J}}.$$

Likewise, for any Hamiltonian $H(\mathbf{J})$ depending only on the angular momentum, one finds by a similar calculation that

$$\dot{\mathbf{J}} = \{\mathbf{J}, H\} = -\mathbf{J} \times \frac{\partial H}{\partial \mathbf{J}}.$$

Thus, by the *product rule*, any function of the angular momentum $F(\mathbf{J})$ satisfies the Poisson bracket relation

$$\frac{dF}{dt} = \{F, H\}(\mathbf{J}) = \mathbf{J} \cdot \frac{\partial F}{\partial \mathbf{J}} \times \frac{\partial H}{\partial \mathbf{J}} \qquad (LPB).$$

▲

Remark 3.3.2 Any function $C(|\mathbf{J}|)$ of the magnitude $|\mathbf{J}|$ satisfies

$$\{C, H\}(\mathbf{J}) = C'(|\mathbf{J}|)\, \mathbf{J} \cdot \frac{\partial |\mathbf{J}|}{\partial \mathbf{J}} \times \frac{\partial H}{\partial \mathbf{J}}$$

$$= C'(|\mathbf{J}|)\, \mathbf{J} \cdot \frac{\mathbf{J}}{|\mathbf{J}|} \times \frac{\partial H}{\partial \mathbf{J}} = 0, \quad \forall H.$$

Functions $C(|\mathbf{J}|)$ that are distinguished by Poisson-commuting with all others under the Poisson bracket are called *Casimirs*.   □

Remark 3.3.3 The map $J^{\xi} = \boldsymbol{\xi} \cdot (\mathbf{q} \times \mathbf{p})$ is an example of a *Poisson map*. This means the Lie–Poisson bracket in (LPB) satisfies the defining properties of a Poisson bracket, including the Jacobi identity. This may be recognised by computing the Jacobi identity for the corresponding Hamiltonian vector fields $X_{\mathbf{J}} = \{\cdot, \mathbf{J}\} = -\mathbf{J} \times \frac{\partial}{\partial \mathbf{J}}$. Another proof is to note that the Lie–Poisson bracket $\{J_a, J_b\} = \epsilon_{abc} J_c$ may be expressed as a linear functional on a Lie algebra with basis elements e_a and dual basis elements e^d as

$$\{F, H\}(\mathbf{J}) = \mathbf{J} \cdot \frac{\partial F}{\partial \mathbf{J}} \times \frac{\partial H}{\partial \mathbf{J}}$$

$$= \left\langle J_d\, e^d, \frac{\partial F}{\partial J_a} \Big[e_a, e_b \Big] \frac{\partial H}{\partial J_b} \right\rangle$$

with pairing $\langle e^d, e_c \rangle = \delta_c^d$.

In this case, the Lie algebra bracket $[\cdot, \cdot]$ with structure constants $C^c_{ab} = \epsilon_{abc}$ in

$$[e_a, e_b] = C^c_{ab} e_c$$

corresponds to $\mathfrak{so}(3)$, the Lie algebra for the rotation group in three dimensions.   □

Exercise. Write hat maps for all three-dimensional matrix Lie algebras.   ★

3.4   Phase-space action principle

Hamilton's principle for stationary variations $\delta S = 0$ of the action integral $S = \int_{t_a}^{t_b} L(q, \dot{q}) dt$ on the tangent space TM of a manifold M may be augmented for clarity by imposing the relation $\dot{q} = dq/dt$ as an *additional constraint* in terms of generalised coordinates $(q, \dot{q}) \in T_q M$. The *Hamilton–Pontryagin constrained action* is given by

$$S = \int_{t_a}^{t_b} L(q, \dot{q}) + \left\langle p, \left(\frac{dq}{dt} - \dot{q} \right) \right\rangle dt, \qquad (3.4.1)$$

where p is a *Lagrange multiplier* for the *constraint*, $\dot{q} = dq/dt$. The variations of this action result in

$$\delta S = \int_{t_a}^{t_b} \left\langle \left(\frac{\partial L}{\partial q} - \frac{dp}{dt} \right), \delta q \right\rangle + \left\langle \left(\frac{\partial L}{\partial \dot{q}} - p \right), \delta \dot{q} \right\rangle$$
$$+ \left\langle \left(\frac{dq}{dt} - \dot{q} \right), \delta p \right\rangle dt + \langle p, \delta q \rangle \Big|_{t_a}^{t_b}. \qquad (3.4.2)$$

The contributions at the endpoints t_a and t_b in time vanish, provided the variations δq are assumed to vanish then.

Thus, stationarity of this action under these variations imposes the three relations

$$\delta q : \quad \frac{\partial L}{\partial q} = \frac{dp}{dt},$$

$$\delta \dot{q} : \quad \frac{\partial L}{\partial \dot{q}} = p, \tag{3.4.3}$$

$$\delta p : \quad \dot{q} = \frac{dq}{dt}.$$

- Combining the first and second of these relations recovers the Euler–Lagrange equations: $[L]_q = 0$.

- The third relation constrains the variable $\dot{q}$ to be the time derivative of the trajectory $q(t)$ at any time t after the variations have been taken.

Substituting the Legendre-transform relation in (3.4.3) into the constrained action (3.4.1) yields the *phase-space action*

$$S = \int_{t_a}^{t_b} \left(p \frac{dq}{dt} - H(q, p) \right) dt. \tag{3.4.4}$$

Varying the phase-space action in (3.4.4) yields

$$\delta S = \int_{t_a}^{t_b} \left(\frac{dq}{dt} - \frac{\partial H}{\partial p} \right) \delta p - \left(\frac{dp}{dt} + \frac{\partial H}{\partial q} \right) \delta q \, dt + \left[p \, \delta q \right]_{t_a}^{t_b}.$$

Provided the variations δq vanish at the endpoints t_a and t_b in time, then the last term vanishes. Thus, stationary variations of the phase-space action in (3.4.4) recover Hamilton's canonical equations (3.2.14) directly.

Hamiltonian evolution along a curve $(q(t), p(t)) \in T^*M$ satisfying equations (3.2.14) induces the evolution of a given function $F(q, p) : T^*M \to \mathbb{R}$ on the phase space T^*M of a manifold M as

$$\frac{dF}{dt} = \frac{\partial F}{\partial q} \frac{dq}{dt} + \frac{\partial H}{\partial q} \frac{dp}{dt}$$

$$= \frac{\partial F}{\partial q} \frac{\partial H}{\partial p} - \frac{\partial H}{\partial q} \frac{\partial F}{\partial p} =: \{F, H\} \tag{3.4.5}$$

$$= \left(\frac{\partial H}{\partial p} \frac{\partial}{\partial q} - \frac{\partial H}{\partial q} \frac{\partial}{\partial p} \right) F =: X_H F. \qquad (3.4.6)$$

The second and third lines of this calculation introduce notation for two natural operations that will be investigated further in the following few lectures. These are the *canonical Poisson brackets* $\{ \cdot, \cdot \}$ in (3.4.5) and the *Hamiltonian vector field* $X_H = \{ \cdot, H \}$ in (3.4.6).

4

MATRIX LIE GROUPS AND LIE ALGEBRAS

Contents

> **What is this lecture about?** This lecture lays out the definitions and basic properties of matrix Lie groups and Lie algebras, as well as their actions on finite-dimensional manifolds.

4.1 Matrix Lie groups

Definition 4.1.1 *A **group** is a set of elements with:*

1. *a binary product (multiplication), $G \times G \to G$, such that*

    a. *the product of g and h is written gh, and*

    b. *the product is associative: $(gh)k = g(hk)$;*

2. *an identity element e such that $eg = g$ and $ge = g$, $\forall g \in G$;*

3. *an inverse operation $G \to G$, such that $gg^{-1} = g^{-1}g = e$.*

Definition 4.1.2 *A **Lie group** is a smooth manifold G which is a group and for which the group operations of multiplication, $(g, h) \to gh$ for $g, h \in G$, and inversion, $g \to g^{-1}$ with $gg^{-1} = g^{-1}g = e$, are smooth.*

Definition 4.1.3 *A **matrix Lie group** is a set of invertible $n \times n$ matrices which is closed under matrix multiplication and which is a submanifold of $\mathbb{R}^{n \times n}$. The conditions showing that a matrix Lie group is a Lie group are easily checked:*

1. *A matrix Lie group is a manifold because it is a submanifold of $\mathbb{R}^{n \times n}$.*

2. *Its group operations are smooth since they are algebraic operations on the matrix entries.*

Example 4.1.1 (The general linear group $GL(n, \mathbb{R})$) The matrix Lie group $GL(n, \mathbb{R})$ is the group of linear isomorphisms of $\mathbb{R}^n$ to itself. The dimension of the matrices in $GL(n, \mathbb{R})$ is n^2.

Proposition 4.1.1 *Let $K \in GL(n, \mathbb{R})$ be a symmetric matrix, $K^T = K$. Then, the subgroup S of $GL(n, \mathbb{R})$ defined by the mapping*

$$S = \{U \in GL(n, \mathbb{R}) | U^T K U = K\}$$

is a submanifold of $\mathbb{R}^{n \times n}$ of dimension $n(n-1)/2$.

Remark 4.1.1 The subgroup S leaves invariant a certain symmetric quadratic form under linear transformations, $S \times \mathbb{R}^n \to \mathbb{R}^n$, given by $\mathbf{x} \to U\mathbf{x}$, since

$$\mathbf{x}^T K \mathbf{x} = \mathbf{x}^T U^T K U \mathbf{x}.$$

So, the matrices $U \in S$ change the basis for this quadratic form, but they leave its value unchanged. Thus, S is the *isotropy subgroup* of the quadratic form associated with K. □

Proof. Is S a subgroup? We check the following three defining properties:

1. Identity: $I \in S$ because $I^T K I = K$.

2. Inverse: $U \in S \implies U^{-1} \in S$ because

$$K = U^{-T}(U^T K U)U^{-1} = U^{-T}(K)U^{-1}.$$

3. Closed under multiplication: $U, V \in S \implies UV \in S$ because

$$(UV)^T K U V = V^T(U^T K U)V = V^T(K)V = K.$$

Hence, S is a subgroup of $GL(n, \mathbb{R})$.

Is S a submanifold of $\mathbb{R}^{n \times n}$ of dimension $n(n-1)/2$?

Indeed, S is the zero locus of the mapping $UKU^T - K$. This makes it a submanifold because it turns out to be a submersion.

For a submersion, the dimension of the level set is the dimension of the domain minus the dimension of the range space. In this case, this dimension is $n^2 - n(n+1)/2 = n(n-1)/2$.   ∎

Exercise. Explain why one can conclude that the zero locus map for S is a submersion. In particular, pay close attention to establishing the constant rank condition for the linearisation of this map.   ★

Answer. Here is why S is a submanifold of $R^{n \times n}$.

First, S is the zero locus of the mapping

$$U \to U^T KU - K \quad \text{(locus map)}.$$

Let $U \in S$, and let δU be an arbitrary element of $R^{n \times n}$. Then, linearise to find

$$(U + \delta U)^T K(U + \delta U) - K$$
$$= U^T KU - K + \delta U^T KU + U^T K \delta U + O(\delta U)^2.$$

We may conclude that S is a submanifold of $R^{n \times n}$ if we can show that the linearisation of the locus map, namely the linear mapping defined by

$$L \equiv \delta U \to \delta U^T KU + U^T K \delta U, \quad R^{n \times n} \to R^{n \times n},$$

has constant rank for all $U \in S$.   ▲

Lemma 4.1.1 *The linearisation map L is onto the space of $n \times n$ of symmetric matrices, and hence the original map is a submersion. That is, the original map is a surjective differentiable map between two manifolds.*

Proof that L is onto.

Both the original locus map and the image of L lie in the subspace of $n \times n$ symmetric matrices.

Indeed, given U and any symmetric matrix S, we can find δU such that

$$\delta U^T K U + U^T K \delta U = S.$$

Namely,

$$\delta U = K^{-1} U^{-T} S / 2.$$

Thus, the linearisation map L is onto the space of $n \times n$ of symmetric matrices, and the original locus map $U \to U K U^T - K$ to the space of symmetric matrices is a submersion. ∎

For a submersion S, the dimension of the level set is the dimension of the domain minus the dimension of the range space. Here, this dimension is $n^2 - n(n+1)/2 = n(n-1)/2$.

Corollary 4.1.1 (S is a matrix Lie group) *S is both a subgroup and a submanifold of the general linear group $GL(n, \mathbb{R})$. Thus, by Definition 4.1.3, S is a matrix Lie group.*

Exercise. What is the tangent space to S at the identity $T_I S$? ★

Exercise. Show that for any pair of matrices $A, B \in T_I S$, the matrix commutator $[A, B] \equiv AB - BA \in T_I S$. ★

Proposition 4.1.2 *The linear space of matrices A satisfying*

$$A^T K + K A = 0$$

defines $T_I S$, the tangent space at the identity of the matrix Lie group S defined in Definition 4.1.1.

Proof. Near the identity, the defining condition for S expands to

$$(I + \epsilon A^T + O(\epsilon^2))K(I + \epsilon A + O(\epsilon^2)) = K, \quad \text{for } \epsilon \ll 1.$$

At linear order $O(\epsilon)$, one finds

$$A^T K + K A = 0.$$

This relation defines the linear space of matrices $A \in T_I S$. ∎

If $A, B \in T_I S$, does it follow that $[A, B] \in T_I S$?

Using $[A, B]^T = [B^T, A^T]$, we check *closure* through direct computation:

$$[B^T, A^T]K + K[A, B]$$
$$= B^T A^T K - A^T B^T K + KAB - KBA$$
$$= B^T A^T K - A^T B^T K - A^T KB + B^T KA = 0.$$

Hence, the tangent space of S at the identity $T_I S$ is closed under the matrix commutator $[\cdot, \cdot]$.

Remark 4.1.2 In a moment, we will show that the matrix commutator for $T_I S$ also satisfies the Jacobi identity. This will imply that the condition $A^T K + K A = 0$ defines a matrix Lie algebra. □

4.2 Defining matrix Lie algebras

We are ready to prove the following proposition, in preparation for defining matrix Lie algebras.

Proposition 4.2.1 *Let S be a matrix Lie group, and let $A, B \in T_I S$ (the tangent space to S at the identity element). Then, $AB - BA \in T_I S$.*

The proof makes use of a lemma.

Lemma 4.2.1 *Let R be an arbitrary element of a matrix Lie group S, and let $B \in T_I S$. Then, $RBR^{-1} \in T_I S$.*

Proof. Let $R_B(t)$ be a curve in S such that $R_B(0) = I$ and $R'(0) = B$. Define $S(t) = RR_B(t)R^{-1} \in T_I S$ for all t. Then, $S(0) = I$ and $S'(0) = RBR^{-1}$. Hence, $S'(0) \in T_I S$, thereby proving the lemma. ∎

Proof of 4.2.1. Let $R_A(s)$ be a curve in S such that $R_A(0) = I$ and $R'_A(0) = A$. Define $S(t) = R_A(t)BR_A(t)^{-1} \in T_I S$. Then, the lemma implies that $S(t) \in T_I S$ for every t. Hence, $S'(t) \in T_I S$, and in particular, $S'(0) = AB - BA \in T_I S$. ∎

Definition 4.2.1 (Matrix commutator) *For any pair of $n \times n$ matrices A, B, the **matrix commutator** is defined as $[A, B] = AB - BA$.*

Proposition 4.2.2 (Properties of the matrix commutator) *The matrix commutator has the following two properties:*

1. *Any two $n \times n$ matrices A and B satisfy*

$$[B, A] = -[A, B].$$

(This is the property of skew-symmetry.)

2. *Any three $n \times n$ matrices A, B and C satisfy*

$$[[A, B], C] + [[B, C], A] + [[C, A], B] = 0.$$

(This is known as the Jacobi identity.)

Definition 4.2.2 (Matrix Lie algebra) *A matrix Lie algebra $\mathfrak{g}$ is a set of $n \times n$ matrices which is a vector space with respect to the usual operations of matrix addition and multiplication by real numbers (scalars) and which is closed under the matrix commutator $[\cdot, \cdot]$.*

Proposition 4.2.3 *For any matrix Lie group S, the tangent space at the identity $T_I S$ is a matrix Lie algebra.*

Proof. This follows by 4.2.1 and because $T_I S$ is a vector space. ∎

4.3 Examples of matrix Lie groups

Example 4.3.1 (The orthogonal group $O(n)$) The mapping condition $U^T K U = K$ in 4.1.1 specialises for $K = I$ to $U^T U = I$, which defines the orthogonal group. Thus, in this case, S specialises to $O(n)$, the group of $n \times n$ orthogonal matrices. The orthogonal group is of special interest in the dynamics of rotating rigid bodies.

Corollary 4.3.1 ($O(n)$ is a matrix Lie group) *By* 4.1.1, *the orthogonal group $O(n)$ is both a subgroup and a submanifold of the general linear group $GL(n, \mathbb{R})$. Thus, by* 4.1.3, *the orthogonal group $O(n)$ is a matrix Lie group.*

Example 4.3.2 (The special linear group $SL(n,\mathbb{R})$) The subgroup of $GL(n, \mathbb{R})$ with $\det(U) = 1$ is called $SL(n, \mathbb{R})$.

Example 4.3.3 (The special orthogonal group $SO(n)$) The special case of S with $\det(U) = 1$ and $K = I$ is called $SO(n)$. In this case, the mapping condition $U^T K U = K$ specialises to $U^T U = I$ with the extra condition of $\det(U) = 1$.

Example 4.3.4 (The tangent space of $SO(n)$ at the identity $T_I SO(n)$) The special case with $K = I$ of $T_I SO(n)$ yields

$$A^T + A = 0.$$

These are antisymmetric matrices. Lying in the tangent space at the identity of a matrix Lie group, this linear vector space forms a matrix Lie algebra.

Example 4.3.5 (The symplectic group) Suppose $n = 2l$ (that is, let n be even), and consider the nonsingular skew-symmetric matrix

$$J = \begin{bmatrix} 0 & I \\ -I & 0 \end{bmatrix},$$

where I is the $l \times l$ identity matrix. One may verify that

$$Sp(l) = \{U \in GL(2l, \mathbb{R}) | U^T JU = J\}$$

is a group. This is called the symplectic group. Reasoning as before, the matrix algebra $T_I Sp(l)$ is defined as the set of $n \times n$ matrices A satisfying $JA^T + AJ = 0$. This algebra is denoted as $\mathfrak{sp}(l)$.

Example 4.3.6 (The special Euclidean group) Consider the set of 4×4 matrices of the form

$$E(R, v) = \begin{bmatrix} R & v \\ 0 & 1 \end{bmatrix},$$

where $R \in SO(3)$ and $v \in \mathbb{R}^3$. This set of 4×4 matrices forms a *representation* of the special Euclidean group in three dimensions, denoted $SE(3)$. The special Euclidean group $SE(3)$ is of central interest in mechanics since it describes the set of rigid motions and linear coordinate transformations of three-dimensional space.

Exercise. A point P in $\mathbb{R}^3$ undergoes a rigid motion associated with $E(R_1, v_1)$ followed by a rigid motion associated with $E(R_2, v_2)$. What matrix element of $SE(3)$ is associated with the composition of these motions in the given order?   ★

Exercise. Multiply the special Euclidean matrices of $SE(3)$. Investigate their matrix commutators in their tangent space at the identity. (This is an example of a semidirect-product Lie group.)   ★

Tripos question. When does a stone at the equator of the Earth weigh the most?

Two hints: (a) assume the Earth's orbit is a circle around the Sun, and ignore the declination of the Earth's axis of rotation; (b) this is an exercise in using $SE(2)$.   ★

Exercise. Suppose the $n \times n$ matrices A and M satisfy

$$AM + MA^T = 0.$$

Show that $\exp(At)M\exp(A^T t) = M$ for all t.

Hint: $A^n M = M(-A^T)^n$. This direct calculation shows that for $A \in \mathfrak{so}(n)$ or $A \in \mathfrak{sp}(l)$, we have $\exp(At) \in SO(n)$ or $\exp(At) \in Sp(l)$, respectively.   ★

4.4 Lie group actions

The action of a Lie group G on a manifold M is a group of transformations of M associated with the elements of the group G, whose composition acting on M corresponds to group multiplication in G.

Definition 4.4.1 (Left and right actions) *Let M be a manifold and G be a Lie group. A **left action** of a Lie group G on M is a smooth mapping $\Phi \colon G \times M \to M$ such that:*

1. $\Phi(e, x) = x$ *for all $x \in M$,*

2. $\Phi(g, \Phi(h, x)) = \Phi(gh, x)$ *for all $g, h \in G$ and $x \in M$, and*

3. $\Phi(g, \cdot)$ *is a diffeomorphism on M for each $g \in G$.*

We often use the convenient notation gx for $\Phi(g, x)$ and consider the group element g acting on the point $x \in M$. The associativity condition (ii) above then simply reads $(gh)x = g(hx)$.

Similarly, one can define a *right action*, which is a map $\Psi\colon M \times G \to M$ satisfying $\Psi(x, e) = x$ and $\Psi(\Psi(x, g), h) = \Psi(x, gh)$. The convenient notation for right action is xg for $\Psi(x, g)$, the right action of a group element g on the point $x \in M$. Associativity $\Psi(\Psi(x, g), h) = \Psi(x, gh)$ can then be expressed conveniently as $(xg)h = x(gh)$.

Example 4.4.1 (Properties of Lie group actions) The action $\Phi\colon G \times M \to M$ of a group G on a manifold M is said to be:

1. *transitive* if for every $x, y \in M$, there exists a $g \in G$, such that $gx = y$;

2. *free* if it has no fixed points, that is, $\Phi_g(x) = x$ implies $g = e$; and

3. *proper* if whenever a convergent subsequence $\{x_n\}$ in M exists and the mapping $g_n x_n$ converges in M, then $\{g_n\}$ has a convergent subsequence in G.

4.4.1   Orbits

Given a group action of G on M, for a given point $x \in M$, the subset

$$\mathcal{O}(x) = \{gx \mid g \in G\} \subset M,$$

is called the *group orbit* through x. In finite dimensions, it can be shown that group orbits are always smooth (possibly immersed) manifolds. Group orbits generalise the notion of orbits of a dynamical system.

> **Exercise.** The flow of a vector field on M can be thought of as an action of $\mathbb{R}$ on M. Show that, in this case, the general notion of group orbit reduces to the familiar notion of orbit used in dynamical systems. ★

Exercise. Compute the group orbit of the action of unitary transformations $SU(2)$ on complex space $\mathbb{C}^2$. For details of the formulation, see Ref. [Ho2011b].   ★

Theorem 4.4.1 *Orbits of proper group actions are embedded submanifolds.*

This theorem is stated by Marsden and Ratiu [MaRa1994, Chapter 9], who refer in turn to Abraham and Marsden [AbMa1978] for the proof.

Example 4.4.2 (Orbits of $SO(3)$) A simple example of a group orbit is the action of $SO(3)$ on $\mathbb{R}^3$ given by matrix multiplication: the action of $A \in SO(3)$ on a point $\mathbf{x} \in \mathbb{R}^3$ is simply the product $A\mathbf{x}$. In this case, the orbit of the origin is a single point (the origin itself), while the orbit of any other point is the sphere through that point.

Example 4.4.3 (Orbits of a Lie group acting on itself) The   action of a group G on itself from either the left or the right also produces group orbits. This action sets the stage for discussing the tangent-lifted action of a Lie group on its tangent bundle.

Left and right translations on the group are denoted L_g and R_g, respectively. For example, $L_g \colon G \to G$ is the map given by $h \to gh$, while $R_g \colon G \to G$ is the map given by $h \to hg$, for $g, h \in G$.

1. *Left translation* $L_g \colon G \to G$; $h \to gh$ defines a transitive and free action of G on itself. Right multiplication $R_g \colon G \to G$; $h \to hg$ defines a right action, while $h \to hg^{-1}$ defines a left action of G on itself. Thus, a right action by the inverse is a left action.

2. G acts on G by conjugation, $g \to I_g = R_{g^{-1}} \circ L_g$. The map $I_g \colon G \to G$ given by $h \to ghg^{-1}$ is the *inner automorphism* associated with g. Orbits of this action are called *conjugacy classes*.

3. Differentiating conjugation at e gives the *adjoint action* of G on $\mathfrak{g}$:
$$\mathrm{Ad}_g := T_e I_g : T_e G = \mathfrak{g} \to T_e G = \mathfrak{g}.$$

Explicitly, the *adjoint action* of G on $\mathfrak{g}$ is given by

$$\mathrm{Ad}: G \times \mathfrak{g} \to \mathfrak{g}, \quad \mathrm{Ad}_g(\xi) = T_e(R_{g^{-1}} \circ L_g)\xi.$$

We have already seen an example of adjoint action for matrix Lie groups acting on matrix Lie algebras when we defined $S(t) = R_A(t)BR_A(t)^{-1} \in T_I S$ as a key step in the proof of Proposition 4.2.1.

4. The *co-Adjoint action* of G on $\mathfrak{g}^*$, the dual of the Lie algebra $\mathfrak{g}$ of G, is defined as follows. Let $\mathrm{Ad}_g^*: \mathfrak{g}^* \to \mathfrak{g}^*$ be the dual of Ad_g, defined by

$$\langle \mathrm{Ad}_g^* \alpha, \xi \rangle = \langle \alpha, \mathrm{Ad}_g \xi \rangle$$

for $\alpha \in \mathfrak{g}^*$, $\xi \in \mathfrak{g}$ and pairing $\langle \cdot, \cdot \rangle: \mathfrak{g}^* \times \mathfrak{g} \to \mathbb{R}$. Then, the map

$$\Phi^*: G \times \mathfrak{g}^* \to \mathfrak{g}^* \quad \text{given by} \quad (g, \alpha) \mapsto \mathrm{Ad}_{g^{-1}}^* \alpha$$

is the co-Adjoint action of G on $\mathfrak{g}^*$.

4.5   Examples: $SO(3)$, $SE(3)$, etc.

4.5.1   A basis for the matrix Lie algebra $\mathfrak{so}(3)$ and a map to $\mathbb{R}^3$

The Lie algebra of $SO(n)$ is called $\mathfrak{so}(n)$. A basis (e_1, e_2, e_3) for $\mathfrak{so}(3)$ when $n = 3$ is given by

$$\hat{\mathbf{x}} = \begin{bmatrix} 0 & -z & y \\ z & 0 & -x \\ -y & x & 0 \end{bmatrix} = xe_1 + ye_2 + ze_3.$$

Exercise. Show that $[e_1, e_2] = e_3$ and its cyclic permutations hold as well, while all other matrix commutators among the basis elements vanish.   ★

Example 4.5.1 (The isomorphism between $\mathfrak{so}(3)$ and $\mathbb{R}^3$) The previous equation may be written equivalently by defining the hat operation $\widehat{\cdot}$ as

$$\hat{x}_{ij} = -\epsilon_{ijk}x^k, \quad \text{where } (x^1, x^2, x^3) = (x, y, z).$$

Here, $\epsilon_{123} = 1$ and $\epsilon_{213} = -1$, with cyclic permutations. The totally antisymmetric tensor $\epsilon_{ijk} = -\epsilon_{jik} = -\epsilon_{ikj}$ also defines the cross product of vectors in $\mathbb{R}^3$. Consequently, we may write

$$(\mathbf{x} \times \mathbf{y})_i = \epsilon_{ijk}x^j y^k = \hat{x}_{ij}y^j, \quad \text{that is, } \mathbf{x} \times \mathbf{y} = \hat{\mathbf{x}}\mathbf{y}.$$

Exercise. What is the analogue of the hat map $\mathfrak{so}(3) \mapsto \mathbb{R}^3$ for the three-dimensional Lie algebras $\mathfrak{sp}(2, \mathbb{R})$, $\mathfrak{so}(2, 1)$, $\mathfrak{su}(1, 1)$, or $\mathfrak{sl}(2, \mathbb{R})$? ★

Background reading for this lecture is the paper by Marsden and Ratiu [MaRa1994, Chapter 9].

4.5.2 Compute the Adjoint and adjoint operations by differentiation

1. Differentiate $I_g(h)$ with respect to h at $h = e$ to produce the *Adjoint operation*

$$\text{Ad} : G \times \mathfrak{g} \to \mathfrak{g} : \quad \text{Ad}_g \eta = T_e I_g \eta.$$

2. Differentiate $\text{Ad}_g \eta$ with respect to g at $g = e$ in the direction ξ to get the *Lie bracket* $[\xi, \eta] : \mathfrak{g} \times \mathfrak{g} \to \mathfrak{g}$ and thereby produce the *adjoint operation*

$$T_e(\text{Ad}_g \eta)\xi = [\xi, \eta] = \text{ad}_\xi \eta.$$

4.5.3 Compute the co-Adjoint and coadjoint operations by taking duals

1. $\mathrm{Ad}_g^* : \mathfrak{g}^* \to \mathfrak{g}^*$, the dual of Ad_g, is defined by

$$\langle \mathrm{Ad}_g^* \alpha, \xi \rangle = \langle \alpha, \mathrm{Ad}_g \xi \rangle$$

for $\alpha \in \mathfrak{g}^*$, $\xi \in \mathfrak{g}$ and the non-degenerate pairing $\langle \cdot, \cdot \rangle$: $\mathfrak{g}^* \times \mathfrak{g} \to \mathbb{R}$. The map

$$\Phi^* : G \times \mathfrak{g}^* \to \mathfrak{g}^* \quad \text{given by } (g, \alpha) \mapsto \mathrm{Ad}_{g^{-1}}^* \alpha$$

defines the *co-Adjoint action* of G on $\mathfrak{g}^*$.

2. The pairing
$$\langle \mathrm{ad}_\xi^* \alpha, \eta \rangle = \langle \alpha, \mathrm{ad}_\xi \eta \rangle$$

defines the *coadjoint action* of $\mathfrak{g}$ on $\mathfrak{g}^*$, for $\alpha \in \mathfrak{g}^*$ and $\xi, \eta \in \mathfrak{g}$.

See [MaRa1994, Chapter 9] for more discussion of the Ad and ad operations.

4.5.4 The Lie algebra $\mathfrak{so}(3)$ and its dual

The special orthogonal group is defined by

$$SO(3) := \{A \mid A \text{ a } 3 \times 3 \text{ orthogonal matrix}, \det(A) = 1\}.$$

Its Lie algebra $\mathfrak{so}(3)$ is formed by 3×3 skew-symmetric matrices, and its dual is denoted $\mathfrak{so}(3)^*$.

4.5.5 The isomorphism $(\widehat{\cdot}) : (\mathfrak{so}(3), [\cdot, \cdot]) \to (\mathbb{R}^3, \times)$

The Lie algebra $(\mathfrak{so}(3), [\cdot, \cdot])$, where $[\cdot, \cdot]$ is the commutator bracket of matrices, is isomorphic to the Lie algebra $(\mathbb{R}^3, \times)$, where $\times$ denotes the vector product in $\mathbb{R}^3$, by the isomorphism

$$
\mathbf{u} := (u^1, u^2, u^3) \in \mathbb{R}^3 \mapsto \hat{u} :=
\begin{bmatrix}
0 & -u^3 & u^2 \\
u^3 & 0 & -u^1 \\
-u^2 & u^1 & 0
\end{bmatrix}
\in so(3),
$$

that is, $\hat{u}_{ij} := -\epsilon_{ijk}u^k$. Equivalently, this isomorphism is given by

$$
\hat{u}\mathbf{v} = \mathbf{u} \times \mathbf{v} \quad \text{for all } \mathbf{u}, \mathbf{v} \in \mathbb{R}^3.
$$

The following formulas for $\mathbf{u}, \mathbf{v}, \mathbf{w} \in \mathbb{R}^3$ may be easily verified:

$$
\begin{aligned}
(\mathbf{u} \times \mathbf{v})\hat{} &= [\hat{u}, \hat{v}] \\
[\hat{u}, \hat{v}]\mathbf{w} &= (\mathbf{u} \times \mathbf{v}) \times \mathbf{w} \\
\mathbf{u} \cdot \mathbf{v} &= -\tfrac{1}{2}\operatorname{trace}(\hat{u}\hat{v}).
\end{aligned}
$$

4.5.6   The Ad action of $SO(3)$ on $so(3)$

The corresponding adjoint action of $SO(3)$ on $so(3)$ may be obtained as follows. For $SO(3)$, we have $I_A(B) = ABA^{-1}$. Differentiating $B(t)$ at $B(0) = \mathrm{Id}$ gives

$$
\mathrm{Ad}_A \hat{v} = \frac{d}{dt}\Big|_{t=0} AB(t)A^{-1} = A\hat{v}A^{-1}, \quad \text{with } \hat{v} = B'(0).
$$

One calculates the pairing with a vector $\mathbf{w} \in \mathbb{R}^3$ as

$$
\mathrm{Ad}_A \hat{v}(\mathbf{w}) = A\hat{v}(A^{-1}\mathbf{w}) = A(\mathbf{v} \times A^{-1}\mathbf{w}) = A\mathbf{v} \times \mathbf{w} = (A\mathbf{v})\hat{}\mathbf{w},
$$

where we have used the relation

$$
A(\mathbf{u} \times \mathbf{v}) = A\mathbf{u} \times A\mathbf{v},
$$

which holds for any $\mathbf{u}, \mathbf{v} \in \mathbb{R}^3$ and $A \in SO(3)$.

Consequently,

$$
\mathrm{Ad}_A \hat{v} = (A\mathbf{v})\hat{}. \tag{4.5.1}
$$

Identifying $so(3) \simeq \mathbb{R}^3$ then gives

$$
\mathrm{Ad}_A \mathbf{v} = A\mathbf{v}. \tag{4.5.2}
$$

So (speaking prose all our lives), the adjoint action of $SO(3)$ on $\mathfrak{so}(3)$ may be identified with multiplication of a matrix in $SO(3)$ times a vector in $\mathbb{R}^3$.

4.5.7   The ad action of $\mathfrak{so}(3)$ on $\mathfrak{so}(3)$

Differentiating again gives the ad-action of the Lie algebra $\mathfrak{so}(3)$ on itself:

$$[\hat{\mathbf{u}}, \hat{\mathbf{v}}] = \mathrm{ad}_{\hat{\mathbf{u}}}\, \hat{\mathbf{v}} = \left.\frac{d}{dt}\right|_{t=0} \left(e^{t\hat{\mathbf{u}}}\mathbf{v}\right)\widehat{} = (\hat{\mathbf{u}}\mathbf{v})\widehat{} = (\mathbf{u} \times \mathbf{v})\widehat{}.$$

So, in this isomorphism, the vector cross product is identified with the matrix commutator of skew-symmetric matrices.

4.5.8   Infinitesimal generator

Likewise, the *infinitesimal generator* corresponding to $\mathbf{u} \in \mathbb{R}^3$ has the expression

$$\mathbf{u}_{\mathbb{R}^3}(\mathbf{v}) := \left.\frac{d}{dt}\right|_{t=0} e^{t\hat{\mathbf{u}}}\mathbf{v} = \hat{\mathbf{u}}\,\mathbf{v} = \mathbf{u} \times \mathbf{v}.$$

Exercise. What is the analogue of the hat map $\mathfrak{so}(3) \mapsto \mathbb{R}^3$ for the three-dimensional Lie algebras $\mathfrak{sp}(2, \mathbb{R})$, $\mathfrak{so}(2, 1)$, $\mathfrak{su}(1, 1)$, or $\mathfrak{sl}(2, \mathbb{R})$?   ★

The dual Lie algebra isomorphism $\breve{}$: $\mathfrak{so}(3)^* \to \mathbb{R}^3$

4.5.9   Coadjoint actions

The dual $\mathfrak{so}(3)^*$ is identified with $\mathbb{R}^3$ by the breve ($\breve{}$) isomorphism

$$\mathbf{\Pi} \in \mathbb{R}^3 \mapsto \breve{\mathbf{\Pi}} \in so(3)^*\colon \breve{\mathbf{\Pi}}(\hat{\mathbf{u}}) := \mathbf{\Pi} \cdot \mathbf{u} \quad \text{for any } \mathbf{u} \in \mathbb{R}^3.$$

In terms of this isomorphism, the co-Adjoint action of $SO(3)$ on $\mathfrak{so}(3)^*$ is given by

$$\text{Ad}^*_{A^{-1}} \check{\Pi} = (A\Pi)^\vee, \tag{4.5.3}$$

and the coadjoint action of $\mathfrak{so}(3)$ on $\mathfrak{so}(3)^*$ is given by

$$\text{ad}^*_{\hat{u}} \check{\Pi} = (\Pi \times u)^\vee. \tag{4.5.4}$$

4.5.10   Computing the co-Adjoint action of $SO(3)$ on $\mathfrak{so}(3)^*$

For $A \in SO(3)$ and $\hat{u} \in \mathfrak{so}(3)$, the co-Adjoint action of $SO(3)$ on $\mathfrak{so}(3)$ is obtained from

$$\left(\text{Ad}^*_{A^{-1}} \check{\Pi}\right)(\hat{u}) = \check{\Pi}(\text{Ad}_{A^{-1}} \hat{u}) = \check{\Pi}((A^{-1}u)^\wedge) = \Pi \cdot A^T u$$

$$= A\Pi \cdot u = (A\Pi)^\vee (\hat{u}),$$

where the second step applies equation (4.5.1). Finally, the co-Adjoint action of $SO(3)$ on $\mathfrak{so}(3)^*$ is expressed as in (4.5.3)

$$\text{Ad}^*_{A^{-1}} \check{\Pi} = (A\Pi)^\vee.$$

Consequently, the *co-Adjoint orbit* $\mathcal{O} = \{A\Pi \mid A \in SO(3)\} \subset \mathbb{R}^3$ of $SO(3)$ through $\Pi \in \mathbb{R}^3$ is a two-sphere of radius $\|\Pi\|$.

4.5.11   Computing the coadjoint action of $\mathfrak{so}(3)$ on $\mathfrak{so}(3)^*$

Let $u, v \in \mathbb{R}^3$, and note that

$$\langle \text{ad}^*_{\hat{u}} \check{\Pi}, \hat{v} \rangle = \langle \check{\Pi}, [\hat{u}, \hat{v}] \rangle = \langle \check{\Pi}, (u \times v)^\wedge \rangle = \Pi \cdot (u \times v)$$

$$= (\Pi \times u) \cdot v = \langle (\Pi \times u)^\vee, \hat{v} \rangle,$$

which shows that $\text{ad}^*_{\hat{u}} \check{\Pi} = (\Pi \times u)^\vee$, thereby proving (4.5.4).

Consequently, the tangent space of the *co-Adjoint orbit* $\mathcal{O}$

$$T_\Pi \mathcal{O} = \{\Pi \times u \mid u \in \mathbb{R}^3\}$$

since the plane is perpendicular to Π, that is, the tangent space to the sphere centred at the origin of radius $\|\Pi\|$ is given by $\{\Pi \times u \mid u \in \mathbb{R}^3\}$.

5

THE RIGID BODY IN $\mathbb{R}^3$

Contents

What is this lecture about? This lecture reformulates Euler's equations for rigid-body dynamics by using the complementary variational approaches of Lagrange and Hamilton.

5.1   Euler's equations for the rigid body in $\mathbb{R}^3$

In the absence of external torques, Euler's equations for rigid-body motion are

$$I_1\dot{\Omega}_1 = (I_2 - I_3)\Omega_2\Omega_3,$$

$$I_2\dot{\Omega}_2 = (I_3 - I_1)\Omega_3\Omega_1, \qquad (5.1.1)$$

$$I_3\dot{\Omega}_3 = (I_1 - I_2)\Omega_1\Omega_2,$$

or, equivalently,

$$\mathbb{I}\dot{\boldsymbol{\Omega}} = \mathbb{I}\boldsymbol{\Omega} \times \boldsymbol{\Omega},$$

where $\boldsymbol{\Omega} = (\Omega_1, \Omega_2, \Omega_3)$ is the body angular velocity vector and I_1, I_2 and I_3 are the moments of inertia of the rigid body.

> **Exercise.** Can equations (5.1.1) be cast into Lagrangian or Hamiltonian form in any sense? (Since there are an odd number of equations, they cannot be put into canonical Hamiltonian form.)   ★

We could reformulate them as:

- Euler–Lagrange equations on $TSO(3)$ or

- canonical Hamiltonian equations on $T^*SO(3)$

by using Euler angles and their velocities or their conjugate momenta. However, these reformulations on $TSO(3)$ or $T^*SO(3)$ would answer a different question for a *six*-dimensional system. We are interested in these structures for the equations as given above.

> **Answer. (Lagrangian formulation)**   The Lagrangian answer is as follows: these equations may be expressed

in Euler–Poincaré form on the Lie algebra $\mathfrak{so}(3) \simeq \mathbb{R}^3$ using the Lagrangian

$$l(\mathbf{\Omega}) = \tfrac{1}{2}(I_1\Omega_1^2 + I_2\Omega_2^2 + I_3\Omega_3^2) = \tfrac{1}{2}\mathbf{\Omega}^T \cdot \mathbb{I}\mathbf{\Omega}, \qquad (5.1.2)$$

which is the (rotational) kinetic energy of the rigid body.[1]   ▲

Proposition 5.1.1 *The Euler rigid body equations are equivalent to the rigid-body action principle for a reduced action*

$$\delta S_{\mathrm{red}} = \delta \int_a^b l(\mathbf{\Omega})\, dt = \delta \int_a^b \tfrac{1}{2}\mathbf{\Omega}^T \cdot \mathbb{I}\mathbf{\Omega}\, dt = 0, \qquad (5.1.3)$$

where variations of $\mathbf{\Omega}$ are restricted to be of the form

$$\delta\mathbf{\Omega} = \dot{\mathbf{\Sigma}} + \mathbf{\Omega} \times \mathbf{\Sigma}, \qquad (5.1.4)$$

in which $\mathbf{\Sigma}(t)$ is a curve in $\mathbb{R}^3$ that vanishes at the endpoints in time.

Proof. Since $l(\mathbf{\Omega}) = \tfrac{1}{2}\langle \mathbb{I}\mathbf{\Omega}, \mathbf{\Omega}\rangle$, and $\mathbb{I}$ is symmetric, we obtain

$$\delta \int_a^b l(\mathbf{\Omega})\, dt = \int_a^b \langle \mathbb{I}\mathbf{\Omega}, \delta\mathbf{\Omega}\rangle\, dt$$

$$= \int_a^b \langle \mathbb{I}\mathbf{\Omega}, \dot{\mathbf{\Sigma}} + \mathbf{\Omega} \times \mathbf{\Sigma}\rangle\, dt$$

$$= \int_a^b \left[\left\langle -\frac{d}{dt}\mathbb{I}\mathbf{\Omega}, \mathbf{\Sigma}\right\rangle + \left\langle \mathbb{I}\mathbf{\Omega}, \mathbf{\Omega} \times \mathbf{\Sigma}\right\rangle\right] dt + \langle \mathbb{I}\mathbf{\Omega}, \mathbf{\Sigma}\rangle \Big|_a^b$$

$$= \int_a^b \left\langle -\frac{d}{dt}\mathbb{I}\mathbf{\Omega} + \mathbb{I}\mathbf{\Omega} \times \mathbf{\Omega}, \mathbf{\Sigma}\right\rangle dt,$$

upon integrating by parts and using the endpoint conditions, $\mathbf{\Sigma}(b) = \mathbf{\Sigma}(a) = 0$. Since $\mathbf{\Sigma}$ is otherwise arbitrary, (5.2.1) is equivalent to

$$-\frac{d}{dt}(\mathbb{I}\mathbf{\Omega}) + \mathbb{I}\mathbf{\Omega} \times \mathbf{\Omega} = 0,$$

which are Euler's rigid-body equations (5.1.1).   ∎

Let's derive this variational principle from the *standard* Hamilton's principle.

---

[1]The Hamiltonian answer to this question will be discussed later.

5.2 Hamilton's principle for rigid body motion on $TSO(3)$

An element $O \in SO(3)$ gives the configuration of the body as a map of a *reference configuration* $\mathcal{B} \subset \mathbb{R}^3$ to the current configuration $O(\mathcal{B})$. The map O takes a reference or label point $\mathbf{X} \in \mathcal{B}$ to a current point $\mathbf{x} = O(\mathbf{X}) \in O(\mathcal{B})$.

When the matrix O is time dependent, the rigid body undergoes motion relative to the reference configuration. Thus,

$$\mathbf{x}(t) = O(t)\mathbf{X}$$

with $O(t)$ a curve parametrised by time in $SO(3)$. The relative velocity in fixed space of a point of the body is given by

$$\dot{\mathbf{x}}(t) = \dot{O}(t)\mathbf{X} = \dot{O}O^{-1}(t)\mathbf{x}(t).$$

As discussed earlier, since O is an orthogonal matrix, $O^{-1}\dot{O}$ and $\dot{O}O^{-1}$ are skew matrices. Consequently, we can write (recall the hat map in Remark 2.5.1)

$$\dot{\mathbf{x}} = \dot{O}O^{-1}\mathbf{x} = \boldsymbol{\omega} \times \mathbf{x}. \tag{5.2.1}$$

This formula defines the *spatial angular velocity vector* $\boldsymbol{\omega}$. Thus, the vector $\boldsymbol{\omega} = (\dot{O}O^{-1})\hat{}$ is associated with the *right* translation of $\dot{O}$ to the identity. The corresponding *body angular velocity* vector is defined by

$$\boldsymbol{\Omega} = O^{-1}\boldsymbol{\omega}, \tag{5.2.2}$$

so that $\boldsymbol{\Omega} \in \mathbb{R}^3$ is the angular velocity relative to a body fixed frame. Note that

$$O^{-1}\dot{O}\mathbf{X} = O^{-1}\dot{O}O^{-1}\mathbf{x} = O^{-1}(\boldsymbol{\omega} \times \mathbf{x})$$

$$= O^{-1}\boldsymbol{\omega} \times O^{-1}\mathbf{x} = \boldsymbol{\Omega} \times \mathbf{X}, \tag{5.2.3}$$

so that $\boldsymbol{\Omega}\times$ is given by *left* translation of $\dot{O}$ to the identity. That is, the skew-symmetric matrix $\boldsymbol{\Omega}\times$ is given by

$$\boldsymbol{\Omega}\times = (O^{-1}\dot{O})\hat{}.$$

The *kinetic energy* is obtained by summing up $m|\dot{\mathbf{x}}|^2/2$ (where $|\cdot|$ denotes the Euclidean norm) over the body. This yields

$$K = \frac{1}{2}\int_{\mathcal{B}} \rho(\mathbf{X})|\dot{O}\mathbf{X}|^2\, d^3X, \qquad (5.2.4)$$

in which $\rho(\mathbf{X})$ is a given mass density in the reference configuration. Since the action of $SO(3)$ on $\mathbb{R}^3$ preserves vector magnitudes, one finds

$$|\dot{O}\mathbf{X}| = |\boldsymbol{\omega}\times\mathbf{x}| = |O^{-1}(\boldsymbol{\omega}\times\mathbf{x})| = |(O^{-1}\boldsymbol{\omega}\times O^{-1}\mathbf{x})| = |\boldsymbol{\Omega}\times\mathbf{X}|,$$

and one sees that K is a quadratic function of $\boldsymbol{\Omega}$. Writing

$$K = \tfrac{1}{2}\boldsymbol{\Omega}^T\cdot\mathbb{I}\boldsymbol{\Omega} \qquad (5.2.5)$$

defines the *moment of inertia tensor*, $\mathbb{I}$, which is a positive-definite (3×3) matrix (provided the body does not degenerate to a line interval) and the kinetic energy K is a quadratic form. This quadratic form can be diagonalised by a change of basis, thereby defining the principal axes and moments of inertia. In this basis, we write $\mathbb{I} = \mathrm{diag}(I_1, I_2, I_3)$.

The function K is taken to be the Lagrangian of the system on $TSO(3)$. Note that K in equation (5.2.4) is *left* (not right) invariant on $TSO(3)$ since

$$\boldsymbol{\Omega}\times = O^{-1}\dot{O} \quad\Longleftrightarrow\quad \boldsymbol{\Omega} = (O^{-1}\dot{O})\widehat{}.$$

In the framework of Hamilton's principle, the relation between motion in $O \in SO(3)$ space and motion in body angular velocity $O^{-1}\dot{O} \in T_e SO(3)$ (or $\boldsymbol{\Omega}$) space is as follows.

Proposition 5.2.1 *The curve $O(t) \in SO(3)$ satisfies the Euler–Lagrange equations for*

$$L(O, \dot{O}) = \frac{1}{2}\int_{\mathcal{B}} \rho(\mathbf{X})|\dot{O}\mathbf{X}|^2\, d^3\mathbf{X} \qquad (5.2.6)$$

if and only if $\boldsymbol{\Omega}(t)$, defined by $O^{-1}\dot{O}\mathbf{v} = \boldsymbol{\Omega}\times\mathbf{v}$ for all $\mathbf{v}\in\mathbb{R}^3$, satisfies Euler's rigid-body equations,

$$\mathbb{I}\dot{\boldsymbol{\Omega}} = \mathbb{I}\boldsymbol{\Omega}\times\boldsymbol{\Omega}. \qquad (5.2.7)$$

The proof of this relation will illustrate how to reduce variational principles using their symmetry groups. By Hamilton's principle, $O(t)$ satisfies the Euler–Lagrange equations if and only if

$$\delta \int L(O, \dot{O}) \, dt = 0.$$

Let $l(\mathbf{\Omega}) = \frac{1}{2}(\mathbb{I}\mathbf{\Omega}) \cdot \mathbf{\Omega}$, so that $l(\mathbf{\Omega}) = L(O, \dot{O})$, where the matrix O and the vector $\mathbf{\Omega}$ are related by the hat map $\widehat{\mathbf{\Omega}} = (O^{-1}\dot{O})\widehat{}$. Thus, the Lagrangian L is left $SO(3)$-invariant. That is,

$$l(\mathbf{\Omega}) = L(O, \dot{O}) = L(e, O^{-1}\dot{O}).$$

To see how we should use this left invariance to transform Hamilton's principle, define the skew matrix $\widehat{\mathbf{\Omega}}$ by $\widehat{\mathbf{\Omega}}\mathbf{v} = \mathbf{\Omega} \times \mathbf{v}$ for any $\mathbf{v} \in \mathbf{R}^3$.

We differentiate the relation $O^{-1}\dot{O} = \widehat{\mathbf{\Omega}}$ with respect to O to get

$$-O^{-1}(\delta O)O^{-1}\dot{O} + O^{-1}(\delta \dot{O}) = \widehat{\delta \mathbf{\Omega}}. \tag{5.2.8}$$

Let the skew matrix $\widehat{\Sigma}$ be defined by

$$\widehat{\Sigma} = O^{-1}\delta O, \tag{5.2.9}$$

and define the corresponding vector Σ by

$$\widehat{\Sigma}\mathbf{v} = \Sigma \times \mathbf{v}. \tag{5.2.10}$$

Note that

$$\dot{\widehat{\Sigma}} = -O^{-1}\dot{O}O^{-1}\delta O + O^{-1}\delta\dot{O}.$$

Consequently,

$$O^{-1}\delta\dot{O} = \dot{\widehat{\Sigma}} + O^{-1}\dot{O}\widehat{\Sigma}. \tag{5.2.11}$$

Substituting (5.2.11) and (5.2.9) into (5.2.8) gives

$$-\widehat{\Sigma}\widehat{\mathbf{\Omega}} + \dot{\widehat{\Sigma}} + \widehat{\mathbf{\Omega}}\widehat{\Sigma} = \widehat{\delta\mathbf{\Omega}},$$

that is, cf. relation (2.10.4),

$$\widehat{\delta\mathbf{\Omega}} = \dot{\widehat{\Sigma}} + [\widehat{\mathbf{\Omega}}, \widehat{\Sigma}]. \tag{5.2.12}$$

The identity $[\widehat{\Omega}, \widehat{\Sigma}] = (\boldsymbol{\Omega} \times \boldsymbol{\Sigma})\widehat{}$ holds by the exercises in Section 2.2.2. Hence, one has

$$\delta\boldsymbol{\Omega} = \dot{\boldsymbol{\Sigma}} + \boldsymbol{\Omega} \times \boldsymbol{\Sigma}. \tag{5.2.13}$$

These calculations have proved the following.

Theorem 5.2.1 *For a Lagrangian that is left invariant under SO(3), Hamilton's variational principle*

$$\delta S = \delta \int_a^b L(O, \dot{O})\, dt = 0 \tag{5.2.14}$$

on $TSO(3)$ is equivalent to the reduced variational principle

$$\delta S_{\mathrm{red}} = \delta \int_a^b l(\boldsymbol{\Omega})\, dt = 0, \tag{5.2.15}$$

with $\boldsymbol{\Omega} = (O^{-1}\dot{O})\widehat{}$ on $\mathbb{R}^3$, where the variations $\delta\boldsymbol{\Omega}$ are of the form

$$\delta\boldsymbol{\Omega} = \dot{\boldsymbol{\Sigma}} + \boldsymbol{\Omega} \times \boldsymbol{\Sigma},$$

with $\boldsymbol{\Sigma}(a) = \boldsymbol{\Sigma}(b) = 0$.

5.3 Reconstruction of $O(t) \in SO(3)$

In 5.2.1, Euler's equation for the rigid body,

$$\mathbb{I}\dot{\boldsymbol{\Omega}} = \mathbb{I}\boldsymbol{\Omega} \times \boldsymbol{\Omega}, \tag{5.3.1}$$

follows from the reduced variational principle (5.2.15) for the Lagrangian

$$l(\boldsymbol{\Omega}) = \tfrac{1}{2}(\mathbb{I}\boldsymbol{\Omega}) \cdot \boldsymbol{\Omega}, \tag{5.3.2}$$

which is expressed in terms of the left-invariant time-dependent angular velocity in the body, $\boldsymbol{\Omega} \in \mathfrak{so}(3)$. The body angular velocity $\boldsymbol{\Omega}(t)$ yields the tangent vector $\dot{O}(t) \in T_{O(t)}SO(3)$ along the integral curve in the rotation group $O(t) \in SO(3)$ by the relation

$$\dot{O}(t) = O(t)\boldsymbol{\Omega}(t). \tag{5.3.3}$$

This relation provides the *reconstruction formula*. Its solution as a linear differential equation with time-dependent coefficients yields the integral curve $O(t) \in SO(3)$ for the orientation of the rigid body, once the time dependence of $\Omega(t)$ is determined from the Euler equations.

5.4   Hamilton–Pontryagin constrained rigid-body variational principle

Formula (5.2.12) for the variation $\widehat{\Omega}$ of the skew-symmetric matrix

$$\widehat{\Omega} = O^{-1}\dot{O}$$

may be imposed as a constraint in Hamilton's principle and thereby provide a variational derivation of Euler's equations (5.3.1) for rigid-body motion in principal axis coordinates. This constraint is incorporated into the matrix Euler equations as follows.

Proposition 5.4.1 (Matrix Euler equations) *Euler's rigid-body equation may be written in matrix form as*

$$\frac{d\Pi}{dt} = -\left[\widehat{\Omega}, \Pi\right] \quad with \ \Pi = \mathbb{I}\widehat{\Omega} = \frac{\delta l}{\delta\widehat{\Omega}}, \tag{5.4.1}$$

for the Lagrangian $l(\widehat{\Omega})$ given by

$$l = \frac{1}{2}\left\langle \mathbb{I}\widehat{\Omega}, \widehat{\Omega} \right\rangle, \tag{5.4.2}$$

in which the bracket

$$\left[\widehat{\Omega}, \Pi\right] := \widehat{\Omega}\Pi - \Pi\widehat{\Omega} \tag{5.4.3}$$

denotes the commutator and $\langle \cdot, \cdot \rangle$ denotes the trace pairing, e.g.,

$$\left\langle \Pi, \widehat{\Omega} \right\rangle =: \frac{1}{2}\,\mathrm{trace}\left(\Pi^T \widehat{\Omega}\right). \tag{5.4.4}$$

Remark 5.4.1 Note that the symmetric part of Π does not contribute in the pairing, and if set equal to zero initially, it will remain zero.

□

Proposition 5.4.2 (Constrained variational principle) *The matrix Euler equations (5.4.1) are equivalent to stationarity $\delta S = 0$ of the following constrained action:*

$$S(\widehat{\Omega}, O, \dot{O}, \Pi) \;=\; \int_a^b l(\widehat{\Omega}, O, \dot{O}, \Pi)\, dt \tag{5.4.5}$$

$$=\; \int_a^b \left[l(\widehat{\Omega}) + \langle\, \Pi,\, O^{-1}\dot{O} - \widehat{\Omega} \,\rangle \right] dt.$$

Remark 5.4.2 The integrand of the constrained action in (5.7.4) is similar to the formula for the Legendre transform, but its functional dependence is different. This variational approach is related to the classic *Hamilton–Pontryagin principle*, which is used in control theory.   □

Proof. The variations of S in formula (5.7.4) are given by

$$\delta S \;=\; \int_a^b \Bigg\{ \Big\langle \frac{\delta l}{\delta\widehat{\Omega}} - \Pi,\, \delta\widehat{\Omega} \Big\rangle$$

$$+ \Big\langle \delta\Pi,\, (O^{-1}\dot{O} - \widehat{\Omega}) \Big\rangle + \Big\langle \Pi,\, \delta(O^{-1}\dot{O}) \Big\rangle \Bigg\}\, dt,$$

where

$$\delta(O^{-1}\dot{O}) = \widehat{\Xi}^{\,\cdot} + [\widehat{\Omega}, \widehat{\Xi}\,], \tag{5.4.6}$$

and $\widehat{\Xi} = (O^{-1}\delta O)$ from equation (5.2.9).

Substituting for $\delta(O^{-1}\dot{O})$ in (5.4.6) into the last term of δS produces

$$\int_a^b \Big\langle \Pi,\, \delta(O^{-1}\dot{O}) \Big\rangle dt \;=\; \int_a^b \Big\langle \Pi,\, \widehat{\Xi}^{\,\cdot} + [\widehat{\Omega}, \widehat{\Xi}\,] \Big\rangle dt$$

$$=\; \int_a^b \Big\langle -\Pi^{\,\cdot} - [\widehat{\Omega}, \Pi],\, \widehat{\Xi} \Big\rangle dt$$

$$+ \Big\langle \Pi,\, \widehat{\Xi} \Big\rangle \Big|_a^b, \tag{5.4.7}$$

where one uses the cyclic properties of the trace operation for matrices,

$$\text{trace}\left(\Pi^T \widehat{\Xi} \widehat{\Omega}\right) = \text{trace}\left(\widehat{\Omega} \Pi^T \widehat{\Xi}\right). \tag{5.4.8}$$

Thus, stationarity of the Hamilton–Pontryagin variational principle for vanishing endpoint conditions $\widehat{\Xi}(a) = 0 = \widehat{\Xi}(b)$ implies the following set of equations:

$$\frac{\delta l}{\delta \widehat{\Omega}} = \Pi, \quad O^{-1}\dot{O} = \widehat{\Omega}, \quad \frac{d\Pi}{dt} = -[\widehat{\Omega}, \Pi]. \tag{5.4.9}$$

■

Remark 5.4.3 (Interpreting the formulas in (5.4.9)) The first formula in (5.4.9) defines the angular momentum matrix Π as the *fibre derivative* of the Lagrangian with respect to the angular velocity matrix $\widehat{\Omega}$. The second formula is the reconstruction formula (5.3.3) for the solution curve $O(t) \in SO(3)$, given the solution $\widehat{\Omega}(t) = O^{-1}\dot{O}$. Finally, the third formula is Euler's equation for rigid-body motion in matrix form. □

We transform the endpoint terms in (5.4.10), arising on integrating the variation δS by parts in the proof of Theorem 5.4.2, into the spatial representation by setting $\widehat{\Xi}(t) =: O(t)\widehat{\xi}O^{-1}(t)$ and $\widehat{\Pi}(t) =: O(t)\widehat{\pi}(t)O^{-1}(t)$ as follows:

$$\left\langle \Pi, \widehat{\Xi} \right\rangle = \text{trace}\left(\Pi^T \widehat{\Xi}\right) = \text{trace}\left(\pi^T \widehat{\xi}\right) = \left\langle \pi, \widehat{\xi} \right\rangle. \tag{5.4.10}$$

Thus, the vanishing of both endpoints for an arbitrary *constant* infinitesimal spatial rotation $\widehat{\xi} = (\delta O O^{-1}) = \text{const}$ implies constancy of the spatial angular momentum,

$$\pi(a) = \pi(b). \tag{5.4.11}$$

This is Noether's theorem for the rigid body.

Theorem 5.4.1 (Noether's theorem for the rigid body) *Invariance of the constrained Hamilton–Pontryagin action under spatial rotations* *implies conservation of spatial angular momentum,*

$$\pi = O^{-1}(t)\Pi(t)O(t) =: \mathrm{Ad}^*_{O^{-1}(t)}\Pi(t).$$ (5.4.12)

Proof.

$$
\begin{aligned}
\frac{d}{dt}\left\langle \pi, \widehat{\xi} \right\rangle &= \frac{d}{dt}\left\langle O^{-1}\Pi O, \widehat{\xi} \right\rangle = \frac{d}{dt}\,\mathrm{trace}\left(\Pi^T O^{-1}\widehat{\xi}O \right) \\
&= \left\langle \frac{d}{dt}\Pi + [\widehat{\Omega},\, \Pi],\, O^{-1}\widehat{\xi}O \right\rangle = 0 \\
&=: \left\langle \frac{d}{dt}\Pi - \mathrm{ad}^*_{\widehat{\Omega}}\Pi,\, \mathrm{Ad}_{O^{-1}}\widehat{\xi} \right\rangle,
\end{aligned}
$$

$$
\frac{d}{dt}\left\langle \mathrm{Ad}^*_{O^{-1}}\Pi, \widehat{\xi} \right\rangle = \left\langle \mathrm{Ad}^*_{O^{-1}}\left(\frac{d}{dt}\Pi - \mathrm{ad}^*_{\widehat{\Omega}}\Pi \right), \widehat{\xi} \right\rangle.
$$ (5.4.13)

The proof of Noether's theorem for the rigid body is already on the second line. However, the last line gives a general result.   ∎

Remark 5.4.4 The proof of Noether's theorem for the rigid body when the constrained Hamilton–Pontryagin action is invariant under spatial rotations also proves a general result in equation (5.4.13), with $\widehat{\Omega} = O^{-1}\dot{O}$ for a Lie group O, that, as in the calculation in (2.10.1),

$$\frac{d}{dt}\left(\mathrm{Ad}^*_{O^{-1}}\Pi \right) = \mathrm{Ad}^*_{O^{-1}}\left(\frac{d}{dt}\Pi - \mathrm{ad}^*_{\widehat{\Omega}}\Pi \right).$$ (5.4.14)

This equation will be useful in the remainder of the text. In particular, it provides the solution of a differential equation defined on the dual of a Lie algebra. Namely, for a Lie group O with Lie algebra $\mathfrak{o}$, the equation for $\Pi \in \mathfrak{o}^*$ and $\widehat{\Omega} = O^{-1}\dot{O} \in \mathfrak{o}$,

$$\frac{d}{dt}\Pi - \mathrm{ad}^*_{\widehat{\Omega}}\Pi = 0 \quad \text{has solution } \Pi(t) = \mathrm{Ad}^*_{O(t)}\pi,$$ (5.4.15)

in which the constant $\pi \in \mathfrak{o}^*$ is obtained from the initial conditions.

   □

5.5   Hamiltonian form of rigid body motion in $\mathbb{R}^3$

A dynamical system on a manifold M,

$$\dot{\mathbf{x}}(t) = \mathbf{F}(\mathbf{x}), \quad \mathbf{x} \in M,$$

is said to be in *Hamiltonian form* if it can be expressed as

$$\dot{\mathbf{x}}(t) = \{\mathbf{x}, H\}, \quad \text{for } H \colon M \mapsto \mathbb{R},$$

in terms of a Poisson bracket operation,

$$\{\cdot, \cdot\} \colon \mathcal{F}(M) \times \mathcal{F}(M) \mapsto \mathcal{F}(M),$$

that is bilinear, skew symmetric and satisfies the Jacobi identity and (usually) the Leibniz rule.

As we shall explain, reduced equations arising from Lie group invariant Hamilton's principles are naturally Hamiltonian. If we *Legendre-transform* our reduced Lagrangian for the $SO(3)$ left-invariant variational principle (5.2.15) for rigid-body dynamics, then its simple, beautiful and well-known Hamiltonian formulation emerges.

Definition 5.5.1 *The Legendre transformation* $\mathbb{F}l \colon \mathfrak{so}(3) \to \mathfrak{so}(3)^*$ *is defined by the fibre derivative*

$$\mathbb{F}l(\Omega) = \frac{\delta l}{\delta \Omega} = \Pi.$$

The Legendre transformation defines the *body angular momentum* by the variations of the rigid body's reduced Lagrangian with respect to the body angular velocity. For the Lagrangian in (5.3.2), the $\mathbb{R}^3$ components of the body angular momentum are

$$\Pi_i = I_i \Omega_i = \frac{\partial l}{\partial \Omega_i}, \quad i = 1, 2, 3. \tag{5.5.1}$$

5.6   Lie–Poisson Hamiltonian formulation of rigid-body dynamics

The rigid-body Hamiltonian is obtained from the *reduced Legendre transform*,

$$h(\Pi) := \langle \Pi, \Omega \rangle - l(\Omega), \tag{5.6.1}$$

where the pairing $\langle \cdot, \cdot \rangle : \mathfrak{so}(3)^* \times \mathfrak{so}(3) \to \mathbb{R}$ is understood via the hat map in components as the vector dot product on $\mathbb{R}^3$

$$\langle \Pi, \Omega \rangle := \mathbf{\Pi} \cdot \mathbf{\Omega}.$$

Hence, one finds the expected expression for the rigid-body Hamiltonian:

$$h = \tfrac{1}{2}\mathbf{\Pi} \cdot \mathbb{I}^{-1}\mathbf{\Pi} := \frac{\Pi_1^2}{2I_1} + \frac{\Pi_2^2}{2I_2} + \frac{\Pi_3^2}{2I_3}. \tag{5.6.2}$$

The reduced Legendre transform $\mathbb{F}l$ for this case in (5.6.1) is a diffeomorphism, so we may take its differential to find

$$dh(\Pi) = \left\langle \frac{\partial h}{\partial \Pi}, d\Pi \right\rangle = \langle \Omega, d\Pi \rangle + \left\langle \Pi - \frac{\partial L}{\partial \Omega}, d\Omega \right\rangle.$$

In $\mathbb{R}^3$ coordinates, this relation expresses the body angular velocity as the derivative of the reduced Hamiltonian with respect to the body angular momentum, namely (introducing grad-notation),

$$\nabla_\Pi h := \frac{dh}{d\Pi} = \mathbf{\Omega}.$$

Hence, the reduced Euler–Lagrange equations for the Lagrangian $l(\Omega)$ in (5.3.2) may be expressed equivalently in angular momentum vector components in $\mathbb{R}^3$ and the Hamiltonian $h(\Pi)$ as

$$\frac{d}{dt}(\mathbb{I}\Omega) = \mathbb{I}\Omega \times \Omega \iff \dot{\mathbf{\Pi}} = \mathbf{\Pi} \times \frac{dh}{d\Pi} := \{\Pi, h\}.$$

This expression suggests we introduce the following rigid-body Poisson bracket on the functions of $\mathbf{\Pi}$'s as

$$\{f,h\}(\mathbf{\Pi}) = -\Pi_i \frac{\partial f}{\partial \Pi_j} \epsilon_{ijk} \frac{\partial h}{\partial \Pi_k} =: -\mathbf{\Pi} \cdot \frac{df}{d\mathbf{\Pi}} \times \frac{dh}{d\mathbf{\Pi}}. \qquad (5.6.3)$$

For the Hamiltonian (5.6.2), one checks that the Euler equations in terms of the rigid-body angular momenta follow as

$$\dot{\Pi}_1 = \frac{I_2 - I_3}{I_2 I_3} \Pi_2 \Pi_3, \quad \dot{\Pi}_2 = \frac{I_3 - I_1}{I_3 I_1} \Pi_3 \Pi_1, \quad \dot{\Pi}_3 = \frac{I_1 - I_2}{I_1 I_2} \Pi_1 \Pi_2.$$
$$(5.6.4)$$

That is, the equation

$$\dot{\mathbf{\Pi}} = \mathbf{\Pi} \times \frac{dh}{d\mathbf{\Pi}} \qquad (5.6.5)$$

implies

$$\frac{df}{dt} = \frac{df}{d\mathbf{\Pi}} \cdot \frac{d\mathbf{\Pi}}{dt} = \frac{df}{d\mathbf{\Pi}} \cdot \mathbf{\Pi} \times \frac{dh}{d\mathbf{\Pi}} = -\mathbf{\Pi} \cdot \frac{df}{d\mathbf{\Pi}} \times \frac{dh}{d\mathbf{\Pi}} = \{f,h\}.$$

The Poisson bracket proposed in (5.6.3) is an example of a *Lie Poisson bracket*, which we will show later satisfies the defining relations to be a Poisson bracket.

Remark 5.6.1 The bracket in (5.6.3) may also be written equivalently as a differential form relation:

$$\{f,h\}d^3\mathbf{\Pi} := -d(|\mathbf{\Pi}|^2/2) \wedge df \wedge dh. \qquad (5.6.6)$$

Note that this bracket is symplectic on the level sets of $|\mathbf{\Pi}|^2$. It also vanishes if either f or h is a function of $|\mathbf{\Pi}|^2$. Hence, for $f(|\mathbf{\Pi}|^2)$, say, the Poisson bracket proposed in (5.6.3) vanishes for every function $h(\mathbf{\Pi})$. Also, the flow of either $f(\mathbf{\Pi})$ or $h(\mathbf{\Pi})$ takes place on the level sets of $|\mathbf{\Pi}|^2$, where it reduces to a canonical Poisson bracket. As we shall see, this is the hallmark of *coadjoint motion*. $\qquad \square$

5.7   Relation to the $\mathbb{R}^3$ Poisson bracket

The rigid-body Poisson bracket (5.6.3) is a special case of the Poisson bracket for functions on $\mathbb{R}^3$,

$$\{f, h\} = -\nabla c \cdot \nabla f \times \nabla h. \tag{5.7.1}$$

This bracket generates the motion

$$\dot{\mathbf{x}} = \{\mathbf{x}, h\} = \nabla c \times \nabla h. \tag{5.7.2}$$

For this bracket, the motion takes place along the intersections of level surfaces of the functions c and h in $\mathbb{R}^3$. In particular, for the rigid body, the motion takes place along intersections of angular momentum spheres $c = \|\mathbf{x}\|^2/2$ and energy ellipsoids $h = \mathbf{x} \cdot \mathbb{I}\mathbf{x}$. (See the cover illustration given by Marsden and Ratiu [MaRa2003].)

Exercise. Consider the $\mathbb{R}^3$ Poisson bracket

$$\{f, h\} = -\nabla c \cdot \nabla f \times \nabla h. \tag{5.7.3}$$

Let $c = \mathbf{x}^T \cdot \mathbb{C}\mathbf{x}$ be a quadratic form on $\mathbb{R}^3$, and let $\mathbb{C}$ be the associated symmetric 3×3 matrix. Determine the conditions on the quadratic function $c(\mathbf{x})$ so that this Poisson bracket will satisfy the Jacobi identity.   ★

Exercise. Find the general conditions on the function $c(\mathbf{x})$ so that the $\mathbb{R}^3$ bracket

$$\{f, h\} = -\nabla c \cdot \nabla f \times \nabla h$$

satisfies the defining properties of a Poisson bracket. Is this $\mathbb{R}^3$ bracket also a derivation satisfying the Leibniz relation for a product of functions on $\mathbb{R}^3$? If so, why?   ★

Exercise. How is the $\mathbb{R}^3$ bracket related to the canonical Poisson bracket?

Hint: Restrict to level surfaces of the function $c(\mathbf{x})$. ★

Exercise. (Casimirs of the $\mathbb{R}^3$ bracket) The Casimirs (or distinguished functions, as Lie called them) of a Poisson bracket satisfy

$$\{c, h\}(\mathbf{x}) = 0, \quad \forall h(\mathbf{x}).$$

Suppose the function $c(\mathbf{x})$ is chosen so that the $\mathbb{R}^3$ bracket (5.7.1) satisfies the defining properties of a Poisson bracket. What are the Casimirs for the $\mathbb{R}^3$ bracket (5.7.1)? Why? ★

Exercise. Show that the motion equation

$$\dot{\mathbf{x}} = \{\mathbf{x}, h\}$$

for the $\mathbb{R}^3$ bracket (5.7.1) is invariant under a certain linear combination of the functions c and h. Interpret this invariance geometrically. ★

Exercise. Use the Hamilton–Pontryagin approach to compute the dynamics arising from stationarity $\delta S = 0$ of the following *right-invariant* constrained action, which is discussed in detail in Ref. [GaMaRa2012]:

$$S(\widehat{\omega}, O, \dot{O}, \pi) = \int_a^b l(\widehat{\omega}, O, \dot{O}, \pi) \, dt \qquad (5.7.4)$$

$$= \int_a^b \left[l(\widehat{\omega}) + \langle \pi, \dot{O}O^{-1} - \widehat{\omega} \rangle \right] dt.$$

★

Answer. The variations of S in formula (5.7.4) are given by

$$\delta S = \int_a^b \left\{ \left\langle \frac{\delta l}{\delta \widehat{\omega}} - \pi, \delta \widehat{\omega} \right\rangle \right.$$

$$\left. + \left\langle \delta \pi, (\dot{O}O^{-1} - \widehat{\omega}) \right\rangle + \left\langle \pi, \delta(\dot{O}O^{-1}) \right\rangle \right\} dt,$$

$$(5.7.5)$$

where

$$\delta(\dot{O}O^{-1}) = \widehat{\dot{\xi}} - [\widehat{\omega}, \widehat{\xi}], \qquad (5.7.6)$$

and one defines $\widehat{\xi} := \delta O O^{-1}$.

Substituting for $\delta(\dot{O}O^{-1})$ in (5.4.6) into the last term of δS produces

$$\int_a^b \left\langle \pi, \delta(\dot{O}O^{-1}) \right\rangle dt = \int_a^b \left\langle \pi, \widehat{\dot{\xi}} - [\widehat{\omega}, \widehat{\xi}] \right\rangle dt$$

$$= \int_a^b \left\langle -\dot{\pi} - [\widehat{\omega}, \pi], \widehat{\xi} \right\rangle dt$$

$$+ \left\langle \pi, \widehat{\xi} \right\rangle \Big|_a^b. \qquad (5.7.7)$$

Here, the square brackets $[\cdot, \cdot]$ denote matrix commutator and the second line in (5.7.7) applies integration by parts and uses the cyclic property of the trace operation for matrices,

$$\text{trace}\left(\pi^T \widehat{\xi}\,\widehat{\omega}\right) = \text{trace}\left(\widehat{\omega}\,\pi^T\widehat{\xi}\right).$$ (5.7.8)

Thus, the stationarity of the Hamilton–Pontryagin variational principle in (5.7.5) for vanishing endpoint conditions $\widehat{\xi}(a) = 0 = \widehat{\xi}(b)$ implies the following set of equations:

$$\frac{\delta l}{\delta\widehat{\omega}} = \pi, \quad \dot{O}O^{-1} = \widehat{\omega}, \quad \frac{d\pi}{dt} = [\widehat{\omega}, \pi].$$ (5.7.9)

▲

Exercise. Recall that the Lagrangian $l(\widehat{\Omega})$ for $\widehat{\Omega} = O^{-1}\dot{O}$ is given in equation (5.4.2) for rigid-body motion in the body frame. Derive the equations of rigid-body dynamics in the spatial frame by using the Hamilton–Pontryagin approach after writing the Lagrangian from (5.4.2) in terms of the spatial angular velocity $\widehat{\omega} = \dot{O}O^{-1}$.

★

Answer. The Euler's rigid-body Lagrangian from (5.4.2) is written in terms of the spatial angular velocity $\widehat{\omega} = \dot{O}O^{-1}$ by substituting the relation between body and spatial angular velocities in (2.9.1). Hence, the rigid-body Lagrangian in the spatial frame transforms into

$$l(\widehat{\Omega}) := \tfrac{1}{2}\Big\langle \widehat{\Omega}, \mathbb{I}\widehat{\Omega} \Big\rangle = \tfrac{1}{2}\Big\langle \widehat{\omega}, \mathbb{I}_{\text{spat}}(t)\,\widehat{\omega} \Big\rangle =: \ell(\widehat{\omega}, \mathbb{I}_{\text{spat}}).$$ (5.7.10)

In terms of the *trace pairing* in (5.4.4), one may verify this formula directly:

$$\frac{-1}{2}\text{tr}\left(\widehat{\Omega}\,\mathbb{I}\,\widehat{\Omega}\right) = \frac{-1}{2}\text{tr}\left(\widehat{\omega}\,O(t)\mathbb{I}O^{-1}(t)\,\widehat{\omega}\right)$$

$$=: \frac{-1}{2}\text{tr}\left(\widehat{\omega}\,\mathbb{I}_{\text{spat}}(t)\,\widehat{\omega}\right).$$ (5.7.11)

Thus, in the spatial frame, the moment of inertia becomes a dynamical variable as

$$\mathbb{I}_{\text{spat}}(t) := O(t)\mathbb{I}O^{-1}(t). \qquad (5.7.12)$$

From its definition, the time-dependent moment of inertia $\mathbb{I}_{\text{spat}}(t)$ satisfies the auxiliary equations:

$$\frac{d\mathbb{I}_{\text{spat}}}{dt} = \widehat{\omega}\,\mathbb{I}_{\text{spat}} - \mathbb{I}_{\text{spat}}\,\widehat{\omega} =: [\widehat{\omega}, \mathbb{I}_{\text{spat}}],$$

$$\delta\mathbb{I}_{\text{spat}} = [\widehat{\xi}, \mathbb{I}_{\text{spat}}] \quad \text{with } \widehat{\xi} := \delta O\, O^{-1}. \qquad (5.7.13)$$

Likewise, the variational relation for the spatial angular velocity becomes

$$\delta(\dot{O}O^{-1}) = \widehat{\xi}\,\dot{} - [\widehat{\omega}, \widehat{\xi}] \quad \text{with } \widehat{\xi} := \delta O\, O^{-1}. \qquad (5.7.14)$$

Thus, in the spatial frame, one deals with a rigid-body Lagrangian that is not right invariant and depends on both $\widehat{\omega}(t)$ and $\mathbb{I}_{\text{spat}}(t)$:

$$S(\widehat{\omega}, O, \dot{O}, \pi) = \int_a^b \left[\ell(\widehat{\omega}, \mathbb{I}_{\text{spat}}) + \langle \pi, \dot{O}O^{-1} - \widehat{\omega} \rangle \right] dt. \qquad (5.7.15)$$

The variations of the action S in formula (5.7.15) are now given by

$$\delta S = \int_a^b \left\{ \left\langle \frac{\delta\ell}{\delta\widehat{\omega}} - \pi, \delta\widehat{\omega} \right\rangle + \left\langle \delta\pi, (\dot{O}O^{-1} - \widehat{\omega}) \right\rangle \right.$$

$$\left. + \left\langle \pi, \delta(\dot{O}O^{-1}) \right\rangle + \left\langle \frac{\delta\ell}{\delta\mathbb{I}_{\text{spat}}}, \delta\mathbb{I}_{\text{spat}} \right\rangle \right\} dt$$

$$= \int_a^b \left\{ \left\langle \frac{\delta\ell}{\delta\widehat{\omega}} - \pi, \delta\widehat{\omega} \right\rangle + \left\langle \delta\pi, (\dot{O}O^{-1} - \widehat{\omega}) \right\rangle \right.$$

$$\left. + \left\langle -\dot{\pi} - [\pi, \widehat{\omega}] + \left[\mathbb{I}_{\text{spat}}, \frac{\delta\ell}{\delta\mathbb{I}_{\text{spat}}}\right], \widehat{\xi} \right\rangle \right\} dt, \qquad (5.7.16)$$

where we have applied formula (5.7.15) for $\delta(\dot{O}O^{-1})$ and then integrated by parts in time. A direct calculation with the Lagrangian $\ell(\widehat{\omega}, \mathbb{I}_{\text{spat}})$ in equation (5.7.10) reveals that the last two terms in the variations paired with $\widehat{\xi}$ in (5.7.16) cancel each other because for this Lagrangian, the equality holds that

$$\left[\pi, \widehat{\omega}\right] = \left[\mathbb{I}_{\text{spat}}, \frac{\delta\ell}{\delta\mathbb{I}_{\text{spat}}}\right].$$

Consequently, the dynamics of the free rigid body in spatial variables satisfies

$$\frac{d\pi}{dt} = 0 \quad \text{and} \quad \frac{d\mathbb{I}_{\text{spat}}}{dt} = \left[\widehat{\omega}, \mathbb{I}_{\text{spat}}\right] \quad \text{with} \quad \pi = \mathbb{I}_{\text{spat}}(t)\,\widehat{\omega}$$

for spatial angular velocity $\widehat{\omega} = \dot{O}O^{-1}$ and dynamical moment of inertia $\mathbb{I}_{\text{spat}}(t) := O(t)\mathbb{I}O^{-1}(t)$.

▲

Exercise. Show that the Hamiltonian equations for the spatial dynamics of the rigid body in (5.7.16) may be expressed via the following Lie–Poisson matrix operator:

$$\frac{d}{dt}\begin{pmatrix} \pi \\ \mathbb{I}_{\text{spat}} \end{pmatrix} = -\begin{pmatrix} [\pi, \square] & [\mathbb{I}_{\text{spat}}, \square] \\ [\mathbb{I}_{\text{spat}}, \square] & 0 \end{pmatrix}$$
$$\times \begin{pmatrix} \delta h/\delta\pi = \widehat{\omega} \\ \delta h/\delta\mathbb{I}_{\text{spat}} = -\delta\ell/\delta\mathbb{I}_{\text{spat}} \end{pmatrix}, \quad (5.7.17)$$

with $(\pi, \mathbb{I}_{\text{spat}}) \in [\mathfrak{so}(3)\,\text{ⓈSym}^3_+(\mathbb{R})]^*$ where Ⓢ denotes the semidirect-product action of the Lie algebra $\mathfrak{so}(3)$ represented as 3×3 antisymmetric real matrices acting on the vector space of 3×3 positive symmetric real matrices, $\text{Sym}^3_+(\mathbb{R})$, and superscript $*$ denotes the dual semidirect-product Lie algebra under matrix trace pairing. ★

Remark 5.7.1 Semidirect-product action by matrix multiplication will be discussed further for the example of the heavy top in body coordinates in the following lecture. For now, one should keep in mind that semidirect-product action arose here when the left-invariant angular velocity producing $SO(3)$ rotation symmetry of the rigid-body Lagrangian in body coordinates was broken by transforming into the right-invariant angular velocity in spatial coordinates.   □

Exercise. Formulate and analyse the equations of motion on $\mathfrak{so}()3^* \times T^*Q$ for a rigid body that has a flywheel attached whose axis of rotation is aligned with the intermediate principal axis of the rigid body. The flywheel's rotation angle α has a harmonic restoring force with spring constant k. The kinetic energy of this system is given by

$$KE = \frac{1}{2}I_1\Omega_1^2 + \frac{1}{2}I_2\Omega_2^2 + \frac{1}{2}I_3\Omega_3^2 + \frac{1}{2}J_2(\dot{\alpha}+\Omega_2)^2,$$

where $\Omega = (\Omega_1, \Omega_2, \Omega_3)$ is the angular velocity vector, $\dot{\alpha}$ is the angular rotational rate of the flywheel about the intermediate principal axis of the rigid body and I_1, I_2, I_3 and J_2 are positive constants.

Hamilton's principle for this system may be written as

$$0 = \delta S = \delta \int_0^T L(\Omega, \alpha, p_\alpha)\, dt$$

$$= \delta \int_0^T \frac{1}{2}\Omega \cdot \mathbb{I}\Omega + p_\alpha(\dot{\alpha}+\Omega_2) \qquad (5.7.18)$$

$$- \left(\frac{1}{2J_2}p_\alpha^2 + \frac{1}{2}k\alpha^2\right) dt.$$

1. Find the angular momenta $\boldsymbol{\Pi} \in \mathbb{R}^3$ and $p_\alpha \in \mathbb{R}^1$.

2. Legendre-transform to obtain the Hamiltonian in the variables $\boldsymbol{\Pi}, p_\alpha, \alpha$.

3. Write the equations of motion for this system in Lie–Poisson bracket form.   ★

Answer.

1. The variations of the Lagrangian in Hamilton's principle (5.7.18) with respect to $\boldsymbol{\Omega}$ and p_α yield

$$\delta\boldsymbol{\Omega}: \quad \boldsymbol{\Pi} = \left(I_1\Omega_1, I_2\Omega_2 + p_\alpha, I_3\Omega_3\right)^T,$$
$$\delta p_\alpha: \quad p_\alpha = J_2(\dot\alpha + \Omega_2). \tag{5.7.19}$$

2. The Legendre transform

$$H(\boldsymbol{\Pi}, \alpha, p_\alpha) = \boldsymbol{\Pi} \cdot \boldsymbol{\Omega} + p_\alpha\dot\alpha - L(\boldsymbol{\Omega}, \alpha, p_\alpha), \tag{5.7.20}$$

produces the Hamiltonian for this system:

$$H(\boldsymbol{\Pi}, \alpha, p_\alpha) = \frac{\Pi_1^2}{2I_1} + \frac{1}{2I_2}(\Pi_2 - p_\alpha)^2$$
$$+ \frac{\Pi_3^2}{2I_3} + \frac{1}{2J_2}p_\alpha^2 + \frac{1}{2}k\alpha^2. \tag{5.7.21}$$

Note that the angular momentum shift $(\Pi_2 - p_\alpha)$ along the intermediate axis couples the dynamics of the two degrees of freedom.

3. The equations of motion for this system in Lie–Poisson bracket form are

$$\frac{d}{dt}\begin{pmatrix} \boldsymbol{\Pi} \\ p_\alpha \\ \alpha \end{pmatrix} = \begin{bmatrix} \boldsymbol{\Pi}\times & 0 & 0 \\ 0 & 0 & -1 \\ 0 & 1 & 0 \end{bmatrix}$$
$$\times \begin{pmatrix} \partial H/\partial\boldsymbol{\Pi} = \boldsymbol{\Omega} \\ \partial H/\partial p_\alpha = p_\alpha/J_2 - \Omega_2 = \dot\alpha \\ \partial H/\partial\alpha = k\alpha \end{pmatrix}.$$
$$\tag{5.7.22}$$

▲

Exercise. Formulate and analyse the equations of motion on $\mathfrak{so}(3)^* \times T^*\mathbb{R}^3$ for a rigid body whose angular velocity $\Omega \in \mathfrak{so}(3)$ acts on $q \in \mathbb{R}^3$ as

$$\dot{q} + (\Omega + \Upsilon) \times q = 0,$$

where $\Upsilon \in \mathbb{R}^3$ is constant. Hamilton's principle for the corresponding constrained rigid-body motion is given via the Clebsch approach as

$$0 = \delta S = \delta \int_0^T L(\Omega) + p \cdot (\dot{q} + (\Omega + \Upsilon) \times q) \, dt$$

$$= \delta \int_0^T \frac{1}{2} \Omega \cdot \mathbb{I}\Omega + p \cdot (\dot{q} + (\Omega + \Upsilon) \times q) \, dt.$$

$$(5.7.23)$$

1. Take the variation of the action integral S in Ω and determine the Euler–Poincaré equation for $\Pi := \partial L / \partial \Omega$ by using the equations arising from the other variations in q and p as constraints. This is the Clebsch variational approach.

2. Write the equations of motion for this system in Lie–Poisson bracket form.   ★

Answer.

1. The stationary variation in Ω of the action integral S in (5.7.23) implies that

$$\frac{\partial L}{\partial \Omega} - (q \times p) = 0,$$

and the variations of the Clebsch constraint in q and p imply that

$$\frac{d}{dt}(q \times p) = (q \times p) \times (\Omega + \Upsilon).$$

Hence,

$$\frac{d}{dt}\frac{\partial L}{\partial \mathbf{\Omega}} = \frac{\partial L}{\partial \mathbf{\Omega}} \times (\mathbf{\Omega} + \mathbf{\Upsilon}). \qquad (5.7.24)$$

2. To write the equations of motion for this system in Lie–Poisson bracket form, we Legendre-transform to derive the Hamiltonian as

$$H(\mathbf{\Pi}) = \mathbf{\Pi} \cdot \mathbf{\Omega} - L(\mathbf{\Omega}),$$

$$dH(\mathbf{\Pi}) = \frac{\partial H}{\partial \mathbf{\Pi}} \cdot d\mathbf{\Pi} = \mathbf{\Omega} \cdot d\mathbf{\Pi} + \left(\mathbf{\Pi} - \frac{\partial L}{\partial \mathbf{\Omega}}\right) \cdot d\mathbf{\Omega}. \qquad (5.7.25)$$

Then, upon rewriting equation (5.7.24) by using (5.7.25), we find

$$\frac{d}{dt}\mathbf{\Pi} = \mathbf{\Pi} \times (\mathbf{\Omega} + \mathbf{\Upsilon}) = \mathbf{\Pi} \times \frac{\partial H}{\partial \mathbf{\Pi}}, \qquad (5.7.26)$$

$$\implies \quad H(\mathbf{\Pi}) = \frac{1}{2}\mathbf{\Pi} \cdot \mathbb{I}^{-1}\mathbf{\Pi} + \mathbf{\Pi} \cdot \mathbf{\Upsilon},$$

where the Lie–Poisson bracket is the same as for standard rigid-body dynamics. However, the rigid-body Hamiltonian has acquired an additional term which, in the motion equation (5.7.26), shifts the angular velocity as $\mathbf{\Omega} \rightarrow (\mathbf{\Omega} + \mathbf{\Upsilon})$. For more discussion, see, e.g., Ref. [ArDeHo2018]. ▲

6

BROKEN SYMMETRY: HEAVY TOP EQUATIONS

Contents

What is this lecture about? This lecture reformulates Euler's equations for heavy top dynamics as breaking of the $SO(3)$ symmetry of the invariant variational principles of Lagrange and Hamilton for rigid-body dynamics.

6.1 Introduction and definitions

A top is a rigid body of mass m rotating with a fixed point of support in a constant gravitational field of acceleration $-g\hat{z}$ pointing vertically downward. The orientation of the body relative to the vertical axis $\hat{z}$ is defined by the unit vector $\Gamma = O^{-1}(t)\hat{z}$ for a curve $O(t) \in SO(3)$.

According to its definition, the unit vector Γ represents the motion of the vertical direction as seen from the rotating body. Consequently, it satisfies the auxiliary motion equation,

$$\dot{\Gamma} = -(O^{-1}\dot{O}(t))\Gamma = \Gamma \times \Omega.$$

Here, the rotation matrix $O(t) \in SO(3)$, the skew matrix $\widehat{\Omega} = O^{-1}\dot{O} \in \mathfrak{so}(3)$ and the body angular frequency vector $\Omega \in \mathbb{R}^3$ are related by the hat map, $\Omega = (O^{-1}\dot{O})\widehat{}$, where $\widehat{} : (\mathfrak{so}(3), [\cdot, \cdot]) \to (\mathbb{R}^3, \times)$, with $\widehat{\Omega}v = \Omega \times v$ for any $v \in \mathbb{R}^3$.

The motion of a heavy top of weight mg is given by Euler's equations in vector form:

$$\mathbb{I}\dot{\Omega} = \mathbb{I}\Omega \times \Omega + mg\,\Gamma \times \chi,$$
$$\dot{\Gamma} = \Gamma \times \Omega,$$

(6.1.1)

where $\Omega, \Gamma, \chi \in \mathbb{R}^3$ are vectors in the rotating body frame.

Here:

- $\Omega = (\Omega_1, \Omega_2, \Omega_3)$ is the body angular velocity vector;

- $\mathbb{I} = \text{diag}(I_1, I_2, I_3)$ is the moment of inertia tensor, diagonalised in the body principle axes;

- $\Gamma = O^{-1}(t)\hat{z}$ represents the motion of the unit vector along the vertical axis, as seen from the body;

- χ is the constant vector in the body from the point of support to the body's centre of mass;

- m is the total mass of the body and g is the constant acceleration of gravity.

6.2   Heavy top action principles

6.2.1   Euler–Poincaré reduced heavy top action integral

Proposition 6.2.1 *The heavy top equations are equivalent to the heavy top action principle for an Euler–Poincaré reduced action integral,*

$$
\delta S_{\mathrm{red}} = 0, \quad \text{with} \quad S_{\mathrm{red}} = \int_a^b l(\mathbf{\Omega}, \mathbf{\Gamma})\, dt = \int_a^b \tfrac{1}{2}\langle \mathbb{I}\mathbf{\Omega}, \mathbf{\Omega}\rangle - \langle mg\chi, \mathbf{\Gamma}\rangle\, dt,
$$
(6.2.1)

where variations of $\mathbf{\Omega}$ and $\mathbf{\Gamma}$ are restricted to be of the form

$$
\delta\mathbf{\Omega} = \dot{\mathbf{\Sigma}} + \mathbf{\Omega} \times \mathbf{\Sigma} \quad \text{and} \quad \delta\mathbf{\Gamma} = \mathbf{\Gamma} \times \mathbf{\Sigma}, \tag{6.2.2}
$$

arising from variations of the definitions $\mathbf{\Omega} = (O^{-1}\dot{O})^\widehat{}$ and $\mathbf{\Gamma} = O^{-1}(t)\hat{\mathbf{z}}$, in which $\mathbf{\Sigma}(t) = (O^{-1}\delta O)^\widehat{}$ is a curve in $\mathbb{R}^3$ that vanishes at the endpoints in time.

Proof. Since $\mathbb{I}$ is symmetric and χ is constant, we obtain the variation

$$
0 = \delta \int_a^b l(\mathbf{\Omega}, \mathbf{\Gamma})\, dt
$$

$$
= \int_a^b \langle \mathbb{I}\mathbf{\Omega}, \delta\mathbf{\Omega}\rangle - \langle mg\,\chi, \delta\mathbf{\Gamma}\rangle\, dt
$$

$$
= \int_a^b \langle \mathbb{I}\mathbf{\Omega}, \dot{\mathbf{\Sigma}} + \mathbf{\Omega} \times \mathbf{\Sigma}\rangle - \langle mg\,\chi, \mathbf{\Gamma} \times \mathbf{\Sigma}\rangle\, dt
$$

$$
= \int_a^b \left\langle -\frac{d}{dt}\mathbb{I}\mathbf{\Omega}, \mathbf{\Sigma}\right\rangle + \langle \mathbb{I}\mathbf{\Omega}, \mathbf{\Omega} \times \mathbf{\Sigma}\rangle - \langle mg\,\chi, \mathbf{\Gamma} \times \mathbf{\Sigma}\rangle\, dt
$$

$$
= \int_a^b \left\langle -\frac{d}{dt}\mathbb{I}\mathbf{\Omega} + \mathbb{I}\mathbf{\Omega} \times \mathbf{\Omega} + mg\,\mathbf{\Gamma} \times \chi, \mathbf{\Sigma}\right\rangle dt + \langle \mathbb{I}\mathbf{\Omega}, \mathbf{\Sigma}\rangle \Big|_a^b
$$

upon integrating by parts and using the endpoint conditions, $\mathbf{\Sigma}(b) = \mathbf{\Sigma}(a) = 0$.

Since Σ is otherwise arbitrary, vanishing in the variation of the reduced action integral of S_{red} in (6.2.1) is equivalent to

$$-\frac{d}{dt}\mathbb{I}\boldsymbol{\Omega} + \mathbb{I}\boldsymbol{\Omega} \times \boldsymbol{\Omega} + mg\,\boldsymbol{\Gamma} \times \boldsymbol{\chi} = 0,$$

which is Euler's motion equation for the heavy top (6.1.1). This motion equation is completed by the auxiliary equation $\dot{\boldsymbol{\Gamma}} = \boldsymbol{\Gamma} \times \boldsymbol{\Omega}$ in (6.1.1) arising from the definition of $\boldsymbol{\Gamma}$. ∎

6.2.2 Hamilton–Pontryagin constrained heavy top action integral

Heavy top dynamics can also be derived by using Lagrange multipliers to impose the constraints on the reduced Lagrangian, just as was done in the Hamilton–Pontryagin principle for the matrix Euler equations in Section 5.4. In particular, we apply the Hamilton–Pontryagin approach to the following class of constrained action integrals:

$$S = \int L(g, \dot{g}, \hat{e}_3)\, dt \tag{6.2.3}$$

$$= \int \left\{ l(\boldsymbol{\Omega}, \boldsymbol{\Gamma}) + \left\langle \boldsymbol{\Pi}, \, g^{-1}\dot{g} - \boldsymbol{\Omega} \right\rangle + \left\langle \boldsymbol{\chi}, \, g^{-1}\hat{e}_3 - \boldsymbol{\Gamma} \right\rangle \right\} dt,$$

$$\tag{6.2.4}$$

where one identifies

$$l(\boldsymbol{\Omega}, \boldsymbol{\Gamma}) = L(e, \, g^{-1}\dot{g}, \, g^{-1}\hat{e}_3). \tag{6.2.5}$$

This class of action integrals produces constrained equations of motion determined from the Hamilton–Pontryagin principle.

Remark 6.2.1 A feature of the Hamilton–Pontryagin principle is the freedom to modify the constraint relations in (6.2.4) and (6.2.5) and thereby accommodate alternative types of dynamical interpretations between the physical quantities $(\boldsymbol{\Omega}, \boldsymbol{\Gamma})$ and the transport maps $(g^{-1}\dot{g}, g^{-1}\hat{e}_3)$. For an example of the use of this feature, see Lecture 28 on nonlinear shallow water equations. □

Theorem 6.2.1 (Hamilton–Pontryagin action principle) *The stationarity condition for the constrained Hamilton–Pontryagin principle defined in Equations (6.2.4) and (6.2.5) implies the following equation of motion:*

$$\left(\frac{d}{dt} - \mathrm{ad}^*_\Omega\right)\Pi = \chi \diamond \Gamma, \tag{6.2.6}$$

with $\Pi = \delta l/\delta\Omega$ *and* $\chi = \delta l/\delta\Gamma$, *where* χ *defines the vector in the body directed from the point of support to the centre of mass.*

Exercise. Prove that Theorem 6.2.1 for the Hamilton–Pontryagin action principle yields the same $\mathbb{R}^3$ vector equations (6.1.1) as determined from the Euler–Poincaré action principle when one sets $g = O \in SO(3)$, $\Gamma = \Gamma \in \mathbb{R}^3$ and $\chi = mg\chi \in \mathbb{R}^3$. ★

Answer. The result here hinges on the variational relations with $\xi := g^{-1}\delta g \in \mathfrak{g}$,

$$\delta(g^{-1}\dot{g}) = \frac{d\xi}{dt} + \mathrm{ad}_{g^{-1}\dot{g}}\xi \quad \text{and} \quad \delta(g^{-1}\hat{e}_3) = -\pounds_\xi\hat{e}_3. \tag{6.2.7}$$

The variations then yield the constraints $g^{-1}\dot{g} = \Omega$ and $g^{-1}\hat{e}_3 = \Gamma$ as well as

$$\frac{\delta l}{\delta\Omega} = \Pi, \quad \frac{\delta l}{\delta\Gamma} = \chi, \tag{6.2.8}$$

and after defining dual actions and the diamond operator and then integrating by parts in time, one has the Hamilton–Pontryagin equations,

$$\left(\frac{d}{dt} - \mathrm{ad}^*_{g^{-1}\dot{g}}\right)\frac{\delta l}{\delta\Omega} = \frac{\delta l}{\delta\Gamma} \diamond \Gamma \quad \text{and} \quad \frac{d\Gamma}{dt} = -\pounds_{g^{-1}\dot{g}}\Gamma, \tag{6.2.9}$$

which completes the proof of the motion equation in (6.2.6) and the advection equation for $\Gamma := g^{-1}\hat{e}_3$. ▲

6.2.3 Legendre transformation for the heavy top

The Legendre transformation for $l(\Omega, \Gamma)$ gives the body angular momenta:

$$\Pi = \frac{\partial l}{\partial \Omega} = \mathbb{I}\Omega.$$

The well-known energy Hamiltonian for the heavy top then emerges from the *reduced Legendre transformation*:

$$h(\Pi, \Gamma) = \Pi \cdot \Omega - l(\Omega, \Gamma) = \tfrac{1}{2}\langle \Pi, \mathbb{I}^{-1}\Pi \rangle + \langle mg\, \chi, \Gamma \rangle, \quad (6.2.10)$$

which is the sum of the kinetic and potential energies of the heavy top.

Definition 6.2.1 (Functional variational derivative) *Let $f, h: \mathfrak{g}^* \to \mathbb{R}$ be two real-valued functions on the dual space $\mathfrak{g}^*$. Upon denoting the elements of $\mathfrak{g}^*$ by μ, the functional variational derivative of f at μ is defined as the unique element $\delta f/\delta\mu$ of $\mathfrak{g}$ emerging in the following limit:*

$$\lim_{\varepsilon \to 0} \frac{1}{\varepsilon}[f(\mu + \varepsilon\delta\mu) - f(\mu)] = \left\langle \delta\mu, \frac{\delta f}{\delta\mu} \right\rangle, \quad (6.2.11)$$

for all $\delta\mu \in \mathfrak{g}^*$, where $\langle \cdot, \cdot \rangle$ denotes the pairing between $\mathfrak{g}^*$ and $\mathfrak{g}$.[1]

Definition 6.2.2 (Lie–Poisson brackets and Lie–Poisson equations) *The $(\pm)$ **Lie–Poisson brackets** for the Euler–Poincaré equations for a kinetic-energy Lagrangian are defined by*

$$\{f, h\}_{\pm}(\mu) = \pm\left\langle \mu, \left[\frac{\delta f}{\delta\mu}, \frac{\delta h}{\delta\mu}\right] \right\rangle = \mp\left\langle \mu, \mathrm{ad}_{\delta h/\delta\mu} \frac{\delta f}{\delta\mu} \right\rangle. \quad (6.2.12)$$

*The corresponding **Lie–Poisson equations**, determined by $\dot{f} = \{f, h\}$, read*

$$\dot{\mu} = \{\mu, h\} = \mp\, \mathrm{ad}^*_{\delta h/\delta\mu}\, \mu, \quad (6.2.13)$$

---

[1]For fluid dynamics, $\langle \cdot, \cdot \rangle$ is the L^2 integral pairing between two functions. See equation (25.2.2).

where one defines the ad^* *operation in terms of the pairing* $\langle \cdot, \cdot \rangle$
by

$$\{f, h\} = \left\langle \mu, \mathrm{ad}_{\delta h/\delta \mu} \frac{\delta f}{\delta \mu} \right\rangle = \left\langle \mathrm{ad}^*_{\delta h/\delta \mu} \mu, \frac{\delta f}{\delta \mu} \right\rangle.$$

The Lie–Poisson setting of mechanics is a special case of the general theory of systems on Poisson manifolds, for which there is now an extensive theoretical development. (See Marsden and Ratiu [MaRa1994] for an introduction to this literature.)

6.3   Lie–Poisson brackets and momentum maps

An important feature of the rigid-body bracket carries over to general Lie algebras. Namely, *Lie–Poisson brackets on* $\mathfrak{g}^*$ *arise from canonical brackets on the cotangent bundle* (phase space) T^*G associated with a Lie group G which has $\mathfrak{g}$ as its associated Lie algebra. Thus, the process by which the Lie–Poisson brackets arise is the *momentum map*

$$T^*G \mapsto \mathfrak{g}^*.$$

For example, a rigid body is free to rotate about its centre of mass, and G is the (proper) rotation group $SO(3)$. The choice of T^*G as the primitive phase space is made according to the classical procedures of mechanics described earlier. For a description using Lagrangian mechanics, one forms the velocity phase space TG. The Hamiltonian description on T^*G is then obtained through standard procedures, such as Legendre transforms.

The passage from T^*G to the space of Π's (body angular momentum space) is determined by *left* translation on the group. The mapping $T^*SO(3) \rightarrow \Pi$ is an example of a *momentum map*, that is, a mapping whose components are the "Noether quantities" associated with a symmetry group. The map from T^*G to $\mathfrak{g}^*$ being a Poisson map *is a general fact about momentum maps*. The Hamiltonian point of view of all this is a standard subject.

Remark 6.3.1 (Lie–Poisson description of the heavy top) As it turns out, the underlying Lie algebra for the Lie–Poisson description of the heavy top consists of the Lie algebra $\mathfrak{se}(3)$ of infinitesimal Euclidean motions in $\mathbb{R}^3$. This is a bit surprising because heavy top motion itself does *not* actually arise through the actions of the Euclidean group of rotations and translations on the body since the body has a fixed point!

Instead, the Lie algebra $\mathfrak{se}(3)$ arises for another reason associated with the *breaking* of the $SO(3)$ isotropy by the presence of the gravitational field. This symmetry breaking introduces a semidirect-product Lie–Poisson structure which happens to coincide with the dual of the Lie algebra $\mathfrak{se}(3)$ in the case of the heavy top. As we shall see later, a close parallel exists between this case and the Lie–Poisson structure for compressible fluids.   □

6.4   Lie–Poisson brackets for the heavy top

The *Lie algebra of the special Euclidean group in 3D* is $\mathfrak{se}(3) = \mathbb{R}^3 \times \mathbb{R}^3$, with the Lie bracket

$$\big[(\boldsymbol{\xi}, \mathbf{u}), (\boldsymbol{\eta}, \mathbf{v})\big] = (\boldsymbol{\xi} \times \boldsymbol{\eta},\ \boldsymbol{\xi} \times \mathbf{v} - \boldsymbol{\eta} \times \mathbf{u}). \tag{6.4.1}$$

We identify the dual space of $\mathfrak{se}(3)$ with pairs $(\boldsymbol{\Pi}, \boldsymbol{\Gamma})$; the corresponding $(-)$ Lie–Poisson bracket, called the *heavy top bracket*, is

$$\begin{aligned} \{f, h\}(\boldsymbol{\Pi}, \boldsymbol{\Gamma}) = {}&-\boldsymbol{\Pi} \cdot \nabla_{\boldsymbol{\Pi}} f \times \nabla_{\boldsymbol{\Pi}} h \\ &- \boldsymbol{\Gamma} \cdot \big(\nabla_{\boldsymbol{\Pi}} f \times \nabla_{\boldsymbol{\Gamma}} h - \nabla_{\boldsymbol{\Pi}} h \times \nabla_{\boldsymbol{\Gamma}} f\big). \end{aligned} \tag{6.4.2}$$

This Lie–Poisson bracket and the Hamiltonian (6.2.10) recover equations (6.1.1) for the heavy top as

$$\dot{\boldsymbol{\Pi}} = \{\boldsymbol{\Pi}, h\} \quad = \quad \boldsymbol{\Pi} \times \nabla_{\boldsymbol{\Pi}} h + \boldsymbol{\Gamma} \times \nabla_{\boldsymbol{\Gamma}} h = \boldsymbol{\Pi} \times \mathbb{I}^{-1}\boldsymbol{\Pi} + \boldsymbol{\Gamma} \times mg\,\boldsymbol{\chi},$$

$$\tag{6.4.3}$$

$$\dot{\boldsymbol{\Gamma}} = \{\boldsymbol{\Gamma}, h\} \quad = \quad \boldsymbol{\Gamma} \times \nabla_{\boldsymbol{\Pi}} h = \boldsymbol{\Gamma} \times \mathbb{I}^{-1}\boldsymbol{\Pi}. \tag{6.4.4}$$

Remark 6.4.1 (Semidirect products and symmetry breaking) The Lie algebra of the Euclidean group has a structure which is a special case of what is called a *semidirect product*, denoted Ⓢ. Here, it is the semidirect-product action $\mathfrak{so}(3)$Ⓢ$\mathbb{R}^3$ of the Lie algebra of rotations $\mathfrak{so}(3)$ acting on the infinitesimal translations $\mathbb{R}^3$, which happens to coincide with $\mathfrak{se}(3, \mathbb{R})$. In general, the Lie bracket for the semidirect-product action $\mathfrak{g}$ⓈV of the Lie algebra $\mathfrak{g}$ on vector space V is given by

$$[(X, a), (\overline{X}, \overline{a})] = ([X, \overline{X}], \overline{X}(a) - X(\overline{a})), \tag{6.4.5}$$

in which $X, \overline{X} \in \mathfrak{g}$ and $a, \overline{a} \in V$. The action of the Lie algebra $\mathfrak{g}$ on the vector space V is denoted, for example, $X(\overline{a})$. Usually, the action $\mathfrak{g} \times V \to V$ would be the Lie derivative, $\pounds$ or its transpose $\pounds^T$. Let variables $\mu \in \mathfrak{g}^*$ and $b \in V^*$ be dual, respectively, to $X, \overline{X} \in \mathfrak{g}$ and $a, \overline{a} \in V$ as

$$\left[\left(\frac{\partial F}{\partial \mu}, \frac{\partial F}{\partial b} \right), \left(\frac{\partial H}{\partial \mu}, \frac{\partial H}{\partial b} \right) \right]$$
$$= \mp \left(\left[\frac{\partial F}{\partial \mu}, \frac{\partial H}{\partial \mu} \right], \pounds^T_{\frac{\partial H}{\partial \mu}} \frac{\partial F}{\partial b} - \pounds^T_{\frac{\partial F}{\partial \mu}} \frac{\partial H}{\partial b} \right), \tag{6.4.6}$$

where in the Lie symmetry reduction of the Lagrangian in Hamilton's principle, one chooses the $(-)$ (resp., $(+)$) sign for Lie algebras which are invariant under a right (resp., left) Lie-group action.

Consequently, one may write the Lie–Poisson bracket dual to the semidirect-product action of vector fields on vector spaces as

$$\{F, H\} := \mp \left\langle (\mu, b), \left[\left(\frac{\partial F}{\partial \mu}, \frac{\partial F}{\partial b} \right), \left(\frac{\partial H}{\partial \mu}, \frac{\partial H}{\partial b} \right) \right] \right\rangle$$
$$= \mp \left\langle (\mu, b), \left(\left[\frac{\partial F}{\partial \mu}, \frac{\partial H}{\partial \mu} \right], \pounds^T_{\frac{\partial H}{\partial \mu}} \frac{\partial F}{\partial b} - \pounds^T_{\frac{\partial F}{\partial \mu}} \frac{\partial H}{\partial b} \right) \right\rangle$$
$$= \pm \left\langle \mu, \text{ad}_{\frac{\partial F}{\partial \mu}} \frac{\partial H}{\partial \mu} \right\rangle \pm \left\langle \pounds^T_{\frac{\partial H}{\partial \mu}} \frac{\partial F}{\partial b} - \pounds^T_{\frac{\partial F}{\partial \mu}} \frac{\partial H}{\partial b}, b \right\rangle$$

$$= \pm \left\langle \mathrm{ad}^*_{\frac{\partial H}{\partial \mu}} \mu , \frac{\partial F}{\partial \mu} \right\rangle \pm \left\langle \frac{\partial F}{\partial b} , -\mathcal{L}_{\frac{\partial H}{\partial \mu}} b \right\rangle \mp \left\langle \frac{\partial H}{\partial b} , -\mathcal{L}_{\frac{\partial F}{\partial \mu}} b \right\rangle$$

$$= \mp \left\langle \mathrm{ad}^*_{\frac{\partial H}{\partial \mu}} \mu , \frac{\partial F}{\partial \mu} \right\rangle \mp \left\langle \mathcal{L}_{\frac{\partial H}{\partial \mu}} b , \frac{\partial F}{\partial b} \right\rangle \mp \left\langle \frac{\partial H}{\partial b} \diamond b , \frac{\partial F}{\partial \mu} \right\rangle .$$

$$(6.4.7)$$

Hence, the Lie–Poisson equations for semidirect-product action may be written in matrix operator form as

$$\frac{d}{dt} \begin{bmatrix} \mu \\ b \end{bmatrix} = \mp \begin{bmatrix} \mathrm{ad}^*_{\square} \mu & \square \diamond b \\ \mathcal{L}_{\square} b & 0 \end{bmatrix} \begin{bmatrix} \partial H / \partial \mu \\ \partial H / \partial b \end{bmatrix} = \mp \begin{bmatrix} \mathrm{ad}^*_{\frac{\partial H}{\partial \mu}} \mu + \frac{\partial H}{\partial b} \diamond b \\ \mathcal{L}_{\frac{\partial H}{\partial \mu}} b \end{bmatrix} ,$$

$$(6.4.8)$$

with the $(-)$ (resp., $(+)$) sign for Lie algebras invariant under a right (resp., left) Lie-group action.

For the left-invariant variables in the heavy top Hamiltonian in (6.2.10), the motion equations in (6.4.8) with the $(+)$ sign become

$$\frac{d}{dt} \begin{bmatrix} \mathbf{\Pi} \\ \mathbf{\Gamma} \end{bmatrix} = + \begin{bmatrix} \mathbf{\Pi} \times & \mathbf{\Gamma} \times \\ \mathbf{\Gamma} \times & 0 \end{bmatrix} \begin{bmatrix} \partial H / \partial \mathbf{\Pi} = \mathbb{I}^{-1} \mathbf{\Pi} \\ \partial H / \partial \mathbf{\Gamma} = mg\,\chi \end{bmatrix}$$

$$= + \begin{bmatrix} \mathbf{\Pi} \times \frac{\partial H}{\partial \mathbf{\Pi}} + \mathbf{\Gamma} \times \frac{\partial H}{\partial \mathbf{\Gamma}} \\ \mathbf{\Gamma} \times \frac{\partial H}{\partial \mathbf{\Pi}} \end{bmatrix} .$$

$$(6.4.9)$$

This recovers the heavy top equations in (6.4.3) for the Hamiltonian in (6.2.10):

$$h(\mathbf{\Pi}, \mathbf{\Gamma}) = \mathbf{\Pi} \cdot \mathbf{\Omega} - l(\mathbf{\Omega}, \mathbf{\Gamma}) = \tfrac{1}{2} \langle \mathbf{\Pi}, \mathbb{I}^{-1} \mathbf{\Pi} \rangle + \langle mg\,\chi, \mathbf{\Gamma} \rangle, \quad (6.4.10)$$

which is the sum of the kinetic and potential energies of the heavy top. ☐

Exercise. Determine the Hamiltonian equations via Hamilton's principle for a heavy top whose body variables and parameters are made *complex*: $\mathbf{\Omega}, \mathbf{\Gamma}, \chi \in \mathbb{C}^3$,

$$\Omega = \Omega_\Re + i\Omega_\Im, \quad \Gamma = \Gamma_\Re + i\Gamma_\Im, \quad \chi = \chi_\Re + i\chi_\Im.$$

The imaginary part of its Lagrangian would be

$$\Im\ell(\Omega_\Re, \Omega_\Im) = \Omega_\Re \cdot \mathbb{I}\Omega_\Im - mg(\chi_\Re\Gamma_\Im + \chi_\Im\Gamma_\Re)$$

$$(6.4.11)$$

for a real moment of inertia $\mathbb{I} = \mathrm{diag}(I_1, I_2, I_3)$.

As one would expect from equation (6.4.11), the corresponding Hamiltonian arising after a Legendre transform is

$$\Im h(\Pi_\Re, \Gamma_\Re, \Pi_\Im, \Gamma_\Im)$$

$$= \Pi_\Re \cdot \mathbb{I}^{-1}\Pi_\Im + mg(\chi_\Re\Gamma_\Im + \chi_\Im\Gamma_\Re).$$

$$(6.4.12)$$

Derive the equations of motion for this complex heavy top via its Euler–Poincaré equations, and then find the $\mathfrak{se}(3)^* \times \mathfrak{se}(3)^*$ Lie–Poisson bracket for this system. What are the Casimirs for this Lie–Poisson system? ★

Answer. The $\mathfrak{se}(3)^* \times \mathfrak{se}(3)^*$ block diagonal Lie–Poisson bracket for this system is given by

$$\frac{d}{dt}\begin{pmatrix} \Pi_\Re \\ \Gamma_\Re \\ \Pi_\Im \\ \Gamma_\Im \end{pmatrix} = \begin{pmatrix} \Pi_\Re\times & \Gamma_\Re\times & 0 & 0 \\ \Gamma_\Re\times & 0 & 0 & 0 \\ 0 & 0 & \Pi_\Im\times & \Gamma_\Im\times \\ 0 & 0 & \Gamma_\Im\times & 0 \end{pmatrix}$$

$$\times \begin{pmatrix} \partial H/\partial\Pi_\Re := \mathbb{I}^{-1}\Pi_\Im \\ \partial H/\partial\Gamma_\Re := mg\chi_\Im \\ \partial H/\partial\Pi_\Im := \mathbb{I}^{-1}\Pi_\Re \\ \partial H/\partial\Gamma_\Im := mg\chi_\Re \end{pmatrix}. \quad (6.4.13)$$

with intertwined equations of motion

$$\frac{d}{dt}\begin{pmatrix} \Pi_{\Re} \\ \Gamma_{\Re} \\ \Pi_{\Im} \\ \Gamma_{\Im} \end{pmatrix} = \begin{pmatrix} \Pi_{\Re} \times \Omega_{\Im} + mg\Gamma_{\Re} \times \chi_{\Im} \\ \Gamma_{\Re} \times \Omega_{\Im} \\ \Pi_{\Im} \times \Omega_{\Re} + mg\Gamma_{\Im} \times \chi_{\Re} \\ \Gamma_{\Im} \times \Omega_{\Re} \end{pmatrix} \qquad (6.4.14)$$

whose Casimirs are $\Pi_{\Re} \cdot \Gamma_{\Re}$, $|\Gamma_{\Re}|^2$ and $\Pi_{\Im} \cdot \Gamma_{\Im}$, $|\Gamma_{\Im}|^2$.

▲

6.5   The heavy top equations via the Kaluza–Klein construction

The Lagrangian in the heavy top action principle (6.2.1) may be transformed into a quadratic form. This is accomplished by suspending the system in a higher-dimensional space via the *Kaluza–Klein construction*. This construction proceeds for the heavy top as a modification of the well-known Kaluza–Klein construction for a charged particle in a prescribed magnetic field [HoMaRa1998a].

Let Q_{KK} be the manifold $SO(3) \times \mathbb{R}^3$ with variables $(O, \mathbf{q})$. On Q_{KK}, introduce the *Kaluza–Klein Lagrangian* $L_{KK} : TQ_{KK} \simeq TSO(3) \times T\mathbb{R}^3 \mapsto \mathbb{R}$, in which one observes that the coordinate $\mathbf{q}, \in \mathbb{R}^3$ is absent, as

$$L_{KK}(O, \dot{O}, \dot{\mathbf{q}}; \hat{\mathbf{z}}) = L_{KK}(\mathbf{\Omega}, \mathbf{\Gamma}, \dot{\mathbf{q}}) = \tfrac{1}{2}\langle \mathbb{I}\mathbf{\Omega}, \mathbf{\Omega} \rangle + \tfrac{1}{2}|\mathbf{\Gamma} + \dot{\mathbf{q}}|^2, \quad (6.5.1)$$

with $\mathbf{\Omega} = (O^{-1}\dot{O})^\smallfrown$ and $\mathbf{\Gamma} = O^{-1}\hat{\mathbf{z}}$. The Lagrangian L_{KK} is positive definite in $(\mathbf{\Omega}, \mathbf{\Gamma}, \dot{\mathbf{q}})$; therefore, it may be regarded as the kinetic energy of a metric, the *Kaluza–Klein metric* on TQ_{KK}.

The Legendre transformation for L_{KK} gives the momenta

$$\frac{\partial L_{KK}}{\partial \mathbf{\Omega}} = \mathbb{I}\mathbf{\Omega} =: \mathbf{\Pi} \quad \text{and} \quad \frac{\partial L_{KK}}{\partial \dot{\mathbf{q}}} = \mathbf{\Gamma} + \dot{\mathbf{q}} =: \mathbf{p}. \qquad (6.5.2)$$

Since L_{KK} does not depend on $\mathbf{q}$, the Euler–Lagrange equation

$$\frac{d}{dt}\frac{\partial L_{KK}}{\partial \dot{\mathbf{q}}} = \frac{\partial L_{KK}}{\partial \mathbf{q}} = 0,$$

shows that the q-momentum given by $\mathbf{p} = \partial L_{KK}/\partial \dot{\mathbf{q}}$ is conserved. We now identify the *constant vector* $\mathbf{p}$ as the following constant vector in the body:

$$\mathbf{p} = \mathbf{\Gamma} + \dot{\mathbf{q}} = -mg\chi.$$

After this identification, the heavy top action principle in (5.5.1) with the Kaluza–Klein Lagrangian returns Euler's motion equation for the heavy top (6.1.1).

The Hamiltonian H_{KK} associated with L_{KK} by the Legendre transformation (6.5.2) is

$$
\begin{aligned}
H_{KK}(\mathbf{\Pi}, \mathbf{\Gamma}, \mathbf{p}) &= \mathbf{\Pi} \cdot \mathbf{\Omega} + \mathbf{p} \cdot \dot{\mathbf{q}} - L_{KK}(\mathbf{\Omega}, \mathbf{\Gamma}, \dot{\mathbf{q}}) \\
&= \tfrac{1}{2}\mathbf{\Pi} \cdot \mathbb{I}^{-1}\mathbf{\Pi} - \mathbf{p} \cdot \mathbf{\Gamma} + \tfrac{1}{2}|\mathbf{p}|^2 \\
&= \tfrac{1}{2}\mathbf{\Pi} \cdot \mathbb{I}^{-1}\mathbf{\Pi} + \tfrac{1}{2}|\mathbf{p} - \mathbf{\Gamma}|^2 - \tfrac{1}{2}|\mathbf{\Gamma}|^2.
\end{aligned}
$$

Recall that $\mathbf{\Gamma}$ is a unit vector. On the constant level set $|\mathbf{\Gamma}|^2 = 1$, the Kaluza–Klein Hamiltonian H_{KK} is a positive quadratic function, shifted by a constant. Likewise, on the constant level set $\mathbf{p} = -mg\chi$, the Kaluza–Klein Hamiltonian H_{KK} is a function of only the variables $(\mathbf{\Pi}, \mathbf{\Gamma})$ and is equal to the Hamiltonian (6.2.10) for the heavy top up to an additive constant. Consequently, the Lie–Poisson equations for the Kaluza–Klein Hamiltonian H_{KK} now reproduce Euler's motion equation for the heavy top (6.1.1).

Exercise. Use the Hamilton–Pontryagin principle to calculate the equations for heavy top dynamics in the spatial frame. ★

Answer. In the spatial frame, one deals with a rigid-body Lagrangian that is not right invariant and depends on both $\widehat{\omega}(t)$, $\mathbb{I}_{\text{spat}}(t)$, and $\chi_{\text{spat}}(t) := \chi O^{-1}(t)$, cf. equation (6.2.4), where its definition implies the dynamics of

χ_{spat}, namely,

$$\frac{\chi_{\text{spat}}}{dt} = -[\chi_{\text{spat}}, \widehat{\omega}].$$

$$S(\widehat{\omega}, O, \dot{O}, \pi) = \int_a^b \Big[\ell(\widehat{\omega}, \mathbb{I}_{\text{spat}}, \chi_{\text{spat}})$$

$$+ \langle \pi, \dot{O}O^{-1} - \widehat{\omega} \rangle \Big] dt. \qquad (6.5.3)$$

The variations of the action S in formula (6.5.3) are now given by

$$\delta S = \int_a^b \Big\{ \Big\langle \frac{\delta\ell}{\delta\widehat{\omega}} - \pi, \delta\widehat{\omega} \Big\rangle + \Big\langle \delta\pi, (\dot{O}O^{-1} - \widehat{\omega}) \Big\rangle$$

$$+ \Big\langle \pi, \delta(\dot{O}O^{-1}) \Big\rangle + \Big\langle \frac{\delta\ell}{\delta\mathbb{I}_{\text{spat}}}, \delta\mathbb{I}_{\text{spat}} \Big\rangle$$

$$+ \Big\langle \frac{\delta\ell}{\delta\chi_{\text{spat}}}, \delta\chi_{\text{spat}} \Big\rangle \Big\} dt$$

$$= \int_a^b \Big\{ \Big\langle \frac{\delta\ell}{\delta\widehat{\omega}} - \pi, \delta\widehat{\omega} \Big\rangle + \Big\langle \delta\pi, (\dot{O}O^{-1} - \widehat{\omega}) \Big\rangle$$

$$+ \Big\langle -\dot{\pi} - [\pi, \widehat{\omega}] + \Big[\mathbb{I}_{\text{spat}}, \frac{\delta\ell}{\delta\mathbb{I}_{\text{spat}}}\Big]$$

$$+ \Big[\chi_{\text{spat}}, \frac{\delta\ell}{\delta\chi_{\text{spat}}}\Big], \widehat{\xi} \Big\rangle \Big\} dt,$$

(6.5.4)

where we have applied formula (6.5.3) for $\delta(\dot{O}O^{-1})$ and then integrated by parts in time. ▲

Exercise. Show that the Hamiltonian equations for the spatial dynamics of the heavy top in (6.5.4) may be expressed via the following Lie–Poisson matrix operator:

$$\frac{d}{dt}\begin{pmatrix} \pi \\ \mathbb{I}_{\text{spat}} \\ \chi_{\text{spat}} \end{pmatrix} = -\begin{pmatrix} [\pi, \square] & [\mathbb{I}_{\text{spat}}, \square] & [\chi_{\text{spat}}, \square] \\ [\mathbb{I}_{\text{spat}}, \square] & 0 & 0 \\ [\chi_{\text{spat}}, \square] & 0 & 0 \end{pmatrix}$$

$$\times \begin{pmatrix} \delta h/\delta \pi = \widehat{\omega} \\ \delta h/\delta \mathbb{I}_{\text{spat}} = -\delta \ell/\delta \mathbb{I}_{\text{spat}} \\ \delta h/\delta \chi_{\text{spat}} = -\delta \ell/\delta \chi_{\text{spat}} \end{pmatrix} \qquad (6.5.5)$$

with $(\pi, \mathbb{I}_{\text{spat}}, \chi_{\text{spat}}) \in [\mathfrak{so}(3)\,\circledS\,(\text{Sym}^3_+(\mathbb{R}) \oplus \mathbb{R}^3_+)]^*$, where $\circledS$ denotes the semidirect-product action of the Lie algebra $\mathfrak{so}(3)$, represented as 3×3 antisymmetric real matrices acting on the direct sum of vector spaces comprising 3×3 positive, symmetric real matrices and 3D vectors, $(\text{Sym}^3_+(\mathbb{R}) \oplus \mathbb{R}^3_+)$, and the superscript * denotes the dual semidirect-product Lie algebra under matrix trace pairing. See Ref. [LeRaSiMa1992] for a complete geometric treatment of the heavy top in both the body and spatial representations in *The Heavy Top: A Geometric Treatment*.

★

6.5.1 Summary

Geometric mechanics deals with dynamical systems defined by Lie group invariant variational principles (VP/G), such as geodesic motion on a Lie group G whose metric is invariant under the action of the Lie group G. An example is the variational formulation of Euler's rigid-body equations in three dimensions, whose solutions are then seen to be geodesics on the rotation group $SO(3)$.

Geometric mechanics also deals with the effects of breaking the symmetry of VP/G from G to a subgroup $G_0 \subset G$ on whose coset G/G_0 the Lie group G may act by the semidirect product, denoted as $G \circledS (G/G_0)$. An example is Euler's heavy top equations, in which the fixed vertical direction $\hat{z}$ of gravity reduces the Lie symmetry of the VP/G for the rigid-body motion from $SO(3)$ to $SO(2)$. Namely, the remaining Lie group symmetry $SO(2)$ corresponds only to rotations which leave the direction of gravity $\hat{z}$ invariant. For the heavy top, the (left) action $O(t)^{-1}\hat{z}$ of $O(t) \in SO(3)$ on its coset $SO(3)/SO(2)$ comprises the motion in time t of the fixed direction of gravity $\hat{z}$ as seen by an observer in the reference frame of the rotating top. The variational formulation of Euler's heavy top equations then leads to motion on the semidirect-product Lie group $SO(3)\circledS(SO(3)/SO(2))$, which is isomorphic to the Euclidean group of rotations and translations in three dimensions, denoted $SE(3)$.

Remark 6.5.1 Lie–Poisson brackets defined on the dual spaces of semidirect-product Lie algebras tend to occur under rather general circumstances when the symmetry in T^*G is broken, for example, when reduced to an isotropy subgroup of a set of parameters. In particular, there are similarities in structure between the Poisson bracket for compressible flow and that for the heavy top. In the latter case, the vertical direction of gravity breaks the isotropy of $\mathbb{R}^3$ from $SO(3)$ to $SO(2)$, and the dynamics of the $SO(3)$ flow acts on $\Gamma \in SO(3)/SO(2)$. In the case of compressible fluid flow, the initial density configuration breaks the allowed diffeomorphisms from $\mathrm{Diff}(M)$ to $\mathrm{Diff}_{\rho_0}(M)$, the isotropy transformations of the initial density distribution, ρ_0. Thus, the dynamics of the $\mathrm{Diff}(M)$ flow acts on $\rho \in \mathrm{Diff}(M)/\mathrm{Diff}_{\rho_0}(M)$.

The general theory for semidirect products has been reviewed by a number of authors, including Marsden, Ratiu and Weinstein [MaRaWe1984b, MaRaWe1984a]. Many interesting examples of Lie–Poisson brackets on semidirect products exist for fluid dynamics.

These semidirect-product Lie–Poisson Hamiltonian fluid theories range from simple fluids to charged fluid plasmas, magnetised fluids, multiphase fluids, super fluids, Yang–Mills plasmas (relativistic or non-relativistic) and liquid crystals, as well as to other complex fluids. See, for example, the papers by Gibbons, Holm and Kupershmidt [GiHoKu1982, HoKu1982, HoKu1983, HoKu1988]. For discussions of many of these theories from the Euler–Poincaré viewpoint, see the works of Holm, Marsden and Ratiu [HoMaRa1998a] and Holm [Ho2002a].   □

7

LAGRANGIAN AND HAMILTONIAN METHODS FOR GEOMETRIC RAY OPTICS IN TRANSLATION-INVARIANT, AXISYMMETRIC MATERIAL

Contents

What is this lecture about? This lecture reformulates Fermat's principle for geometrical optics with axial symmetry to illustrate the modern geometrical mechanics versions of the variational methods of Lagrange and Hamilton.

7.1   Fermat's principle: Rays take paths of least optical length

In geometrical optics, the ray path is determined by Fermat's principle of least optical length:

$$\delta \int n(x, y, z)\, ds = 0.$$

Here, $n(x, y, z)$ is the index of refraction at the spatial point (x, y, z) and ds is the element of arc length along the ray path through that point. Choosing coordinates so that the z-axis coincides with the optical axis (the general direction of propagation) gives

$$ds = [(dx)^2 + (dy)^2 + (dz)^2]^{1/2} = [1 + \dot{x}^2 + \dot{y}^2]^{1/2}\, dz,$$

with $\dot{x} = dx/dz$ and $\dot{y} = dy/dz$. Thus, Fermat's principle can be written in Lagrangian form, with z playing the role of time:

$$\delta \int L(x, y, \dot{x}, \dot{y}, z)\, dz = 0.$$

Here, the optical Lagrangian is

$$L(x, y, \dot{x}, \dot{y}, z) = n(x, y, z)[1 + \dot{x}^2 + \dot{y}^2]^{1/2} =: n/\gamma,$$

or, equivalently, in two-dimensional vector notation with $\mathbf{q} = (x, y)$,

$$L(\mathbf{q}, \dot{\mathbf{q}}, z) = n(\mathbf{q}, z)[1 + |\dot{\mathbf{q}}|^2]^{1/2} =: n/\gamma \ \text{ with } \ \gamma = [1 + |\dot{\mathbf{q}}|^2]^{-1/2} \le 1.$$

Consequently, the vector Euler–Lagrange equation of the light rays is

$$\frac{d}{ds}\left(n\frac{d\mathbf{q}}{ds}\right) = \gamma\frac{d}{dz}\left(n\gamma\frac{d\mathbf{q}}{dz}\right) = \frac{\partial n}{\partial \mathbf{q}}.$$

The momentum p canonically conjugate to the ray path position q in an "image plane", or on an "image screen", at a fixed value of z is given by

$$\mathbf{p} = \frac{\partial L}{\partial \dot{\mathbf{q}}} = n\gamma\dot{\mathbf{q}},$$

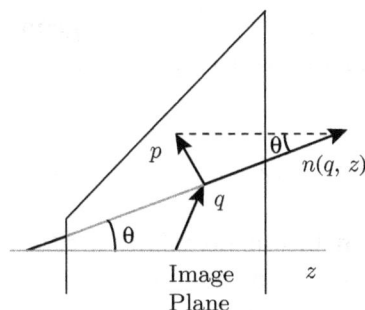

Figure 7.1.   The canonical momentum **p** associated with the coordinate **q** on the image plane at z has magnitude $|\mathbf{p}| = n(\mathbf{q}, z) \sin \theta$, where $\cos \theta = dz/ds$ is the direction cosine of the ray with respect to the optical z-axis.

which satisfies $|\mathbf{p}|^2 = n^2(1 - \gamma^2)$. This implies the velocity

$$\dot{\mathbf{q}} = \mathbf{p}/(n^2 - |\mathbf{p}|^2)^{1/2}.$$

Hence, the momentum is real-valued and the Lagrangian is hyperregular, provided $n^2 - |\mathbf{p}|^2 > 0$. When $n^2 = |\mathbf{p}|^2$, the ray trajectory is vertical and has *grazing incidence* with the image screen.

Defining $\sin \theta = dz/ds = \gamma$ leads to $|\mathbf{p}| = n \cos \theta$ and gives the following geometrical picture of the ray path. Along the optical axis (the z-axis), each image plane normal to the axis is pierced at a point $\mathbf{q} = (x, y)$ by a vector of magnitude $n(\mathbf{q}, z)$ tangent to the ray path. In Figure 7.1, this vector makes an angle θ to the plane. The projection of this vector onto the image plane is the canonical momentum **p**. This picture of the ray paths captures all but the rays of grazing incidence to the image planes. Such grazing rays are ignored in what follows.

Passing now via the usual Legendre transformation from the Lagrangian to the Hamiltonian description gives

$$H = \mathbf{p} \cdot \dot{\mathbf{q}} - L = n\gamma|\dot{\mathbf{q}}|^2 - n/\gamma = -n\gamma = -\left[n(\mathbf{q}, z)^2 - |\mathbf{p}|^2\right]^{1/2}.$$

Thus, in the geometrical picture, the component of the tangent vector of the ray path along the optical axis is (minus) the Hamiltonian, that is, $n(\mathbf{q}, z) \sin \theta = -H$.

The phase-space description of the geometric optics ray paths now follows from Hamilton's equations:

$$\dot{\mathbf{q}} = \frac{\partial H}{\partial \mathbf{p}} = \frac{-1}{H}\mathbf{p}, \qquad \dot{\mathbf{p}} = -\frac{\partial H}{\partial \mathbf{q}} = \frac{-1}{2H}\frac{\partial n^2}{\partial \mathbf{q}}.$$

Remark 7.1.1 (Translation-invariant media) If $n = n(\mathbf{q})$, so that the medium is translation-invariant along the optical axis z, then $H = -n\sin\theta$ is conserved. (The conservation of H at an interface corresponds to Snell's law.) For translation-invariant media, the vector ray path equation simplifies to

$$\ddot{\mathbf{q}} = -\frac{1}{2H^2}\frac{\partial n^2}{\partial \mathbf{q}}.$$

This is the Newtonian dynamics for $\mathbf{q} \in \mathbb{R}^2$.

Thus, in this case, geometrical ray tracing reduces to "Newtonian dynamics" in z, with potential $-n^2(\mathbf{q})$ and with "time" rescaled along each path by the constant value of $\sqrt{2H}$ determined from the initial conditions for each ray.   □

7.2 Axisymmetric, translation-invariant optical materials

In axisymmetric, translation-invariant media, the index of refraction is a function of the radius alone. Axisymmetry implies an additional constant of motion and, hence, reduction of the Hamiltonian system for the light rays to phase-plane analysis. For such media, the index of refraction satisfies

$$n(\mathbf{q}, z) = n(r), \qquad r = |\mathbf{q}|.$$

Passing to polar coordinates (r, ϕ) with $\mathbf{q} = (x, y) = r(\cos\phi, \sin\phi)$ leads, in the usual way, to

$$|\mathbf{p}|^2 = p_r^2 + p_\phi^2/r^2.$$

Consequently, the optical Hamiltonian

$$H = -\left[n(r)^2 - p_r^2 - p_\phi^2/r^2\right]^{1/2}$$

is independent of the azimuthal angle ϕ; therefore, its canonically conjugate "angular momentum" p_ϕ is conserved.

Using the relation $\mathbf{q} \cdot \mathbf{p} = rp_r$ leads to an interpretation of p_ϕ in terms of the image-screen phase-space variables $\mathbf{p}$ and $\mathbf{q}$. Namely,

$$|\mathbf{p} \times \mathbf{q}|^2 = |\mathbf{p}|^2|\mathbf{q}|^2 - (\mathbf{p} \cdot \mathbf{q})^2 = p_\phi^2.$$

The conserved quantity $p_\phi = \mathbf{p} \times \mathbf{q} = yp_x - xp_y$ is called the skewness function or the *Petzval invariant* for axisymmetric media.

Vanishing of p_ϕ occurs for *meridional rays*, for which $\mathbf{p}$ and $\mathbf{q}$ are collinear in the image plane. On the other hand, p_ϕ takes its maximum value for *sagittal rays*, for which $\mathbf{p} \cdot \mathbf{q} = 0$, so that $\mathbf{p}$ and $\mathbf{q}$ are orthogonal in the image plane.

Exercise. (Axisymmetric, translation-invariant materials) Write Hamilton's canonical equations for axisymmetric, translation-invariant media.

Solve these equations for the case of an optical fibre with a radially graded index of refraction in the following form by reducing the problem to phase-plane analysis:

$$n^2(r) = \lambda^2 + (\mu - \nu r^2)^2, \quad \lambda, \mu, \nu = \text{constants}.$$

How does the corresponding phase-space portrait differ between $p_\phi = 0$ and $p_\phi \neq 0$?

Show that for $p_\phi \neq 0$, the problem reduces to a Duffing oscillator in a rotating frame up to a rescaling of time by the value of the Hamiltonian on each ray "orbit".   ★

7.3 The Petzval invariant and its Poisson bracket relations

The skewness function in the image plane

$$S = p_\phi = \mathbf{p} \times \mathbf{q} \cdot \hat{\mathbf{z}} = y p_x - x p_y$$

generates rotations of phase space, rotating $\mathbf{q}$ and $\mathbf{p}$ jointly, each in its plane, around the optical axis $\hat{\mathbf{z}}$. Its square, S^2 (called the Petzval invariant), is conserved for ray optics in axisymmetric media. That is, $\{S^2, H\} = 0$ for optical Hamiltonians of the form

$$H = -\left[n(|\mathbf{q}|^2)^2 - |\mathbf{p}|^2\right]^{1/2}.$$

We define the axisymmetric invariant coordinates by the map $T^*\mathbb{R}^2 \mapsto \mathbb{R}^3$ $(\mathbf{q}, \mathbf{p}) \mapsto (X_1, X_2, X_3)$,

$$X_1 = |\mathbf{q}|^2 \geq 0, \quad X_2 = |\mathbf{p}|^2 \geq 0, \quad X_3 = \mathbf{p} \cdot \mathbf{q}. \tag{7.3.1}$$

The following Poisson bracket relations hold since rotations preserve dot products:

$$\{S^2, X_1\} = 0, \quad \{S^2, X_2\} = 0, \quad \{S^2, X_3\} = 0.$$

In terms of these invariant coordinates, the Petzval invariant and optical Hamiltonian satisfy

$$S^2 = X_1 X_2 - X_3^2 \geq 0 \quad \text{and} \quad H^2 = n^2(X_1) - X_2 \geq 0. \tag{7.3.2}$$

The level sets of S^2 are hyperboloids of revolution around the $X_1 = X_2$ axis, extending through the interior of the $S = 0$ cone, and lying between the X_1- and X_2-axes. The level sets of H^2 in (7.3.2) depend on the functional form of the index of refraction, but they are X_3-independent.

7.4 Hamilton's characteristic function for optics in 3D

The tangents to Fermat's light rays in an isotropic medium are normal to Huygens wave fronts. The phase of such a wave front is given by [BoWo2013]

$$\phi = \int \mathbf{k} \cdot d\mathbf{r} - \omega(\mathbf{k}, \mathbf{r}) \, dt. \tag{7.4.1}$$

The Huygens wave front is a travelling wave, for which the phase ϕ is constant. Consequently, the phase shift

$$\int \mathbf{k} \cdot d\mathbf{r} = \int \frac{d\phi}{d\mathbf{r}} \cdot d\mathbf{r} = \int d\phi = \int \omega(\mathbf{k}, \mathbf{r}) \, dt$$

along a ray trajectory for a travelling wave is given by the integral $\int \omega(\mathbf{k}, \mathbf{r}) \, dt$.

Physically, the index of refraction $n(\mathbf{r})$ of the medium at position $\mathbf{r}$ relates to the travelling wave phase speed ω/k as

$$\frac{\omega}{k} = \frac{c}{n(\mathbf{r})}, \qquad k = |\mathbf{k}|,$$

where c is the speed of light in a vacuum. The index of refraction in a material medium always satisfies $n(\mathbf{r}) > 1$.

As it turns out, the frequency ω of the travelling wave plays the role of the Hamiltonian and the wave vector $\mathbf{k}$ corresponds to the canonical momentum. Consequently, we may write Hamilton's canonical equations for a wave front as

$$\frac{d\mathbf{r}}{dt} = \frac{\partial \omega}{\partial \mathbf{k}} = \frac{c}{n(\mathbf{r})} \frac{\mathbf{k}}{k} = \frac{c^2}{n^2 \omega} \mathbf{k}, \tag{7.4.2}$$

$$\frac{d\mathbf{k}}{dt} = -\frac{\partial \omega}{\partial \mathbf{r}} = \frac{ck}{2n^3} \frac{\partial n^2}{\partial \mathbf{r}} = \frac{\omega}{n} \frac{\partial n}{\partial \mathbf{r}}. \tag{7.4.3}$$

After a short manipulation, these canonical equations combine into

$$\frac{n^2}{c} \frac{d}{dt} \left(\frac{n^2}{c} \frac{d\mathbf{r}}{dt} \right) = \frac{1}{2} \frac{\partial n^2}{\partial \mathbf{r}}. \tag{7.4.4}$$

Equation (7.4.4) may also be expressed in terms of a *different* variable time increment $cdt = n^2 d\tau$ in the form of Newton's second law:

$$\frac{d^2\mathbf{r}}{d\tau^2} = \frac{1}{2}\frac{\partial n^2}{\partial \mathbf{r}} \qquad \text{(Newton's second law)}. \qquad (7.4.5)$$

If instead of τ, we define the variable time increment $cdt = nd\sigma$, then equation (7.4.4) yields the following *eikonal equation* for the paths of light rays in geometric optics, $\mathbf{r}(\sigma) \in \mathbb{R}^3$:

$$\frac{d}{d\sigma}\left(n(\mathbf{r})\frac{d\mathbf{r}}{d\sigma}\right) = \frac{\partial n}{\partial \mathbf{r}} \qquad \text{(Eikonal equation)}. \qquad (7.4.6)$$

This equation also follows from Fermat's principle of stationarity of the optical length under variations of the ray paths,

$$\delta \int_A^B n(\mathbf{r}(\sigma))\, d\sigma = 0 \qquad \text{(Fermat's principle)}, \qquad (7.4.7)$$

with arc-length parameter σ satisfying $d\sigma^2 = d\mathbf{r}(\sigma) \cdot d\mathbf{r}(\sigma)$ and, hence, $|d\mathbf{r}/d\sigma| = 1$.

From this vantage point, one sees that replacing $\mathbf{k} \to \frac{\omega}{c}\nabla S$ in the first Hamilton equation in (7.4.2) yields

$$n(\mathbf{r})\frac{d\mathbf{r}}{d\sigma} = \nabla S(\mathbf{r}) \qquad \text{(Huygens equation)}, \qquad (7.4.8)$$

from which the eikonal equation (7.4.6) may be recovered by differentiating and using

$$d/d\sigma = n^{-1}\nabla S \cdot \nabla \quad \text{and} \quad |\nabla S|^2 = n^2. \qquad (7.4.9)$$

Exercise. Derive the eikonal equation (7.4.6) by differentiating Huygens equation (7.4.8) and using the transformation relations in Equation (7.4.9). ★

7.5 $\mathbb{R}^3$ Poisson bracket for ray optics

The Poisson brackets among the axisymmetric variables X_1, X_2 and X_3 in (7.3.1) close among themselves:

$$\{X_1, X_2\} = 4X_3, \quad \{X_2, X_3\} = -2X_2, \quad \{X_3, X_1\} = -2X_1. \quad (7.5.1)$$

These Poisson brackets derive from a single $\mathbb{R}^3$ Poisson bracket for $\mathbf{X} = (X_1, X_2, X_3)$ given by

$$\{F, H\} = -\nabla S^2 \cdot \nabla F \times \nabla H.$$

Consequently, we may re-express the equations of Hamiltonian ray optics in axisymmetric media with $H = H(X_1, X_2)$ as

$$\dot{\mathbf{X}} = \nabla S^2 \times \nabla H,$$

with Casimir S^2, for which $\{S^2, H\} = 0$, for every H. Thus, the flow preserves volume (div$\dot{\mathbf{X}} = 0$), and the evolution occurs on the intersections of level surfaces of the axisymmetric media invariants S^2 and $H(X_1, X_2)$.

7.6   Poisson bracket properties for ray optics

The Casimir invariant $S^2 = X_1 X_2 - X_3^2$ is quadratic. In such cases, one may write the $\mathbb{R}^3$ Poisson bracket in the suggestive form

$$\{F, H\} = -C_{ij}^k X_k \frac{\partial F}{\partial X_i} \frac{\partial H}{\partial X_j}.$$

In this particular case, $C_{12}^3 = 4$, $C_{23}^2 = 2$ and $C_{31}^1 = 2$, and the rest either vanish or are obtained from the antisymmetry of C_{ij}^k under the exchange of any pair of its indices. These values are the structure constants of any of the Lie algebras $\mathfrak{sp}(2, \mathbb{R})$, $\mathfrak{so}(2, 1)$, $\mathfrak{su}(1, 1)$ or $\mathfrak{sl}(2, \mathbb{R})$. Thus, the reduced description of Hamiltonian ray optics in terms of axisymmetric $\mathbb{R}^3$ variables is said to be "Lie–Poisson" on the dual space of any of these Lie algebras, say, $\mathfrak{sp}(2, \mathbb{R})^*$ for definiteness. We will have more to say about Lie–Poisson brackets later when we discuss the Euler–Poincaré reduction theorem.

Remark 7.6.1 (Coadjoint orbits) As one might expect, the coadjoint orbits of the symplectic group $SP(2, \mathbb{R})$ lie on the level sets of the Petzval invariant $S^2 = X_1X_2 - X_3^2$ in $\mathbb{R}^3$, in contrast to the level sets of spheres $X_1^2 + X_2^2 + X_3^2$ for the orthogonal group $SO(3)$.

The level sets of the Petzval invariant $S^2 = X_1X_2 - X_3^2$ comprise hyperboloids of revolution $H^2 \otimes S^1$ around the $X_1 = X_2$ axis in the $\pi/4$ direction on the horizontal plane $X_3 = 0$. The level sets of the Hamiltonian H in (7.3.2) are independent of the vertical coordinate. The axisymmetric invariants $\mathbf{X} = (X_1, X_2, X_3) \in \mathbb{R}^3$ evolve along the intersections of these level sets by $\dot{\mathbf{X}} = \nabla S^2 \times \nabla H$, as the vertical Hamiltonian knife $H = constant$ slices through the hyperbolic onion of level sets of the Petzval invariant S^2. To visualise the hyperbolic onion in $\mathbb{R}^3$, note that in the coordinates $Y_1 = (X_1+X_2)/2, Y_2 = (X_2 - X_1)/2, Y_3 = X_3$, one has $S^2 = Y_1^2 - Y_2^2 - Y_3^2$.

Being invariant under the flow (integral curves on H^2) of the Hamiltonian vector field given by

$$X_S = \{\cdot, S\} = \frac{\partial S}{\partial p}\frac{\partial}{\partial q} - \frac{\partial S}{\partial q}\frac{\partial}{\partial p}, \tag{7.6.1}$$

each point on any layer H^2 of the *hyperbolic onion* $H^2 \otimes S^1$ in Figure 7.2 comprises an S^1 orbit in phase space under rotation by p_ϕ since the quantities (X_1, X_2, X_3) are all invariant under S^1 rotations about the optical axis. In phase space, this orbit is a circular rotation by an angle ϕ about the optical axis of both $\mathbf{q}$ and $\mathbf{p}$ on an $\mathbb{R}^2$ image screen at position z. □

Exercise. (Potential relation to the Hopf fibration) The Petzval fibration $R^4 \rightarrow H^2 \otimes S^1$ (locally) seems to be a hyperbolic analogue of the spherical Hopf fibration: $S^3 \rightarrow S^2 \otimes S^1$ (locally). This observation may be worth further investigation. ★

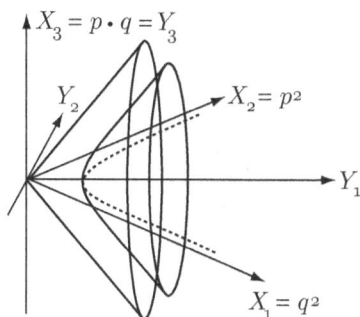

Figure 7.2. The hyperbolic onion of level sets of the Petzval invariant H^2 given in $\mathbb{R}^3$ coordinates as $S^2 = X_1 X_2 - X_3^2 = \mathrm{const}$.

Definition 7.6.1 Hamiltonian matrices m_i *with* $i = 1, 2, 3$ *each satisfy*

$$J m_i + m_i^T J = 0, \quad \text{where } J = \begin{pmatrix} 0 & -1 \\ 1 & 0 \end{pmatrix}. \qquad (7.6.2)$$

That is, $J m_i = (J m_i)^T$ *is a symmetric matrix.*

The following traceless constant Hamiltonian matrices provide a representation of the symplectic Lie algebra $sp(2, \mathbb{R})$:

$$m_1 = \begin{pmatrix} 0 & 0 \\ -2 & 0 \end{pmatrix}, \quad m_2 = \begin{pmatrix} 0 & 2 \\ 0 & 0 \end{pmatrix}, \quad m_3 = \begin{pmatrix} 1 & 0 \\ 0 & -1 \end{pmatrix}. \qquad (7.6.3)$$

Exercise. Show that the corresponding matrix commutation relations of the Hamiltonian matrices are

$$[m_1, m_2] = 4m_3, \quad [m_2, m_3] = -2m_2, \quad [m_3, m_1] = -2m_1.$$

★

Definition 7.6.2 *The map* $\mathcal{J} : T^*\mathbb{R}^2 \simeq \mathbb{R}^2 \times \mathbb{R}^2 \rightarrow sp(2,\mathbb{R})^*$ *is defined by*

$$
\begin{aligned}
\mathcal{J}^\xi(\mathbf{z}) &:= \Big\langle \mathcal{J}(\mathbf{z}), \xi \Big\rangle_{sp(2,\mathbb{R})^* \times sp(2,\mathbb{R})} \\
&= \Big(\mathbf{z}, J\xi\mathbf{z} \Big)_{\mathbb{R}^2 \times \mathbb{R}^2} \\
&:= z_k (J\xi)_{kl} z_l \\
&= \mathbf{z}^T \cdot J\xi\mathbf{z} \\
&= \mathrm{tr}\Big((\mathbf{z} \otimes \mathbf{z}^T J)\xi \Big),
\end{aligned}
\tag{7.6.4}
$$

where $\xi \in sp(2,\mathbb{R})$ *and* $\mathbf{z} = (\mathbf{q}, \mathbf{p})^T \in \mathbb{R}^2 \times \mathbb{R}^2$.

Remark 7.6.2 The map $\mathcal{J}(\mathbf{z})$ obtained in the last line of (7.6.4),

$$
\mathcal{J}(\mathbf{z}) = (\mathbf{z} \otimes \mathbf{z}^T J) \in sp(2,\mathbb{R})^*,
\tag{7.6.5}
$$

sends $\mathbf{z} = (\mathbf{q}, \mathbf{p})^T \in \mathbb{R}^2 \times \mathbb{R}^2$ to $\mathcal{J}(\mathbf{z}) = (\mathbf{z} \otimes \mathbf{z}^T J)$, which is an element of $sp(2,\mathbb{R})^*$, the dual space to $sp(2,\mathbb{R})$. Under the pairing $\langle \cdot, \cdot \rangle : sp(2,\mathbb{R})^* \times sp(2,\mathbb{R}) \rightarrow \mathbb{R}$ given by the trace of the matrix product, one finds the Hamiltonian, or phase-space function,

$$
\Big\langle \mathcal{J}(\mathbf{z}), \xi \Big\rangle = \mathrm{tr}\left(\mathcal{J}(\mathbf{z})\,\xi \right),
\tag{7.6.6}
$$

in which $\xi \in sp(2,\mathbb{R})$ and $\mathcal{J}(\mathbf{z}) = (\mathbf{z}\otimes\mathbf{z}^T J) \in sp(2,\mathbb{R})^*$ is a cotangent lift *momentum map.* $\square$

Remark 7.6.3 (Map to axisymmetric invariant variables) The map

$$
\mathcal{J} : T^*\mathbb{R}^2 \simeq \mathbb{R}^2 \times \mathbb{R}^2 \rightarrow sp(2,\mathbb{R})^*
$$

in (7.6.4) for $Sp(2,\mathbb{R})$ acting diagonally on $\mathbb{R}^2 \times \mathbb{R}^2$ in Equation (7.6.5) may be expressed in matrix form as

$$
\begin{aligned}
\mathcal{J} &= (\mathbf{z} \otimes \mathbf{z}^T J) \\
&= 2 \begin{pmatrix} \mathbf{p} \cdot \mathbf{q} & -|\mathbf{q}|^2 \\ |\mathbf{p}|^2 & -\mathbf{p} \cdot \mathbf{q} \end{pmatrix} \\
&= 2 \begin{pmatrix} X_3 & -X_1 \\ X_2 & -X_3 \end{pmatrix}.
\end{aligned}
\tag{7.6.7}
$$

This is none other than the matrix form of the map (7.3.1) to axisymmetric invariant variables,

$$T^*\mathbb{R}^2 \to \mathbb{R}^3 : (\mathbf{q}, \mathbf{p})^T \to \mathbf{X} = (X_1, X_2, X_3),$$

defined as

$$X_1 = |\mathbf{q}|^2 \geq 0, \quad X_2 = |\mathbf{p}|^2 \geq 0, \quad X_3 = \mathbf{p} \cdot \mathbf{q}. \quad (7.6.8)$$

Applying the momentum map $\mathcal{J}$ to the vector of Hamiltonian matrices $\mathbf{m} = (m_1, m_2, m_3)$ in Equation (7.6.3) yields the individual components

$$\mathcal{J} \cdot \mathbf{m} = 4\mathbf{X} \quad \Longleftrightarrow \quad \mathbf{X} = \frac{1}{4} z_k (\mathcal{J}\mathbf{m})_{kl} z_l. \quad (7.6.9)$$

Thus, the momentum map $\mathcal{J} : T^*\mathbb{R}^2 \simeq \mathbb{R}^2 \times \mathbb{R}^2 \to sp(2, \mathbb{R})^*$ recovers the components of the vector $\mathbf{X} = (X_1, X_2, X_3)$ at any point on a level set of the Petzval invariant $S^2 = X_1 X_2 - X_3^2$.   □

Exercise. Consider the $\mathbb{R}^3$ Poisson bracket

$$\{f, h\} = -\nabla c \cdot \nabla f \times \nabla h. \quad (7.6.10)$$

Let $c = \mathbf{x}^T \cdot \mathbb{C}\mathbf{x}$ be a quadratic form on $\mathbb{R}^3$, and let $\mathbb{C}$ be the associated symmetric 3×3 matrix. Show that this is the Lie–Poisson bracket for the Lie algebra structure

$$[\mathbf{u}, \mathbf{v}]_\mathbb{C} = \mathbb{C}(\mathbf{u} \times \mathbf{v}).$$

What is the underlying matrix Lie algebra? What are the coadjoint orbits of this Lie algebra? What are its structure constants? What quadratic form provides the pairing of this Lie algebra with its dual Lie algebra?   ★

For more discussion of Lagrangian and Hamiltonian methods for geometric ray optics, see [Ho2011b, Ho2011c].

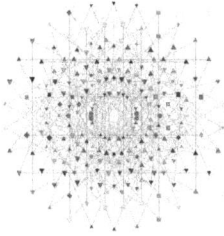

8

RIGID BODY EQUATIONS ON $SO(n)$

Contents

What is this lecture about? This lecture outlines Manakov's approach to proving algebraic integrability of the $SO(n)$ rigid body dynamics via its isospectral eigenvalue problem.

8.1 Manakov's formulation of the $SO(n)$ rigid body

Proposition 8.1.1 (Manakov [Ma1976]) *Euler's equations for a rigid body on $SO(n)$ take the matrix commutator form:*

$$\frac{dM}{dt} = [M, \Omega] \quad \textit{with } M = \mathbb{A}\Omega + \Omega\mathbb{A}, \qquad (8.1.1)$$

where the $n \times n$ matrices M, Ω are skew symmetric (forgoing superfluous hats) and $\mathbb{A}$ is symmetric.

Proof. Manakov's commutator form of the $SO(n)$ rigid-body equations (8.1.1) follows from the Euler–Lagrange equations for Hamilton's principle $\delta S = 0$, with $S = \int l \, dt$ for the Lagrangian

$$l(\Omega) = -\frac{1}{2}\mathrm{tr}(\Omega \mathbb{A} \Omega),$$

where $\Omega = O^{-1}\dot{O} \in \mathfrak{so}(n)$ and the $n \times n$ matrix $\mathbb{A}$ is symmetric. Taking matrix variations in Hamilton's principle yields

$$\delta S = -\frac{1}{2}\int_a^b \mathrm{tr}\big(\delta\Omega\,(\mathbb{A}\Omega + \Omega\mathbb{A})\big)\,dt = -\frac{1}{2}\int_a^b \mathrm{tr}\big(\delta\Omega\,M\big)\,dt$$

after cyclically permuting the order of matrix multiplication under the trace and substituting $M := \mathbb{A}\Omega + \Omega\mathbb{A}$. Using the Euler–Poincaré variational formula for $\delta\Omega$ now leads to

$$\delta S = -\frac{1}{2}\int_a^b \mathrm{tr}\big((\Xi^{\cdot} + \Omega\Xi - \Xi\Omega)M\big)\,dt.$$

Integrating by parts and permuting under the trace then yields the equation

$$\delta S = \frac{1}{2}\int_a^b \mathrm{tr}\big(\Xi\,(\dot{M} + \Omega M - M\Omega)\big)\,dt.$$

Finally, invoking stationarity for arbitrary Ξ implies the commutator form (8.1.1). ∎

8.2   Matrix Euler–Poincaré equations

Manakov's commutator form of the rigid-body equations recalls a much earlier work by Poincaré [Po1901], who also observed that the matrix commutator form of Euler's rigid-body equations suggests an additional mathematical structure going back to Lie's theory of groups of transformations depending continuously on parameters. In particular, Poincaré [Po1901] remarked that the commutator form of Euler's rigid-body equations would make sense for any Lie algebra, not just for $\mathfrak{so}(3)$. The proof of Manakov's commutator form (8.1.1) by Hamilton's principle is essentially the same as Poincaré's proof in Ref. [Po1901].

Theorem 8.2.1 (Matrix Euler–Poincaré equations)

The Euler–Lagrange equations for Hamilton's principle $\delta S = 0$, with $S = \int l(\Omega)\, dt$, may be expressed in matrix commutator form:

$$\frac{dM}{dt} = [M, \Omega] \quad \textit{with } M = \frac{\delta l}{\delta \Omega}, \tag{8.2.1}$$

for any Lagrangian $l(\Omega)$, where $\Omega = g^{-1}\dot{g} \in \mathfrak{g}$ and $\mathfrak{g}$ is the matrix Lie algebra of any matrix Lie group G.

Proof. The proof here is the same as the proof of Manakov's commutator formula via Hamilton's principle, modulo replacing $O^{-1}\dot{O} \in so(n)$ with $g^{-1}\dot{g} \in \mathfrak{g}$. ∎

Remark 8.2.1 Poincaré's observation leading to the matrix Euler–Poincaré equation (8.2.1) was reported in two pages with no references while using a variational approach [Po1901]. Hence the name Euler–Poincaré equations. Note that if $\Omega = g^{-1}\dot{g} \in \mathfrak{g}$, then $M = \delta l/\delta \Omega \in \mathfrak{g}^*$, where the dual is defined in terms of the matrix trace pairing. □

Exercise. Retrace the proof of the variational principle for the Euler–Poincaré equation, replacing the left-invariant quantity $g^{-1}\dot{g}$ with the right-invariant quantity $\dot{g}g^{-1}$. ★

8.3   An isospectral eigenvalue problem for the $SO(n)$ rigid body

The solution of the $SO(n)$ rigid-body dynamics

$$\frac{dM}{dt} = [M, \Omega] \quad \text{with } M = \mathbb{A}\Omega + \Omega\mathbb{A},$$

for the evolution of the $n \times n$ skew-symmetric matrices M, Ω, with constant symmetric $\mathbb{A}$, is given by a similarity transformation (later to be identified as coadjoint motion):

$$M(t) = O(t)^{-1}M(0)O(t) =: \mathrm{Ad}^*_{O(t)}M(0),$$

with $O(t) \in SO(n)$ and $\Omega := O^{-1}\dot{O}(t)$. Consequently, the evolution of $M(t)$ is *isospectral*. This means we have the following:

- The initial eigenvalues of the matrix $M(0)$ are preserved by the motion; that is, $d\lambda/dt = 0$ in

$$M(t)\psi(t) = \lambda\psi(t),$$

  provided its eigenvectors $\psi \in \mathbb{R}^n$ evolve according to

$$\psi(t) = O(t)^{-1}\psi(0).$$

  The proof of this statement follows from the corresponding property of similarity transformations.

- Its matrix invariants are preserved:

$$\frac{d}{dt}\mathrm{tr}(M - \lambda\mathrm{Id})^K = 0,$$

  for every non-negative integer power K.

  This is clear because the invariants of the matrix M may be expressed in terms of its eigenvalues; however, these are invariant under a similarity transformation.

Proposition 8.3.1 *Isospectrality allows the quadratic rigid-body dynam-ics (8.1.1) on $SO(n)$ to be rephrased as a system of two coupled linear equations: the eigenvalue problem for M and an evolution equation for its eigenvectors ψ, as follows:*

$$M\psi = \lambda\psi \quad and \quad \dot\psi = -\Omega\psi, \quad with \; \Omega = O^{-1}\dot O(t).$$

Proof. Applying isospectrality in the time derivative of the first equation yields

$$(\dot M + [\Omega, M])\psi + (M - \lambda\mathrm{Id})(\dot\psi + \Omega\psi) = 0.$$

Now, substitute the second equation to recover (8.1.1).   ∎

8.4   Manakov's proof of algebraic integrability of the $SO(4)$ rigid body

The Euler equations on $SO(4)$ are

$$\frac{dM}{dt} = M\Omega - \Omega M = [M, \Omega],$$

where Ω and M are skew-symmetric 4×4 matrices. The angular frequency Ω is a linear function of the angular momentum M. Manakov [Ma1976] "deformed" these equations into

$$\frac{d}{dt}(M + \lambda A) = [(M + \lambda A), (\Omega + \lambda B)],$$

where A and B are also skew-symmetric 4×4 matrices and λ is a scalar constant parameter. For these equations to hold for any value of λ, the coefficient of each power must vanish.

- The coefficient of λ^2 is

$$0 = [A, B].$$

  Therefore, A and B must commute. So, let them be constant and diagonal:

$$A_{ij} = \mathrm{diag}(a_i)\delta_{ij}, \quad B_{ij} = \mathrm{diag}(b_i)\delta_{ij}. \qquad \text{(no sum)}$$

- The coefficient of λ is

$$0 = \frac{dA}{dt} = [A, \Omega] + [M, B].$$

Therefore, by antisymmetry of M and Ω,

$$(a_i - a_j)\Omega_{ij} = (b_i - b_j)M_{ij} \quad \Longleftrightarrow \quad \Omega_{ij} = \frac{b_i - b_j}{a_i - a_j}M_{ij}.$$

(no sum)

- Finally, the coefficient of λ^0 is the Euler equation

$$\frac{dM}{dt} = [M, \Omega],$$

but now with the restriction that the moments of inertia are of the form

$$\Omega_{ij} = \frac{b_i - b_j}{a_i - a_j}M_{ij}, \qquad \text{(no sum)}$$

which turns out to possess only five free parameters.

With these conditions, Manakov's deformation of the $SO(4)$ rigid body implies for every power n that

$$\frac{d}{dt}(M + \lambda A)^n = [(M + \lambda A)^n, (\Omega + \lambda B)].$$

Since the commutator is antisymmetric, its trace vanishes, and one has

$$\frac{d}{dt}\text{tr}(M + \lambda A)^n = 0$$

after commuting the trace operation with time derivative. Consequently,

$$\text{tr}(M + \lambda A)^n = \text{constant}$$

for each power of λ. That is, all the coefficients of each power of λ are constant in time for the $SO(4)$ rigid body. Manakov [Ma1976] proved that these constants of motion are sufficient to completely determine the solution.

Remark 8.4.1 This result generalises considerably. First, it holds for $SO(n)$. Indeed, as proven using the theory of algebraic varieties by Haine [Ha1984], Manakov's method captures all the algebraically integrable rigid bodies on $SO(n)$, and the moments of inertia of these bodies possess only $2n - 3$ parameters. (Recall that in Manakov's case for $SO(4)$, the moment of inertia possesses only five parameters.) Moreover, Miščenko and Fomenko [MiFo1978] prove that every compact Lie group admits a family of left-invariant metrics with completely integrable geodesic flows. ☐

Exercise. Try computing the constants of motion $\mathrm{tr}(M + \lambda A)^n$ for the values $n = 2, 3, 4$. How many additional constants of motion are needed for integrability in these cases? How many for general n?

Hint: Keep in mind that M is a skew-symmetric matrix, $M^T = -M$, so the trace of the product of any diagonal matrix times an odd power of M vanishes. ★

Answer. The traces of the powers $\mathrm{tr}(M + \lambda A)^n$ are given by

$$n{=}2:\quad \mathrm{tr}M^2 + 2\lambda\mathrm{tr}(AM) + \lambda^2\mathrm{tr}A^2$$

$$n{=}3:\quad \mathrm{tr}M^3 + 3\lambda\mathrm{tr}(AM^2) + 3\lambda^2\mathrm{tr}A^2M + \lambda^3\mathrm{tr}A^3$$

$$n{=}4:\quad \mathrm{tr}M^4 + 4\lambda\mathrm{tr}(AM^3) + \lambda^2(2\mathrm{tr}A^2M^2 + 4\mathrm{tr}AMAM)$$
$$+\lambda^3\mathrm{tr}A^3M + \lambda^4\mathrm{tr}A^4.$$

The number of conserved quantities for $n = 2, 3, 4$ are, respectively, one ($C_1 = \mathrm{tr}M^2$), one ($I_1 = \mathrm{tr}AM^2$) and two ($C_2 = \mathrm{tr}M^4$ and $I_2 = 2\mathrm{tr}A^2M^2 + 4\mathrm{tr}AMAM$). The quantities C_1 and C_2 are Casimirs for the Lie–Poisson bracket for the rigid body. Thus, $\{C_1, H\} = 0 = \{C_2, H\}$ for any Hamiltonian $H(M)$, so of course C_1 and C_2 are conserved. However, each Casimir only reduces the dimension of the system by one. The dimension of the original phase space is $\dim T^*SO(n) = n(n-1)$. This is reduced in half by the left invariance of the Hamiltonian to the dimension of the dual Lie algebra $\dim \mathfrak{so}(n)^* = n(n-1)/2$. For $n = 4$, $\dim \mathfrak{so}(4)^* = 6$. One then subtracts the number of Casimirs (two) by passing to their level surfaces, which leaves four dimensions remaining in this case. The other two constants of motion I_1 and I_2 turn out to be sufficient for integrability because they are in involution $\{I_1, I_2\} = 0$ and because the level surfaces of the Casimirs are symplectic manifolds by the Marsden–Weinstein reduction theorem [MaWe1974]. For more details, see the work of Ratiu [Ra1980].   ▲

Exercise. How do the Euler equations look on $\mathfrak{so}(n)^*$ as a matrix equation? Is there an analogue of the hat map for $\mathfrak{so}(3)^*$?

Hint: The Lie algebra $\mathfrak{so}(4)$ is locally isomorphic to $\mathfrak{so}(3) \times \mathfrak{so}(3)$.   ★

8.5   Implications of left invariance

This Hamiltonian $H(M)$ for the $SO(n)$ rigid-body equations is invariant under the action of $SO(n)$ from the left. The corresponding conserved momentum map under this symmetry is known from the previous section as

$$J_L \colon T^*SO(n) \mapsto \mathfrak{so}(n)^* \quad \text{is } J_L(Q, P) = PQ^T$$

On the other hand, we will see in Lecture 17 that the momentum map for right action is

$$J_R \colon T^*SO(n) \mapsto \mathfrak{so}(n)^*, \quad J_R(Q, P) = Q^T P$$

Hence, $M = Q^T P = J_R$. Therefore, one computes

$$H(Q, P) = H(Q, Q \cdot M) = H(\text{Id}, M) \quad \text{(by left invariance)}$$
$$= H(M) = \tfrac{1}{2}\langle M, \mathbb{I}^{-1}(M)\rangle$$
$$= \tfrac{1}{2}\langle Q^T P, \mathbb{I}^{-1}(Q^T P)\rangle.$$

Hence, we may write the $SO(n)$ rigid-body Hamiltonian as

$$H(Q, P) = \tfrac{1}{2}\langle Q^T P, \Omega(Q, P)\rangle.$$

Consequently, the variational derivatives of $H(Q, P) = \tfrac{1}{2}\langle Q^T P, \Omega(Q, P)\rangle$ are

$$\delta H = \langle Q^T \delta P + \delta Q^T P, \Omega(Q, P)\rangle$$
$$= \text{tr}(\delta P^T Q\Omega) + \text{tr}(P^T \delta Q\Omega)$$
$$= \text{tr}(\delta P^T Q\Omega) + \text{tr}(\delta Q\Omega P^T)$$
$$= \text{tr}(\delta P^T Q\Omega) + \text{tr}(\delta Q^T P\Omega^T)$$
$$= \langle \delta P, Q\Omega\rangle - \langle \delta Q, P\Omega\rangle,$$

where the skew symmetry of Ω is used in the last step, that is, $\Omega^T = -\Omega$. Thus, Hamilton's canonical equations take the symmetric form

$$\dot{Q} = \frac{\delta H}{\delta P} = Q\Omega,$$
$$\dot{P} = -\frac{\delta H}{\delta Q} = P\Omega. \tag{8.5.1}$$

Equations (8.5.1) are the *symmetric generalised rigid-body equations*, derived earlier by Bloch, Brockett and Crouch [BlCr1996, BlBrCr1997] from the viewpoint of optimal control. Combining them yields the left-invariant relations,

$$Q^{-1}\dot{Q} = \Omega = P^{-1}\dot{P} \iff \frac{d}{dt}(PQ^T) = 0,$$

in agreement with conservation of the momentum map $J_L(Q, P) = PQ^T$ corresponding to symmetry of the Hamiltonian under the left action of $SO(n)$. This momentum map is the angular momentum in space, which is related to the angular momentum in the body by $PQ^T = m = QMQ^T$. Thus, we recognise the canonical momentum as $P = QM$ and the momentum maps for left and right actions as

$$J_L = m = PQ^T \qquad \text{(spatial angular momentum)},$$

$$J_R = M = Q^T P \qquad \text{(body angular momentum)}.$$

Thus, momentum maps $T^*G \mapsto \mathfrak{g}^*$ corresponding to symmetries of the Hamiltonian produce conservation laws; while momentum maps $T^*G \mapsto \mathfrak{g}^*$ which do *not* correspond to symmetries may be used to re-express the equations on $\mathfrak{g}^*$ in terms of variables on T^*G.

Exercise. Write Manakov's deformation of the rigid-body equations in the symmetric form (8.5.1).   ★

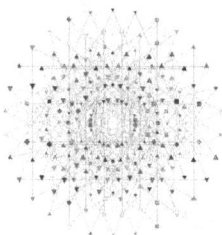

9

EXERCISES: INSIDE THE GEOMETRIC MECHANICS CUBE

Contents

What is this lecture about? This lecture provides a plethora of instructive worked examples of geometric formulations of finite-dimensional dynamics based on the Lie symmetry of Hamilton's principle.

9.1 Introduction

Following Noether [No1918], geometric mechanics deals with Lie group invariant variational principles.

This section applies the framework of geometric mechanics to Hamilton's principle (HP) for a dynamical system with two degrees of freedom (dof$_1$ and dof$_2$), taking values on $(\mathbb{R}_1^n \times \mathbb{R}_2^d)$, respectively. The Lagrangian $L : T(\mathbb{R}_1^n \times \mathbb{R}_2^d) \to \mathbb{R}$ is invariant under the transitive action of a Lie group G on $\mathbb{R}_1^n$, given by $G \times \mathbb{R}_1^n \to \mathbb{R}_1^n$. Geometric mechanics uses this framework to gain insight into the similarities in structure among many different topics, ranging from geometric optics to classical mechanics, quantum mechanics and onwards to ideal fluid mechanics. The main goal of the lecture is to gain insight into how the framework of geometric mechanics applies the transformation theory of smooth invertible flows of Lie groups acting on configuration manifolds to unify our perception of physically disparate topics.

- HP:
  $\delta S = 0$ with $S = \int_a^b L(q, v) + \langle p, \dot{q} - v \rangle_{TM} \, dt$
  and
  $S = \int_a^b \langle p, \dot{q} \rangle_{TM} - H(q, p) dt$

- Reduction by Lie symmetry TM:
  $q_t = g_t q_0$, $\dot{q}_t = \dot{g}_t q_0$.
  Let $L(g, v) = L(kg, kv)$, $k \in G$,
  set $L(e, g^{-1} v) =: l(\xi)$ and apply HP.

- Noether's theorem: one-parameter Lie G-symmetry of HP for $\xi \in \mathfrak{g} \simeq T_e G$ implies the conservation of

$$\left\langle \frac{\partial L}{\partial \dot{q}}, \delta q \right\rangle_{TM} = \langle p, \delta q \rangle_{TM} = \langle p, -\pounds_\xi q \rangle_{TM}$$

$$=: \langle p \diamond q, \xi \rangle_{\mathfrak{g}} = \langle J(q, p), \xi \rangle_{\mathfrak{g}},$$

which is the Hamiltonian defined by the ξ-component of the momentum map $J(q, p) = p \diamond q \in \mathfrak{g}^*$.

- Legendre transformation (LT):
  $p := \partial L / \partial \dot{q}$, $H(q, p) := \langle p, v \rangle_{TM} - L(q, v)$,
  and
  $J := \partial l / \partial \xi$, $h(J) := \langle J, \xi \rangle_{\mathfrak{g}} - l(\xi)$.

- Reduced (left-invariant) HP:
  $S_{\text{red}} = \int_a^b l(\xi) + \langle J, g^{-1} \dot{g} - \xi \rangle_{\mathfrak{g}} \, dt$
  and
  $S_{\text{red}} = \int_a^b \langle J, g^{-1} \dot{g} \rangle_{\mathfrak{g}} - h(J) dt.$

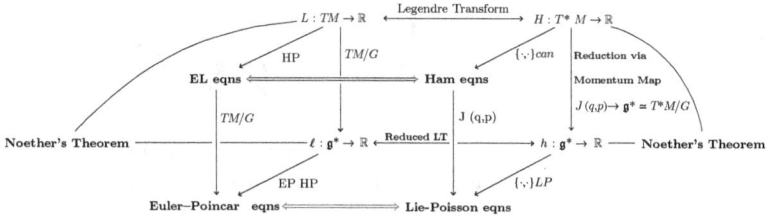

Figure 9.1. Geometric mechanics as a cube of six commuting diagrams.

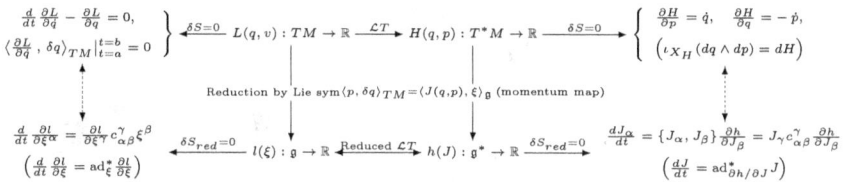

Figure 9.2. Unfolding the cube of commuting diagrams for geometric mechanics.

In studying these notes, the reader might keep in mind the equivariance of the diagrams in Figures 9.1 and 9.2.

Six commuting diagrams comprise the box of geometric mechanics relations.

Exercise. Write HP for two degrees of freedom in the configuration manifold $(q_1, q_2) \in M = \mathbb{R}_1^n \times \mathbb{R}_2^d$.

1. Formulate the Lagrangian $L \in C^\infty(TM)$ as $L(q_1, u_1; q_2, u_2) : T(\mathbb{R}_1^n \times \mathbb{R}_2^d) \to \mathbb{R}$, where, e.g., $u_1 \in T_{q_1}\mathbb{R}_1^n$ lies in the tangent fibre over the point $q_1 \in \mathbb{R}_1^n$.

2. For a family of curves $(q_1(t), q_2(t))$ with tangents $(\dot{q}_1(t), \dot{q}_2(t))$, define Hamilton's variational principle as

$$0 = \delta S = \delta \int_a^b L(q_1, u_1; q_2, u_2) + \langle p_1, \tfrac{dq_1}{dt} - u_1 \rangle$$
$$+ \langle p_2, \tfrac{dq_2}{dt} - u_2 \rangle \, dt,$$

with natural pairing $\langle \cdot, \cdot \rangle : T^*M \times TM \to \mathbb{R}$ for $p_i \in T^*M$ and $u_i \in TM, i = 1, 2$.

3. Set $\delta S = 0$ under variations $\delta q_i, \delta u_i$ and δp_i for $i = 1, 2$ to derive the Euler–Lagrange (EL) equations:

$$p_1 = \frac{\partial L}{\partial u_1}\bigg|_{u_1 = \dot{q}_1}, \quad \frac{dp_1}{dt} = \frac{\partial L}{\partial q_1} \quad \text{and}$$

$$p_2 = \frac{\partial L}{\partial u_2}\bigg|_{u_1 = \dot{q}_2}, \quad \frac{dp_2}{dt} = \frac{\partial L}{\partial q_2},$$

for endpoint conditions $\langle p_1, \delta q_1 \rangle|_a^b = 0$ and $\langle p_2, \delta q_2 \rangle|_a^b = 0$.   ★

Exercise. Recall that a Lie group G is a group whose transformations g_ϵ depend differentiably on a set of parameters, $\epsilon \in \mathbb{R}$.

Show that a Lie group is also a manifold.

Hint: A manifold is a space on which the rules of calculus apply.   ★

Exercise. The vector space $T_eG = \mathfrak{g}$ (i.e., the tangent space of G evaluated at the identity of G) is called the Lie algebra of the Lie group G.

Show that the differential $\frac{d}{d\epsilon}\big|_{\epsilon=0}$ of the composition law of the *left action* of a Lie group G evaluated at the identity $\epsilon = 0$ yields its Lie algebra bracket operation $[\cdot, \cdot] : \mathfrak{g} \times \mathfrak{g} \to \mathfrak{g}$, also written as the adjoint action $\mathrm{ad}_\xi \theta = [\xi, \theta] = -[\theta, \xi]$ for $\xi, \theta \in \mathfrak{g}$.   ★

Answer. Consult Chapters 4–6 of Ref. [Ho2008]. ▲

Exercise. Suppose a Lie group G acts transitively on a submanifold $\mathbb{R}_1^n$ of the configuration manifold $M = \mathbb{R}_1^n \times \mathbb{R}_2^d$, so that $q_1(\epsilon) = g_\epsilon q_1(0)$, with $\epsilon = 0$ at the identity element of G. Assume this G-action only affects the submanifold $\mathbb{R}_1^n$ and does not act on the remaining submanifold $\mathbb{R}_2^d$ of the configuration manifold M.

Prove the Noether theorem [No1918], that each one-parameter Lie symmetry of the Lagrangian in HP implies a conserved quantity under the dynamics of the corresponding EL equations for this composite system on $M = \mathbb{R}_1^n \times \mathbb{R}_2^d$. ★

Answer. Invariance of the Lagrangian $\delta L = \frac{dL}{d\epsilon}\big|_{\epsilon=0} = 0$ for an infinitesimal Lie symmetry transform $(\delta q_1 = \frac{dq_1}{d\epsilon}\big|_{\epsilon=0}$ and $\delta u_1 = \frac{du_1}{d\epsilon}\big|_{\epsilon=0})$ implies (via the endpoint condition $\langle p_1, \delta q_1 \rangle\big|_a^b = 0$) that if the EL equations hold, then the pairing $\langle p_1, \delta q_1 \rangle$ must be conserved. Here, $\delta u_1 \in T_{q_1}\mathbb{R}_1^n$ is the tangent lift at q_1 of the action

$$\delta q_1 = \frac{d}{d\epsilon}\bigg|_{\epsilon=0} q_1(\epsilon) =: -\pounds_\xi q_1$$

of Lie algebra element $\xi \in \mathfrak{g}$ on position $q_1 \in \mathbb{R}_1^n \subset M$. The operation $\pounds_\xi q_1$ is the Lie derivative action of $\xi \in \mathfrak{g}$ on the position $q_1 \in \mathbb{R}_1^n$.

Define the diamond operation to relate vector space pairings

$$\langle p_1, \delta q_1 \rangle = \langle p_1, -\pounds_\xi q_1 \rangle =: \langle p_1 \diamond q_1, \xi \rangle_\mathfrak{g}.$$

In pairing $\langle \, \cdot \, , \, \cdot \, \rangle_\mathfrak{g} : \mathfrak{g}^* \times \mathfrak{g} \to \mathbb{R}$, we have $p_1 \diamond q_1 \in \mathfrak{g}^* := T_e^* G$ and $\xi \in \mathfrak{g} := T_e G$.

The map $\mu = p_1 \diamond q_1$ is called a cotangent-lift momentum map $\mu : T^*\mathbb{R}^n_1 \to \mathfrak{g}^*$.

Cotangent-lift momentum maps also figure in the definitions of Lie–Poisson (LP) brackets in Hamiltonian dynamics.   ▲

9.2   Reduction by Lie symmetry

Exercise. Reduction by Lie symmetry of Lagrangian:

$$T(\mathbb{R}^n_1 \times \mathbb{R}^d_2) \to ((T\mathbb{R}^n_1)/G) \times T\mathbb{R}^d_2 \simeq \mathfrak{g} \times T\mathbb{R}^d_2$$

for a Lagrangian that is left invariant under the Lie group action $G \times T\mathbb{R}^n_1 \to T\mathbb{R}^n_1$.

The left-invariant G-reduced Lagrangian appears in the action integral of HP as

$$S = \int_a^b L(\xi; q_2, u_2) + \langle \mu, g^{-1}\dot{g} - \xi \rangle_\mathfrak{g} + \langle p_2, \dot{q}_2 - u_2 \rangle dt.$$

Assuming that $\dot{q}_2$ (resp., ξ) may be obtained from p_2 and q_2 (resp., from $\mu = \frac{\partial L}{\partial \xi}$) shows that HP $\delta S = 0$ implies

$$\mu = \frac{\partial L}{\partial \xi}, \quad \frac{d\mu}{dt} = \mathrm{ad}^*_\xi \mu \quad \text{and}$$

$$p_2 = \frac{\partial L}{\partial u_2}\bigg|_{u_2 = \dot{q}_2}, \quad \frac{dp_2}{dt} = \frac{\partial L}{\partial q_2},$$

where $\mathrm{ad}^*_\xi \mu$ is defined by $\langle \mathrm{ad}^*_\xi \mu, \theta \rangle_\mathfrak{g} := \langle \mu, \mathrm{ad}_\xi \theta \rangle_\mathfrak{g}$, with $\theta := g^{-1}\delta g$ and $\delta g := \frac{dg}{d\epsilon}\big|_{\epsilon=0}$. These are the

Euler–Poincaré (EP) equations for $\mu = \frac{\partial L}{\partial \xi} \in \mathfrak{g}^*$ and the EL equations for $(q_2, p_2) \in T^* \mathbb{R}^d$.

Hint: The derivation of the EP equations from HP follows by defining $\theta = g^{-1} \delta g$ with $\delta g := \frac{dg}{d\epsilon}|_{\epsilon=0}$ and proving by equality of cross derivatives in $\frac{d}{dt}$ and $\frac{d}{d\epsilon}|_{\epsilon=0}$ that

$$\frac{d\xi}{d\epsilon}\Big|_{\epsilon=0} - \frac{d\theta}{dt} = -\mathrm{ad}_\theta \xi \quad \text{with}$$

$$\mathrm{ad}_\theta \xi = [\theta, \xi] \quad \text{for left Lie group action.} \qquad \bigstar$$

Answer. To show that the combination of EP and EL equations above follows from HP, one first shows by direct computation that

$$0 = \delta S = \int_a^b \left\langle \frac{\partial L}{\partial \xi} - \mu, \delta \xi \right\rangle_{\mathfrak{g}} + \left\langle \mu, \frac{d\theta}{dt} + \mathrm{ad}_\xi \theta \right\rangle_{\mathfrak{g}}$$

$$+ \left\langle \frac{\partial L}{\partial u_2} - p_2, \delta u_2 \right\rangle_{T\mathbb{R}_2} + \left\langle \frac{\partial L}{\partial q_2} - \dot{p}_2, \delta q_2 \right\rangle_{T\mathbb{R}_2} dt,$$

with the vanishing endpoint term $\langle p_2, \delta q_2 \rangle|_a^b = 0$. Then, upon integration by parts in time and use of the definition of $\mathrm{ad}_\xi^* \mu$, show that

$$0 = \delta S = \int_a^b \left\langle \frac{\partial L}{\partial \xi} - \mu, \delta \xi \right\rangle_{\mathfrak{g}} - \left\langle \frac{d\mu}{dt} - \mathrm{ad}_\xi^* \mu, \theta \right\rangle_{\mathfrak{g}}$$

$$+ \left\langle \frac{\partial L}{\partial u_2} - p_2, \delta u_2 \right\rangle_{T\mathbb{R}_2} + \left\langle \frac{\partial L}{\partial q_2} - \dot{p}_2, \delta q_2 \right\rangle_{T\mathbb{R}_2} dt,$$

by imposing another vanishing endpoint term $\langle \mu, \theta \rangle_{\mathfrak{g}}|_a^b = 0$. $\qquad \blacktriangle$

Exercise. Derive canonical Hamiltonian dynamics for two degrees of freedom as follows:

1. For $M = (\mathbb{R}_1^n \times \mathbb{R}_2^d)$, pass from the Lagrangian $L \in C^\infty(TM)$ to the Hamiltonian $H \in C^\infty(T^*M)$ via the Legendre transformation

$$H(q_1, p_1; q_2, p_2) := \langle p_1, u_1 \rangle + \langle p_2, u_2 \rangle$$
$$- L(q_1, u_1; q_2, u_2).$$

   Then, equate the differentials on the left and right sides of the Legendre transformation, and use EL equations to find Hamilton's canonical equations:

$$H_{q_1} = -L_{q_1} = -\dot{p}_1, \quad H_{p_1} = u_1 = \dot{q}_1,$$
$$H_{q_2} = -L_{q_2} = -\dot{p}_2, \quad H_{p_2} = u_2 = \dot{q}_2.$$

2. Using the chain rule for $\frac{dF}{dt}$ with $F(q_1, p_1; q_2, p_2)$, show that Hamilton's equations follow from a bracket operation $\{\cdot, \cdot\}$ by $\{q_1, p_1\} = 1$, $\{q_2, p_2\} = 1$, and zero otherwise.

   Verify that the bracket operation $\{F, H\}$ defined by

$$\frac{dF}{dt} = tr \left(\begin{bmatrix} \frac{\partial F}{\partial q} \\ \frac{\partial F}{\partial p} \end{bmatrix}^T \begin{bmatrix} 0 & Id \\ -Id & 0 \end{bmatrix} \begin{bmatrix} \frac{\partial H}{\partial q} \\ \frac{\partial H}{\partial p} \end{bmatrix} \right)$$

$$= \sum_{i=1}^{2} \frac{\partial H}{\partial p_i} \frac{\partial F}{\partial q_i} - \frac{\partial H}{\partial q_i} \frac{\partial F}{\partial p_i} =: \{F, H\},$$

   is skew and satisfies both the Leibnitz product rule and the Jacobi identity, $\{F, \{G, H\}\} + c.p. = 0$. ★

Exercise. Show that *Hamiltonian vector fields* defined via the canonical Poisson bracket $\{\,\cdot\,,\,\cdot\,\}$ as

$$X_H = \{\cdot, H\} \in \mathfrak{X}(T^*(M))$$

satisfy the commutation relation

$$[X_F, X_H] = -X_{\{F,H\}}$$

for phase-space functions $F, H \in C^\infty(T^*(M))$.   ★

Answer. This follows from Jacobi's identity for the canonical Poisson bracket.   ▲

Exercise. Explain how the Lie derivative of a phase-space function F by a Hamiltonian vector field X_H is related to the canonical Poisson bracket.   ★

Answer.
$$\pounds_{X_F} H = X_F(H) = \{F, H\}.$$ ▲

Exercise. Prove the following:

1. Invariance of a scalar function H under the flow ϕ_ϵ^X of a vector field X means that the Lie derivative $\pounds_X H$ vanishes. That is, $\phi_\epsilon^{X*} H(z) = H(\phi_\epsilon^X(z)) = H(z) \Leftrightarrow \pounds_X H = 0$.
2. The Lie derivative action $\pounds_X F$ on a scalar phase-space function $F(z)$ is the familiar directional derivative.

3. A smooth vector field X generates a flow ϕ_ϵ^X
which takes place along its integral curves.   ★

Answer.

1. Invariance of a scalar function H under the flow ϕ_ϵ^X
   of a vector field X means that the Lie derivative
   $\mathcal{L}_X H$ vanishes. This property follows from the *Lie
   chain rule*, given by

   $$\frac{d}{d\epsilon}\left(\phi_\epsilon^{X*} H(z)\right) = \frac{d}{d\epsilon} H\left(\phi_\epsilon^X(z)\right) = \phi_\epsilon^{X*}\left(\mathcal{L}_X H\right)$$

   $$\Leftrightarrow \mathcal{L}_X H = 0, \tag{9.2.1}$$

   by evaluating at the identity ($\epsilon = 0$) with
   $\phi_0^{X*} = Id$.

2. The directional derivative $X(H) = \frac{\partial H}{\partial z^i} X^i(z)$ governs how a function such as the Hamiltonian $H(z)$
   changes along the flow of the vector field X. The
   directional derivative of a scalar function is a special case of the Lie derivative $\mathcal{L}_X$ since $X(H) =
   \mathcal{L}_X H$ for scalar functions. To see this, one computes
   via the converse of the Lie chain rule,

   $$\mathcal{L}_X H = \frac{d}{d\epsilon}\left(\phi_\epsilon^{X*} H(z)\right)\Big|_{\epsilon=0}$$

   $$= \frac{d}{d\epsilon}\Big|_{\epsilon=0} H(\phi_\epsilon^X z) = \frac{\partial H}{\partial z} \cdot X(z) = X(H).$$

3. The integral curves of a vector field X are given by
   the unique solution of the system of ODEs $\frac{dz^i}{d\epsilon} =
   X^i(z)$ with initial condition $z^i(0)$. Consequently, the
   integral curve ϕ_ϵ^X of a vector field X satisfies the
   flow condition $\phi_\epsilon^X \circ \phi_\tau^X = \phi_{\epsilon+\tau}^X$.   ▲

Exercise. Derive a coordinate-free definition of the Poisson bracket in terms of the operation of insertion

$$\iota : \mathfrak{X} \times \Lambda^n \to \Lambda^{n-1}$$

of Hamiltonian vector fields X_F and X_H into the closed symplectic two-form $\omega \in \Lambda^2$ with $d\omega = 0$.   ★

Answer. The required formula may be verified as

$$\{F, H\} = \omega(X_F, X_H) = \iota_{X_H}(\iota_{X_F}\omega),$$

which is related to the original phase-space coordinates by

$$\omega = \sum_{i=1}^{2} dq_i \wedge dp_i = -\sum_{i=1}^{2} dp_i \wedge dq_i,$$

$$X_H = \{\cdot, H\} = \sum_{i=1}^{2} \frac{\partial H}{\partial p_i}\frac{\partial}{\partial q_i} - \frac{\partial H}{\partial q_i}\frac{\partial}{\partial p_i},$$

which is familiar from earlier lectures.   ▲

Exercise. Explicitly calculate the verification of the coordinate-free definition of the Poisson bracket in terms of the operation of insertion

$$\iota : \mathfrak{X} \times \Lambda^n \to \Lambda^{n-1}$$

for the specific Hamiltonian vector fields X_F and X_H into the exact symplectic two-form $\omega \in \Lambda^2$ with $d\omega = 0$.

The dynamics along each integral curves of X_H is determined by

$$dH = \omega(X_H, \cdot) = \iota_{X_H}\omega,$$

in which the vector field X_H is inserted (ι) into the symplectic two-form ω to create the exact 1-form dH.

In the original coordinates, this is

$$dH = \iota_{\left(\frac{\partial H}{\partial p}\frac{\partial}{\partial q} - \frac{\partial H}{\partial q}\frac{\partial}{\partial p}\right)}(dq \wedge dp) = \frac{\partial H}{\partial q}dq + \frac{\partial H}{\partial p}dp.$$

Show that $\omega(X_F, X_H) = \{F, H\}$ in the original coordinates.   ★

Answer.

$$\frac{dF}{dt} = \omega(X_F, X_H)$$

$$= \iota_{X_H}(\iota_{X_F}\omega) = \iota_{X_H}dF$$

$$= \iota_{\left(\frac{\partial H}{\partial p}\frac{\partial}{\partial q} - \frac{\partial H}{\partial q}\frac{\partial}{\partial p}\right)}\left(\frac{\partial F}{\partial q}dq + \frac{\partial F}{\partial p}dp\right)$$

$$= \frac{\partial H}{\partial p}\frac{\partial F}{\partial q} - \frac{\partial H}{\partial q}\frac{\partial F}{\partial p}$$

$$= \{F, H\}.$$

The opposite calculation was used to motivate the introduction of the insertion operation in previous lectures.

▲

Exercise. Use Cartan's geometric definition of the *Lie derivative* of the symplectic two-form ω with respect to the Hamiltonian vector field $X_F = \{\cdot, F\}$ to show that the symplectic form ω is invariant under the Lie algebra actions of Hamiltonian vector fields. ★

Answer. The coordinate-free expression

$$\pounds_{X_F}\omega = d(\iota_{X_F}\omega) + \iota_{X_F}d\omega = (d\,\iota_{X_F} + \iota_{X_F}d)\,\omega$$

is Cartan's geometric definition of the *Lie derivative* of the symplectic two-form ω with respect to the Hamiltonian vector field $X_F = \{\cdot,\, F\}$.

Since the symplectic form ω is closed ($d\omega = 0$) and $\iota_{X_F}\omega = dF$ for a Hamiltonian vector field X_F, we have

$$\pounds_{X_F}\omega = d(\iota_{X_F}\omega) = d^2 F = 0.$$

The finite transformation ϕ_ϵ generated by the left-invariant Hamiltonian vector field $X_F = \phi_\epsilon^{-1}\phi_\epsilon'|_{\epsilon=0}$ is called a *symplectic flow*. ▲

9.3 Lie chain rule and Noether's theorem

Exercise. Use the Lie chain rule and the equivalence of the dynamic and Cartan definitions of the Lie derivative of a differential form to prove that a smooth symplectic flow $\phi_\epsilon^{X_F}$ generated by a Hamiltonian vector field given by $X_F = \frac{d}{d\epsilon}\phi_\epsilon|_{\epsilon=0} = \{\cdot,\, F\}$, with $\iota_{X_F}\omega = dF$, preserves the symplectic two-form ω under the pull-back $\phi_\epsilon^*\omega(q,p) := \omega(\phi_\epsilon q, \phi_\epsilon p)$. ★

Answer. In the context of flows here, the dynamic definition of the Lie derivative is natural:

$$\pounds_{X_F}\omega = \frac{d}{d\epsilon}(\phi_\epsilon^*\omega)\Big|_{\epsilon=0} \quad \text{with } X_F = \phi_\epsilon^{-1}\phi_\epsilon'\big|_{\epsilon=0}.$$

One immediately finds that

$$\frac{d}{d\epsilon}(\phi_\epsilon^*\omega) = \phi_\epsilon^*(\pounds_{X_F}\omega) \overset{d\omega=0}{=} \phi_\epsilon^* d(\iota_{X_F}\omega)$$

$$\overset{\iota_{X_F}\omega=dF}{=} \phi_\epsilon^* d(dF) = 0$$

since $d^2 F = 0$.

The first step invokes the Lie chain rule. The second step invokes the equivalence of the dynamic and Cartan definitions of the Lie derivative of a differential form.   ▲

Exercise. Show that the diamond operation ($\diamond$) is natural under Ad*. That is,

$$\mathrm{Ad}_g^*(b \diamond a) = (bg) \diamond (ag) = (g^*b) \diamond (g^*a, w)_V, \qquad (9.3.1)$$

for $g \in \mathrm{Diff}(\mathcal{D})$, $b \in V$ and $a \in V^*$.   ★

Answer. Let V be a vector space and V^* be its dual under a pairing $\langle b, a\rangle_V$ for $b \in V$ and $a \in V^*$. Assume that the smooth invertible maps (diffeomorphisms) $\mathrm{Diff}(\mathcal{D})$ defined in domain $\mathcal{D}$ has a right representation on V and an induced right representation on V^*, both of which are denoted by composition. Also, let $w \in \mathfrak{X}(\mathcal{D})$ be an arbitrary fixed vector field, and let $\langle \cdot, \cdot\rangle_{\mathfrak{X}}$ be the L^2 pairing between the Lie algebra of vector fields and its dual $\mathfrak{X}^*(\mathcal{D})$, the 1-form densities. The diamond operation ($\diamond$) in this situation is defined as

$$\langle b \diamond a, w\rangle_{\mathfrak{X}} := \langle b, -\pounds_w a\rangle_V.$$

Next, recall the proof that the Lie derivative by vector fields is natural under Adjoint action of a Lie group element $g \in G$ on its Lie algebra $\mathfrak{g}$, which in this case becomes, for $g \in \mathrm{Diff}(\mathcal{D})$, $w \in \mathfrak{X}(\mathcal{D})$, $b \in V$ and $a \in V^*$,

$$\mathcal{L}_{\mathrm{Ad}_g w} a = g_* \left(\mathcal{L}_w g^* a \right) = \left(\mathcal{L}_w a g \right) g^{-1}. \qquad (9.3.2)$$

Upon recalling that the differential is natural under the push-forward g_*, the identity (9.3.2) can be shown via the Cartan form of the Lie derivative as

$$\mathcal{L}_{\mathrm{Ad}_g w} a = \mathbf{d}\left(g_* w \lrcorner a \right) + g_* w \lrcorner \mathbf{d} a$$
$$= g_* \left(\mathbf{d}(w \lrcorner g^* a) + w \lrcorner \mathbf{d} g^* a \right) = g_* \left(\mathcal{L}_w g^* a \right).$$

Since the vector field w is fixed, equation (9.3.2) implies that the diamond operation ($\diamond$) is natural under Ad^*, for $\mathrm{Diff}(\mathcal{D})$ acting by push-forward on arbitrary vector spaces V and V^* defined in domain $\mathcal{D}$:

$$\left\langle \mathrm{Ad}_g^*(b \diamond a), w \right\rangle_{\mathfrak{X}} = \left\langle b \diamond a, \mathrm{Ad}_g w \right\rangle_{\mathfrak{X}} = \left\langle -b, \mathcal{L}_{(\mathrm{Ad}_g w)} a \right\rangle_V$$
$$= \left\langle -b, \left(\mathcal{L}_w a g \right) g^{-1} \right\rangle_V = \left\langle -bg, \mathcal{L}_w a g \right\rangle_V$$
$$= \left\langle (bg) \diamond (ag), w \right\rangle_V = \left\langle g^* b \diamond g^* a, w \right\rangle_V.$$

$\blacktriangle$

Exercise. State and prove Noether's theorem on the Hamiltonian side. ★

Answer. On the Hamiltonian side, Noether's theorem states that if the one-parameter flow $\phi_\epsilon^{X_F}$ is a symmetry of H, then skew symmetry of the Poisson bracket $\{F, H\} = 0 = -\{H, F\}$ implies that the Hamiltonians $F(z(t))$ and $H(z(t))$ in the notation $z = (q, p) \in T^*M$ are preserved under each other's dynamics.

Proof. For Hamiltonian vector fields $X_F := \{\,\cdot\,, F\}$, the antisymmetry of the Poisson bracket implies

$$\mathcal{L}_{X_F} H = X_F(H) = \{F, H\}$$
$$= -\{H, F\} = -X_H(F) = -\mathcal{L}_{X_H} F.$$

In this situation, if the flow $\phi_\epsilon^{X_F}$ is a symmetry of H, then both $\phi_\epsilon^{X_F*}H(z) = H(\phi_\epsilon^{X_F}z) = H(z)$ and $\phi_\epsilon^{X_H*}F(z) = F(\phi_t^{X_H}z) = F(z)$ are conserved, so that the flow $\phi_t^{X_H}$ is a symmetry of F. Consequently, $\{F, H\} = 0 = -\{H, F\}$ implies that the Hamiltonians $F(z(t))$ and $H(z(t))$ are preserved under each other's dynamics.

This is Noether's theorem on the Hamiltonian side.   ▲

Exercise. What are the Hamiltonian actions of Noether's cotangent-lift momentum map? Provide proofs.   ★

Answer. Suppose the variation δq_1 in Noether's theorem is defined by the infinitesimal action (i.e., the action tangent to the identity) of the Lie group element $g_\epsilon \in G$ parameterised by $\epsilon \in \mathbb{R}$ acting on the configuration space coordinate $q_1 \in \mathbb{R}_1^n$ by push-forward $q_1(\epsilon) = q_1(0)g_\epsilon^{-1}$ (i.e., pull-back by the inverse).

In this case, Noether's theorem on the Hamiltonian side introduces a quantity $N^\xi(q_1, p_1)$ given by

$$N^\xi(q_1, p_1) := \langle p_1, \delta q_1 \rangle_{TM} = \langle p_1, -\pounds_\xi q_1 \rangle_{TM}, \quad (9.3.3)$$

where the variation δq_1 is given by the Lie algebra action

$$\delta q_1 = -\pounds_\xi q_1 = \frac{d}{d\epsilon}\Big|_{\epsilon=0}[q_1(0)g_\epsilon^{-1}] = \frac{d}{d\epsilon}\Big|_{\epsilon=0}q_1(\epsilon),$$
$$(9.3.4)$$

with $q_1(\epsilon) := q_1(0)g_\epsilon^{-1} = g_{\epsilon*}q_1(0)$.   ▲

Exercise. To prove the statement (9.3.4) above, show that the tangent to the push-forward evaluated at the identity ($\epsilon = 0$) is given by (minus) the Lie derivative with respect to $\xi = [\frac{dg}{d\epsilon}g^{-1}]_{\epsilon=0}$ by first proving the Lie chain rule for the push-forward

$$\frac{d}{d\epsilon}(g_{\epsilon\,*}q_1(0)) = -\mathcal{L}_{\frac{dg}{d\epsilon}g_\epsilon^{-1}}(g_{\epsilon\,*}q_1(0)),$$

then evaluating at the identity.   ★

Answer. To prove the Lie chain rule $\frac{d}{d\epsilon}(g_{\epsilon\,*}q_1(0)) = -\mathcal{L}_{\frac{dg}{d\epsilon}g_\epsilon^{-1}}g_{\epsilon\,*}q_1(0)$, one computes

$$\frac{d}{d\epsilon}q_1(0)g_\epsilon^{-1} = -\left[q_1(0)g_\epsilon^{-1}\frac{dg}{d\epsilon}g_\epsilon^{-1}\right] = -\left[\mathcal{L}_{\frac{dg}{d\epsilon}g_\epsilon^{-1}}q_1(0)g_\epsilon^{-1}\right]$$

$$= -\mathcal{L}_{\frac{dg}{d\epsilon}g_\epsilon^{-1}}g_{\epsilon\,*}q_1(0).$$

(The statement above then holds at $\epsilon = 0$.)   ▲

Exercise. Write the Noether Hamiltonian in terms of the diamond operator and identify the momentum map.   ★

Answer. By writing

$$N^\xi(q_1, p_1) = \langle p_1, -\mathcal{L}_\xi q_1\rangle_{TM} = \langle p_1 \diamond q_1, \xi\rangle_{\mathfrak{g}},$$

one identifies the momentum map as $N(q_1, p_1) = p_1 \diamond q_1$.
                                                                    ▲

Exercise. Compute the canonical transformation generated by the corresponding phase-space Hamiltonian $N^\xi(q_1, p_1)$.   ★

Answer. The canonical transformations may be calculated as

$$\delta q_1 = \{q_1, N^\xi\} = \frac{\partial N^\xi}{\partial p_1} = -\pounds_\xi q_1,$$

$$\delta p_1 = \{p_1, N^\xi\} = -\frac{\partial N^\xi}{\partial q_1} = \pounds_\xi^T p_1.$$

This is the infinitesimal action of the Lie group G on the cotangent space $T^*\mathbb{R}_1^n$ for the first degree of freedom with phase-space coordinates (q_1, p_1).   ▲

Exercise. An infinitesimal left action of $SO(3)$ on $(\mathbf{q}, \mathbf{p}) \in \mathbb{R}^3 \times \mathbb{R}^3$ is given by

$$(\delta \mathbf{q}, \delta \mathbf{p}) = (\widehat{\xi}\mathbf{q}, \widehat{\xi}\mathbf{p}),$$

where $\widehat{\xi} \in \mathfrak{so}(3)$ is a 3×3 skew-symmetric matrix. These infinitesimal actions of the Lie algebra $\mathfrak{so}(3)$ are given by the vector cross product as $\mathfrak{so}(3) \simeq \mathbb{R}^3$ via the hat map $\widehat{N}_{ij} = -\epsilon_{ijk}N^k$ for $k = 1, 2, 3$, as

$$(\delta \mathbf{q}, \delta \mathbf{p}) = (\widehat{\xi}_{ij}q_j, \widehat{\xi}_{ij}p_j) = (-\epsilon_{ijk}\xi^k q^j - \epsilon_{ijk}\xi^k p^j)$$
$$= (\boldsymbol{\xi} \times \mathbf{q}, \boldsymbol{\xi} \times \mathbf{p}).$$

Show that the Noether quantity $N^\xi = \mathbf{N} \cdot \boldsymbol{\xi}$ with

$$\mathbf{N} = \mathbf{q} \times \mathbf{p} \in \mathfrak{so}(3)^* \simeq \mathbb{R}^3$$

provides this infinitesimal left action as a symplectic transformation of $(\mathbf{q}, \mathbf{p}) \in T^*\mathbb{R}^3$. Identify the corresponding conserved quantity.   ★

Answer.

$$N^\xi(q,p) := \langle p, \delta q \rangle_{\mathbb{R}^3} = \mathbf{p} \cdot \delta \mathbf{q}$$

$$= \mathbf{p} \cdot \boldsymbol{\xi} \times \mathbf{q} = \mathbf{q} \times \mathbf{p} \cdot \boldsymbol{\xi} =: \mathbf{N}(\mathbf{q}, \mathbf{p}) \cdot \boldsymbol{\xi}.$$

The corresponding conserved quantity is the spatial angular momentum as discussed previously.   ▲

Exercise. More about Noether's theorem and Hamiltonian dynamics for two degrees of freedom:

Previously, in an earlier part of this lecture, we transformed to the reduced Lagrangian $L(\xi, q_2, u_2)$ defined on

$$(T\mathbb{R}^n_1/G) \times T\mathbb{R}^d_2 \simeq \mathfrak{g} \times T\mathbb{R}^d_2$$

and found the EP and EL equations:

$$\frac{d\mu}{dt} = \mathrm{ad}^*_\xi \mu \quad \text{with } \mu := \frac{\partial L}{\partial \xi}$$

and

$$\frac{dp_2}{dt} := \frac{\partial L}{\partial q_2} \quad \text{with } p_2 = \frac{d}{dt}\frac{\partial L}{\partial u_2}\bigg|_{u_2 = \dot{q}_2}. \qquad ★$$

Exercise. Compute the partial derivatives of the corresponding Hamiltonian by Legendre-transforming to $(T^*\mathbb{R}^n_1/G) \times T^*\mathbb{R}^d_2 \simeq \mathfrak{g}^* \times T\mathbb{R}^d_2$ via

$$H(\mu; q_2, p_2) = \langle \mu, \xi \rangle_{\mathfrak{g}} + \langle p_2, u_2 \rangle - L(\xi, q_2, u_2). \qquad ★$$

Answer. The Legendre transform yields the following partial derivatives of the corresponding Hamiltonian:

$$dH(\mu; q_2, p_2) = \left\langle d\mu, \xi \right\rangle_{\mathfrak{g}} + \left\langle \mu - \frac{\partial L}{\partial \xi}, d\xi \right\rangle_{\mathfrak{g}} + \left\langle dp_2, u_2 \right\rangle$$

$$+ \left\langle p_2 - \frac{\partial L}{\partial u_2}, du_2 \right\rangle - \left\langle \frac{\partial L}{\partial q_2}, dq_2 \right\rangle$$

$$= \left\langle d\mu, \frac{\partial H}{\partial \mu} \right\rangle_{\mathfrak{g}} + \left\langle dp_2, \frac{\partial H}{\partial p_2} \right\rangle + \left\langle dq_2, \frac{\partial H}{\partial q_2} \right\rangle.$$

▲

Exercise. Show the following relations by equating like terms in the partial derivatives of the Lagrangian and the corresponding Hamiltonian:

$$\frac{\partial H}{\partial \mu} = \xi, \quad \frac{\partial H}{\partial p_2} = u_2, \quad \frac{\partial H}{\partial q_2} = -\frac{\partial L}{\partial q_2},$$

with

$$p_2 = \frac{\partial L}{\partial u_2} \quad \text{and} \quad \mu = \frac{\partial L}{\partial \xi},$$

so that the EL and EP equations imply

$$EL: \frac{\partial L}{\partial q_2} = \frac{dp_2}{dt} = -\frac{\partial H}{\partial q_2}, \quad u_2 = \frac{dq_2}{dt} = \frac{\partial H}{\partial p_2},$$

and

$$\frac{d\mu}{dt} = \mathrm{ad}^*_{\frac{\partial H}{\partial \mu}} \mu.$$

★

Exercise. Compute the following sum of an LP bracket and a symplectic Poisson bracket by expanding out the time derivative,

$dF(\mu, q_2, p_2)/dt$ as

$$\frac{dF}{dt} = \left\langle \mathrm{ad}^*_{\frac{\partial H}{\partial \mu}} \mu, \frac{\partial F}{\partial \mu} \right\rangle_{\mathfrak{g}} + \left\langle \frac{\partial H}{\partial p_2}, \frac{\partial F}{\partial q_2} \right\rangle_{T\mathbb{R}_2}$$

$$- \left\langle \frac{\partial H}{\partial q_2}, \frac{\partial F}{\partial p_2} \right\rangle_{T\mathbb{R}_2}$$

$$=: \{F, H\}. \qquad \bigstar$$

Exercise. Write the combined Poisson bracket as a block-diagonal combination of LP and symplectic Poisson brackets,

$$\frac{dF}{dt} = tr \left(\begin{bmatrix} \frac{\partial F}{\partial \mu} \\ \frac{\partial F}{\partial q_2} \\ \frac{\partial F}{\partial p_2} \end{bmatrix}^T \begin{bmatrix} \mathrm{ad}^*_\Box \mu & 0 & 0 \\ 0 & 0 & 1 \\ 0 & -1 & 0 \end{bmatrix} \begin{bmatrix} \frac{\partial H}{\partial \mu} \\ \frac{\partial H}{\partial q_2} \\ \frac{\partial H}{\partial p_2} \end{bmatrix} \right)$$

$$= - \left\langle \mu, \left[\frac{\partial F}{\partial \mu}, \frac{\partial H}{\partial \mu} \right] \right\rangle_{\mathfrak{g}} + \frac{\partial H}{\partial p_2} \frac{\partial F}{\partial q_2} - \frac{\partial H}{\partial q_2} \frac{\partial F}{\partial p_2}$$

$$=: \{F, H\}. \qquad \bigstar$$

Exercise. Prove that the Jacobi identity holds for this block-diagonal Poisson operator. $\qquad \bigstar$

Answer. The LP bracket holds for the first term because it is a linear functional of the Lie bracket of Hamiltonian vector fields. The second part of the bracket is canonical and is independent of the first part. Thus, the sum of the two Poisson brackets also satisfies the Jacobi identity. ▲

Exercise. Show by direct calculation that Hamilton's canonical equations yield the block-diagonal Poisson operator above when the Hamiltonian is given by $H(\mu = p_1 \diamond q_1; q_2, p_2)$.

That is, when the $(q_1, p_1) \in T^*\mathbb{R}^n_1$ dependence of the original Hamiltonian depends only on the Noether quantity $\mu = p_1 \diamond q_1 \in \mathfrak{g}^*$, show that its evolution is given by the LP equation,

$$\frac{d\mu}{dt} = \mathrm{ad}^*_\xi \mu \quad \text{with} \quad \mu = p_1 \diamond q_1 \quad \text{and} \quad \xi = \frac{\partial H(\mu; q_2, p_2)}{\partial \mu}.$$

In this case, the Hamiltonian $H(\mu = p_1 \diamond q_1; q_2, p_2) : \mathfrak{g}^* \times T^*\mathbb{R}^d_2$ is said to have *collectivised* in the $T^*\mathbb{R}^n_1$ phase-space variables and remained symplectic in the $T^*\mathbb{R}^d_2$ phase-space variables. ★

Answer. For an arbitrary fixed $\eta \in \mathfrak{X}$, one computes the pairing

$$\langle \partial_t \mu, \eta \rangle_{\mathfrak{X}} = \langle \partial_t p \diamond q + p \diamond \partial_t q, \eta \rangle_{\mathfrak{X}}$$

$$= \langle (\mathcal{L}^T_\xi p) \diamond q - p \diamond \mathcal{L}_\xi q, \eta \rangle_{\mathfrak{X}}$$

$$= \langle p, (-\mathcal{L}_\xi \mathcal{L}_\eta + \mathcal{L}_\eta \mathcal{L}_\xi) q \rangle_V \qquad (9.3.5)$$

$$= \langle p, -(\mathrm{ad}_\xi \eta) q \rangle_V = \langle p \diamond q, \mathrm{ad}_\xi \eta \rangle_{\mathfrak{X}}$$

$$= \langle \mathrm{ad}^*_\xi (p \diamond q), \eta \rangle_{\mathfrak{X}} = \langle \mathcal{L}_\xi \mu, \eta \rangle_{\mathfrak{X}}.$$

Since $\eta \in \mathfrak{X}$ was arbitrary, the last line completes the proof of the lemma. In the last step, we have also used the fact that coadjoint action is identical to the Lie-derivative action for vector fields acting on 1-form densities such as $m = p \diamond q$. ▲

Exercise. ($G = GL(n, \mathbb{R}) \circledS \mathbb{R}^n$ affine invariant motions) Begin with the Lagrangian given by the kinetic energy of the Fisher–Rao metric $g(TG)$ for multivariate Gaussian probability densities: [BaGa2020]

$$L(S, \dot{S}, \dot{\mathbf{q}}) = \frac{1}{2} \operatorname{tr}\left(\dot{S} S^{-1} \dot{S} S^{-1} \right) + \frac{1}{2} \dot{\mathbf{q}}^T S^{-1} \dot{\mathbf{q}}.$$

Here, S is an $n \times n$ symmetric matrix and $\mathbf{q} \in \mathbb{R}^n$ is an n-component column vector. The quantities S and $\mathbf{q}$ represent, respectively, the covariance matrix and the mean of the Gaussian probability distribution. Conveniently, the Lagrangian $L(S, \dot{S}, \dot{\mathbf{q}})$ is independent of the coordinate $\mathbf{q}$.

1. Calculate the EL equations for this Lagrangian.

2. Legendre-transform to find the Hamiltonian for this system and write its *canonical* equations.

3. Show that the Lagrangian and Hamiltonian for this system are both invariant under the group action of the general affine group $G = GL(n) \circledS \mathbb{R}^n$

$$\mathbf{q} \to G\mathbf{q} \quad \text{and} \quad S \to G S G^T$$

for any constant invertible $n \times n$ matrix, G.

4. a. Linearise this group action around the identity in terms of $A = G'G^{-1}$ and construct the infinitesimal transformations $X_A \mathbf{q}$ and $X_A S$ for the linearised action of G on the configuration space $(\mathbf{q}, S)$.

   b. Find the phase-space function (*infinitesimal generator*) whose canonical Poisson brackets produce these infinitesimal transformations by pairing $X_A \mathbf{q}$ and $X_A S$ with the corresponding canonical momenta and summing.

c. Compute the Poisson bracket of the canonical momenta with the infinitesimal generator. (This is the *cotangent lift* to the full phase
space of the infinitesimal action of G on the
configuration space.)

5.   a. Verify directly that the infinitesimal generator of the G-action is a conserved $n \times n$ matrix
quantity by using the equations of motion.

b. Determine whether this Hamiltonian system
has sufficiently many conservation laws in
involution to be completely integrable, for
any dimension n.   ★

Answer.

1. The EL equations for this Lagrangian are

$$\ddot{S} + \mathbf{q} \otimes \mathbf{q}^T - \dot{S}S^{-1}\dot{S} = 0, \quad \text{and} \quad \ddot{\mathbf{q}} + \dot{S}S^{-1}\dot{\mathbf{q}} = 0.$$

2. The Legendre transform yields fibre derivatives,

$$P = \frac{\partial L}{\partial \dot{S}} = S^{-1}\dot{S}S^{-1} \quad \text{and} \quad \mathbf{p} = \frac{\partial L}{\partial \dot{\mathbf{q}}} = S^{-1}\dot{\mathbf{q}}.$$

Thus, the Hamiltonian $H(\mathbf{q}, \mathbf{p}, S, P)$ is

$$H(\mathbf{q}, \mathbf{p}, S, P) = \frac{1}{2}\text{tr}\left(PS \cdot PS\right) + \frac{1}{2}\mathbf{p} \cdot S\mathbf{p},$$

and its canonical Hamilton equations are

$$\dot{S} = \frac{\partial H}{\partial P} = SPS,$$

$$\dot{P} = -\frac{\partial H}{\partial S} = -\left(PSP + \frac{1}{2}\mathbf{p} \otimes \mathbf{p}\right),$$

$$\dot{\mathbf{q}} = \frac{\partial H}{\partial \mathbf{p}} = S\mathbf{p}, \quad \dot{\mathbf{p}} = \frac{\partial H}{\partial \mathbf{q}} = 0.$$

3. Under the $G = GL(n, \mathbb{R}) \circledS \mathbb{R}^n$ affine group action $\mathbf{q} \to G\mathbf{q}$ and $S \to GSG^T$ for any constant invertible $n \times n$ matrix, G, one finds $\dot{S}S^{-1} \to G\dot{S}S^{-1}G^{-1}$ and $\dot{\mathbf{q}} \cdot S^{-1}\dot{\mathbf{q}} \to \dot{\mathbf{q}} \cdot S^{-1}\dot{\mathbf{q}}$. Hence, $L \to L$ and this Lagrangian is invariant.

Likewise, $P \to G^{-T}PG^{-1}$ so $PS \to G^{-T}PSG^T$ and $\mathbf{p} \to G^{-T}\mathbf{p}$ so that $S\mathbf{p} \to GS\mathbf{p}$. Hence, $H \to H$, as well; therefore, both L and H for this system are invariant under the G affine group action.

4. a. The infinitesimal actions for $G(\epsilon) = Id + \epsilon A + O(\epsilon^2)$, where $A \in \mathfrak{g}(n)$, are

$$X_A\mathbf{q} = \frac{d}{d\epsilon}\Big|_{\epsilon=0} G(\epsilon)\mathbf{q} = A\mathbf{q} \quad \text{and}$$

$$X_A S = \frac{d}{d\epsilon}\Big|_{\epsilon=0} \left(G(\epsilon)SG(\epsilon)^T\right) = AS + SA^T.$$

b. Pairing $X_A\mathbf{q}$ and $X_A S$ with their corresponding canonical momenta and summing using $P^T = P$ yields

$$\langle J, A \rangle := \text{tr}\,(PX_A S) + \mathbf{p} \cdot X_A\mathbf{q}$$
$$= \text{tr}\,\left(P(AS + SA^T)\right) + \mathbf{p} \cdot A\mathbf{q}.$$

Hence,

$$\langle J, A \rangle := \text{tr}\,\left(JA^T\right) = \text{tr}\,\left((2SP + \mathbf{q} \otimes \mathbf{p})A\right),$$
$$\text{so} \quad J = (2PS + \mathbf{p} \otimes \mathbf{q}),$$

where $J : T^*(S, \mathbf{q}) \to \mathfrak{g}^*$ is the momentum map of the cotangent-lifted action of $G(n)$ relative to the canonical symplectic form.

c. For any choice of the matrix A, the Poisson bracket with $\langle J, A \rangle$ generates the Hamiltonian vector field

$$\left\{\cdot, \langle J, A \rangle\right\}$$

$$= \text{tr}\,\left(\frac{\partial \langle J, A \rangle}{\partial P}\frac{\partial}{\partial S}\right) + \frac{\partial \langle J, A \rangle}{\partial \mathbf{p}} \cdot \frac{\partial}{\partial \mathbf{q}}$$

$$- \operatorname{tr}\left(\frac{\partial\langle J, A\rangle}{\partial S}\frac{\partial}{\partial P}\right) - \frac{\partial\langle J, A\rangle}{\partial \mathbf{q}}\cdot\frac{\partial}{\partial \mathbf{p}}$$

$$= \operatorname{tr}\left((AS + SA^T)\frac{\partial}{\partial S}\right) + A\mathbf{q}\cdot\frac{\partial}{\partial \mathbf{q}}$$

$$- \operatorname{tr}\left((PA + A^T P)\frac{\partial}{\partial P}\right) - A^T\mathbf{p}\cdot\frac{\partial}{\partial \mathbf{p}},$$

which recovers the infinitesimal action on $(S, \mathbf{q})$ and provides the cotangent-lifted infinitesimal action on the canonical momenta $(P, \mathbf{p})$.

5. a. Conservation of $\langle J, A\rangle$ is verified directly in

$$\frac{d}{dt}\langle J, A\rangle = \langle \dot{J}, A\rangle$$

by computing

$$\dot{J} = \left(2\dot{P}S + 2P\dot{S} + \dot{\mathbf{p}}\otimes\mathbf{q} + \mathbf{p}\otimes\dot{\mathbf{q}}\right)$$

$$= -\left(2PSP + (\mathbf{p}\otimes\mathbf{p})\right)S + 2P\left(SPS\right)$$

$$+ \mathbf{0}\otimes\mathbf{q} + \mathbf{p}\otimes S\mathbf{p}$$

$$= 0.$$

b. The system has $n(n + 1)/2 + n = n(n + 3)/2$ degrees of freedom. It conserves the n components of linear momentum $\mathbf{p}$ and the $n(n + 1)/2$ components of J. Thus, there is one constant of motion for each degree of freedom.

However, these two sets of independent conservation laws do not Poisson commute since

$$\left\{\mathbf{p}, \langle J, A\rangle\right\} = -A^T\mathbf{p}.$$

This means that the naive count of degrees of freedom will not produce complete integrability because the momentum map constants of motion arising from Noether's theorem are not in involution. In general, something more would be needed for complete integrability of this system to hold. This is a potential research question. For more discussion of the geometric dynamics based on the Fisher–Rao metric in probability theory, see Ref. [BaGa2020].   ▲

Exercise. Hamiltonian symmetry reduction by stages.

1. Write Hamilton's equations on $\mathfrak{so}^*(4) \simeq \mathfrak{so}(3)^* \times \mathfrak{so}(3)^*$ using the Poisson brackets

   $$\{M_i, M_j\} = \epsilon_{ijk} M_k, \quad \{N_i, N_j\} = \epsilon_{ijk} N_k,$$
   $$\{M_i, N_j\} = 0,$$

   among the components of the $\mathbb{R}^3$ vectors $\mathbf{M}$ and $\mathbf{N}$.

2. Compute the equations of motion and identify the functionally independent conserved quantities for the following two Hamiltonians:

   $$H_1 = \hat{\mathbf{z}} \cdot (\mathbf{M} \times \mathbf{N}) \quad \text{and} \quad H_2 = \mathbf{M} \cdot \mathbf{N}.$$

   (9.3.6)

3. Determine whether these Hamiltonians have sufficiently many symmetries and associated conservation laws to be completely integrable (i.e., reducible to Hamilton's canonical equations for a single degree of freedom) and explain why.

4. Transform the Hamiltonians in (9.3.6) from Cartesian components of the vectors $(\mathbf{M}, \mathbf{N}) \in \mathbb{R}^3 \times \mathbb{R}^3$ into spherical coordinates, $(\theta, \phi) \in S^2$ and $(\bar{\theta}, \bar{\phi}) \in S^2$, respectively.

5. Use the S^1 symmetries and their associated conservation laws to reduce the dynamics in $\mathbb{R}^3 \times \mathbb{R}^3$ to canonical Hamiltonian equations first on $S^2 \times S^2$ and then on S^2 through a two-stage sequence of canonical transformations.   ★

Answer.

1. The Hamiltonian equations for this system are

$$\dot{\mathbf{M}} = \mathbf{M} \times \frac{\partial H}{\partial \mathbf{M}} \quad \text{and} \quad \dot{\mathbf{N}} = \mathbf{N} \times \frac{\partial H}{\partial \mathbf{N}}.$$

2. The system with Hamiltonian $H_1 = \hat{\mathbf{z}} \cdot (\mathbf{M} \times \mathbf{N})$ conserves

$$|\mathbf{M}|^2, \quad |\mathbf{N}|^2 \quad \text{and} \quad L_3 = M_3 + N_3.$$

The first two conservation laws reduce the problem to $S^2 \times S^2$, and the last one provides a further $SO(2)$ symmetry under simultaneous rotation of each of the spheres about its vertical three-axis. As we shall see, this symmetry and its conservation law are enough to reduce $S^2 \times S^2$ to S^2 and thereby make the system completely integrable.

3. The system with Hamiltonian $H_2 = \mathbf{M} \cdot \mathbf{N}$ conserves $|\mathbf{M}|^2$, $|\mathbf{N}|^2$ and all the components of $\mathbf{L} = \mathbf{M} + \mathbf{N}$. These conserved quantities are *not* all functionally independent since

$$|\mathbf{L}|^2 = |\mathbf{M} + \mathbf{N}|^2 = |\mathbf{M}|^2 + |\mathbf{N}|^2 + 2\,\mathbf{M} \cdot \mathbf{N}.$$

However, enough symmetry still remains for these equations to be integrated by employing $L_3 = M_3 + N_3$ and its associated $SO(2)$ symmetry for simultaneous rotation of each of the spheres about its vertical three-axis. This symmetry reduces its $S^2 \times S^2$ phase space to S^2 and thereby allows it to be integrated as before.

4. The vectors $\mathbf{M}$ and $\mathbf{N}$ may be written in spherical coordinates (θ, ϕ) and $(\bar{\theta}, \bar{\phi})$, respectively, as

$$\mathbf{M} = (M_1, M_2, M_3)^T$$
$$= M(\sin\theta\cos\phi, \sin\theta\sin\phi, \cos\theta)^T,$$
$$\mathbf{N} = (N_1, N_2, N_3)^T$$
$$= N(\sin\bar{\theta}\cos\bar{\phi}, \sin\bar{\theta}\sin\bar{\phi}, \cos\bar{\theta})^T.$$

In terms of these variables, we may write

$$H_1 = M_1 N_2 - M_2 N_1 \quad \text{and}$$
$$H_2 = M_1 N_1 + M_2 N_2 + M_3 N_3.$$

5. Reduction $S^2 \times S^2 \rightarrow S^2$ may be accomplished through a canonical transformation using the conservation of $|\mathbf{M}|^2$, $|\mathbf{N}|^2$ and $L_3 = M_3 + N_3$. The symplectic form on $S^2 \times S^2$ is given in spherical coordinates by

$$\omega = M^2 d\cos\theta \wedge d\phi + N^2 d\cos\bar{\theta} \wedge d\bar{\phi}. \quad (9.3.7)$$

We transform to weighted sum and difference variables by

$$\sqrt{2}\lambda = M_3 + N_3 = M\cos\theta + N\cos\bar{\theta},$$
$$\sqrt{2}\alpha = M\phi + N\bar{\phi},$$
$$\sqrt{2}\kappa = M_3 - N_3 = M\cos\theta - N\cos\bar{\theta},$$
$$\sqrt{2}\beta = M\phi - N\bar{\phi}.$$

This transformation is canonical and yields the new symplectic form:

$$\omega = d\kappa \wedge d\beta + d\lambda \wedge d\alpha. \quad (9.3.8)$$

Expressing the Hamiltonians H_1 and H_2 in terms of these new canonical variables reduces the problem to the (κ, β) phase plane, with motion parameterised by the third component of total angular momentum λ and independent of its canonically conjugate angle α. In each case, Hamilton's canonical equations separate into reduced dynamics on S^2, plus reconstruction of the phase $\alpha \in S^1$:

$$\underbrace{\dot{\kappa} = -\frac{\partial H}{\partial \beta}, \quad \dot{\beta} = \frac{\partial H}{\partial \kappa},}_{\text{Reduced dynamics on } S^2} \quad \underbrace{\dot{\lambda} = -\frac{\partial H}{\partial \alpha} = 0, \quad \dot{\alpha} = \frac{\partial H}{\partial \lambda}}_{\text{Reconstruction of the phase, } \alpha}.$$

▲

9.4   Composition of maps and Lagrangian reduction by stages

Exercise. Reduction by stages of the composition of two non-commutative left group actions $G_1 \times G_2$.

Consider the following HP on the product space $(TG_1 \times TG_2)$ and verify the symmetry reduction by stages $TG_1/G_1 \times (TG_2/G_2)/G_1$:

$$S = \int_{t_1}^{t_2} L(g_1, \dot{g}_1; g_2, \dot{g}_2)\, dt$$

$$= \int_{t_1}^{t_2} L\big(g_1^{-1}\dot{g}_1, \mathrm{Ad}_{g_1^{-1}}(g_2^{-1}\dot{g}_2)\big)\, dt \qquad (9.4.1)$$

$$=: \int_{t_1}^{t_2} L(\Omega_1, \Omega_2)\, dt,$$

where one introduces the following two velocities [HoHuSt2023]:

$$\Omega_1 = g_1^{-1}\dot{g}_1, \quad \Omega_2 = g_1^{-1}(\dot{g}_2 g_2^{-1})g_1 =: \mathrm{Ad}_{g_1^{-1}}(g_2^{-1}\dot{g}_2),$$

for the nested symmetries of $(TG_1 \times TG_2)$ under the respective left actions of $G_1 \times G_2$.   ★

Exercise. After verifying the line of reasoning in deriving the Lagrangian shown above, prove the following equations for iterated left actions leading to semidirect-product Lie algebras, with [HoHuSt2023]

$$\omega_1 = g_1^{-1}\delta g_1 =: g_1^{-1}g_1', \quad \omega_2 = \mathrm{Ad}_{g_1^{-1}}(g_2^{-1}g_2').$$

$$\Omega_1' - \dot\omega_1 = \mathrm{ad}_{\Omega_1}\omega_1 := [\Omega_1, \omega_1],$$
$$\Omega_2' - \dot\omega_2 = \mathrm{ad}_{\Omega_2}\omega_2 + \mathrm{ad}_{\Omega_2}\omega_1 + \mathrm{ad}_{\Omega_1}\omega_2 \qquad (9.4.2)$$
$$= \mathrm{ad}_{\Omega_1+\Omega_2}\omega_2 + \mathrm{ad}_{\Omega_2}\omega_1. \qquad \bigstar$$

Answer. Recall the definitions

$$\Omega_1 = g_1^{-1}\dot g_1, \quad \Omega_2 =: \mathrm{Ad}_{g_1^{-1}}(g_2^{-1}\dot g_2).$$

The first relation has already been derived in equation (2.10.4):

$$\Omega_1' = (g_1'g_1^{-1})' - \mathrm{ad}_{\dot g_1 g_1^{-1}} g_1'g_1^{-1} = \dot\omega_1 - \mathrm{ad}_{\Omega_1}\omega_1.$$

One derives the second formula in (9.4.12) by direct calculation of Ω_2' and $\dot\omega_2$ as follows:

$$\Omega_2' = \mathrm{Ad}_{g_1^{-1}}(g_2^{-1}\dot g_2)' + [\omega_2, \Omega_1]$$
$$\dot\omega_2 = \mathrm{Ad}_{g_1^{-1}}(g_2^{-1}g_2')' + [\omega_1, \Omega_2].$$

Consequently, the difference is given by

$$\Omega_2' - \dot\omega_2 = \mathrm{Ad}_{g_1^{-1}}\left((g_2^{-1}\dot g_2)' - (g_2^{-1}g_2')'\right)$$
$$+ [\omega_2, \Omega_1] - [\omega_1, \Omega_2]$$
$$= [\Omega_2, \omega_2] + [\Omega_2, \omega_1] + [\Omega_1, \omega_2]$$
$$= \mathrm{ad}_{\Omega_1+\Omega_2}\omega_2 + \mathrm{ad}_{\Omega_2}\omega_1.$$

This is how infinitesimal transformations compose under the iteration of left semidirect-product Lie algebra action [BrGaHoRa2011, Ho2008, HoHuSt2023].   ▲

Exercise. Calculate the EP equations of motion resulting from

$$0 = \delta S = \delta \int_{t_1}^{t_2} L(\Omega_1, \Omega_2)\, dt.$$

Write these EP equations of motion in LP form.   ★

Answer. The EP equations of motion resulting from HP can be read off from the following variational calculation:

$$0 = \delta S = \delta \int_{t_1}^{t_2} L(\Omega_1, \Omega_2)\, dt$$

$$= \int_{t_1}^{t_2} \left\langle \frac{\delta L}{\delta \Omega_1},\, \delta\Omega_1 \right\rangle + \left\langle \frac{\delta L}{\delta \Omega_2},\, \delta\Omega_2 \right\rangle dt$$

$$= \int_{t_1}^{t_2} \left\langle \frac{\delta L}{\delta \Omega_1},\, \dot{\omega}_1 + \mathrm{ad}_{\Omega_1}\, \omega_1 \right\rangle$$

$$+ \left\langle \frac{\delta L}{\delta \Omega_2},\, \dot{\omega}_2 + \mathrm{ad}_{\Omega_1 + \Omega_2}\, \omega_2 + \mathrm{ad}_{\Omega_2}\, \omega_1 \right\rangle dt$$

$$= \int_{t_1}^{t_2} \left\langle -(\partial_t - \mathrm{ad}^*_{\Omega_1})\frac{\delta L}{\delta \Omega_1} + \mathrm{ad}^*_{\Omega_2}\frac{\delta L}{\delta \Omega_2},\, \omega_1 \right\rangle$$

$$+ \left\langle -(\partial_t - \mathrm{ad}^*_{\Omega_1 + \Omega_2})\frac{\delta L}{\delta \Omega_2},\, \omega_2 \right\rangle dt.$$

The LP form of these EP equations of motion is given in terms of angular moments defined as

$$\Pi_1 := \frac{\delta L}{\delta \Omega_1} \quad \text{and} \quad \Pi_2 := \frac{\delta L}{\delta \Omega_2}.$$

Namely, the LP form is the following:

$$\partial_t \begin{pmatrix} \Pi_1 \\ \Pi_2 \end{pmatrix} = \begin{bmatrix} \mathrm{ad}^*_\square \Pi_1 & \mathrm{ad}^*_\square \Pi_2 \\ \mathrm{ad}^*_\square \Pi_2 & \mathrm{ad}^*_\square \Pi_2 \end{bmatrix} \begin{pmatrix} \delta h/\delta \Pi_1 = \Omega_1 \\ \delta h/\delta \Pi_2 = \Omega_2 \end{pmatrix}. \quad (9.4.3)$$

If one transforms variables by the linear transformation $(\Pi_1, \Pi_2) \to (\Pi_1 - \Pi_2, \Pi_2)$, them the Poisson matrix in (9.4.3) transforms to

$$\begin{bmatrix} 1 & -1 \\ 0 & 1 \end{bmatrix} \begin{bmatrix} \mathrm{ad}^*_\square \Pi_1 & \mathrm{ad}^*_\square \Pi_2 \\ \mathrm{ad}^*_\square \Pi_2 & \mathrm{ad}^*_\square \Pi_2 \end{bmatrix} \begin{bmatrix} 1 & 0 \\ -1 & 1 \end{bmatrix}$$

$$= \begin{bmatrix} \mathrm{ad}^*_\square (\Pi_1 - \Pi_2) & 0 \\ 0 & \mathrm{ad}^*_\square \Pi_2 \end{bmatrix},$$

which is the same LP bracket as that appearing in Exercise 9.3, upon identifying $G_1 \times G_2 \to SO(3) \times SO(3)$, $(\Pi_1 - \Pi_2) \to M$ and $\Pi_2 \to N$. ▲

Exercise. Reduction by stages of the composition of three non-commutative *left* group actions $G_1 \times G_2 \times G_3$.

Consider the following HP on the product space $(TG_1 \times TG_2 \times TG_3)$ and verify the symmetry reduction by stages $TG_1/G_1 \times (TG_2/G_2)/G_1 \times ((TG_3/G_3)/G_2)/G_1)$:

$$S = \int_{t_1}^{t_2} L(g_1, \dot{g}_1; g_2, \dot{g}_2; g_3, \dot{g}_3)\, dt$$

$$= \int_{t_1}^{t_2} L(g_1^{-1} \dot{g}_1\,; \mathrm{Ad}_{g_1^{-1}} (g_2^{-1} \dot{g}_2)\,; \mathrm{Ad}_{g_1^{-1}} \mathrm{Ad}_{g_2^{-1}} (\dot{g}_3 g_3^{-1}))\, dt$$

$$=: \int_{t_1}^{t_2} L(\Omega_1; \Omega_2; \Omega_3)\, dt, \quad (9.4.4)$$

where one introduces sequential Ad actions to define three velocities:

$$\Omega_1 := g_1^{-1}\dot{g}_1, \quad \Omega_2 := \mathrm{Ad}_{g_1^{-1}}(g_2^{-1}\dot{g}_2),$$

$$\Omega_3 := \mathrm{Ad}_{g_1^{-1}}\mathrm{Ad}_{g_2^{-1}}(\dot{g}_3 g_3^{-1}), \tag{9.4.5}$$

for the nested symmetries of $(TG_1 \times TG_2 \times TG_3)$ under the respective left actions of $G_1 \times G_2 \times G_3$. ★

Exercise. After verifying the line of reasoning in deriving the Lagrangian shown above, introduce variations with sequential Ad actions:

$$\omega_1 := g_1^{-1}g'_1, \quad \omega_2 := \mathrm{Ad}_{g_1^{-1}}(g_2^{-1}g'_2),$$

$$\omega_3 := \mathrm{Ad}_{g_1^{-1}}\mathrm{Ad}_{g_2^{-1}}(g_3^{-1}g'_3). \tag{9.4.6}$$

Then, prove the following equations for iterated left actions leading to semidirect-product Lie algebras:

$$\Omega'_1 - \dot{\omega}_1 = \mathrm{ad}_{\Omega_1}\omega_1,$$

$$\Omega'_2 - \dot{\omega}_2 = \mathrm{ad}_{\Omega_2}\omega_2 + \mathrm{ad}_{\Omega_2}\omega_1 + \mathrm{ad}_{\Omega_1}\omega_2$$

$$= \mathrm{ad}_{\Omega_1+\Omega_2}\omega_2 + \mathrm{ad}_{\Omega_2}\omega_1,$$

$$\Omega'_3 - \dot{\omega}_3 = \mathrm{ad}_{\Omega_3}\omega_3 + \mathrm{ad}_{\Omega_3}(\omega_1 + \omega_2) + \mathrm{ad}_{\Omega_1+\Omega_2}\omega_3$$

$$= \mathrm{ad}_{\Omega_1+\Omega_2+\Omega_3}\omega_3 + \mathrm{ad}_{\Omega_3}(\omega_1 + \omega_2).$$

$$\tag{9.4.7}$$

The sequence of EP variational relations for compositions of group actions follows a clear pattern [HoHuSt2023]. ★

Answer. The definitions in equations (9.4.5) and (9.4.6) have already implied the first two equations in the set (9.4.7) in the previous exercise.

The last equation in (9.4.7) is obtained from the following intermediate relations:

$$
\Omega_3' = -(\omega_1 + \omega_2 + \omega_3)\Omega_3 + \mathrm{Ad}_{g_1^{-1}}\mathrm{Ad}_{g_2^{-1}}(g_3^{-1}\dot{g}_3')
$$
$$
+ (\Omega_1 + \Omega_2)\omega_3,
$$
$$
\dot{\omega}_3 = -(\Omega_1 + \Omega_2 + \Omega_3)\omega_3 + \mathrm{Ad}_{g_1^{-1}}\mathrm{Ad}_{g_2^{-1}}(g_3^{-1}\dot{g}_3')
$$
$$
+ (\omega_1 + \omega_2)\Omega_3 \qquad (9.4.8)
$$

Taking the difference of the previous two equations yields

$$
\Omega_3' - \dot{\omega}_3 = \big[\Omega_1 + \Omega_2 + \Omega_3\big]\omega_3 + \big[\Omega_3, \omega_1 + \omega_2\big]
$$
$$
= \mathrm{ad}_{(\Omega_1+\Omega_2+\Omega_3)}\,\omega_3 + \mathrm{ad}_{\Omega_3}(\omega_1 + \omega_2).
$$
$$
(9.4.9)
$$
▲

Exercise. Reduction by stages of the composition of three non-commutative right group actions $G_1 \times G_2 \times G_3$.

Consider the following HP on the product space $(TG_1 \times TG_2 \times TG_3)$ and verify the symmetry reduction by stages $TG_1/G_1 \times (TG_2/G_2)/G_1 \times ((TG_3/G_3)/G_2)/G_1)$:

$$
S = \int_{t_1}^{t_2} L(g_1, \dot{g}_1; g_2, \dot{g}_2; g_3, \dot{g}_3)\, dt
$$
$$
= \int_{t_1}^{t_2} L(\dot{g}_1 g_1^{-1}; (\dot{g}_2 g_2^{-1})g_1^{-1}; (\dot{g}_3 g_3^{-1})g_2^{-1}g_1^{-1})\, dt
$$
$$
=: \int_{t_1}^{t_2} L(u_1; u_2; u_3)\, dt, \qquad (9.4.10)
$$

where g_{1*} (resp., $(g_1 g_2)_*$) denotes push-forward by g_1 (resp., $g_1 g_2$), and one introduces the following three velocities:

$$u_1 = \dot{g}_1 g_1^{-1}, \quad u_2 = (\dot{g}_2 g_2^{-1}) g_1^{-1} =: g_{1*}(\dot{g}_2 g_2^{-1}),$$

$$u_3 = (\dot{g}_3 g_3^{-1}) g_2^{-1} g_1^{-1} =: g_{1*} g_{2*}(\dot{g}_3 g_3^{-1}),$$

for the nested symmetries of $(TG_1 \times TG_2 \times TG_3)$ under the respective right actions of $G_1 \times G_2 \times G_3$. ★

Exercise. After verifying the line of reasoning in deriving the Lagrangian shown above, prove the following equations for iterated right actions leading to semidirect-product Lie algebras, with $w_1 = \delta g_1 g_1^{-1} =: g_1' g_1^{-1}, w_2 = g_{1*} g_2' g_2^{-1}, w_3 = g_{1*} g_{2*} g_3' g_3^{-1}$:

$$u_1' - \dot{w}_1 = - \operatorname{ad}_{u_1} w_1,$$

$$u_2' - \dot{w}_2 = - \operatorname{ad}_{u_2} w_2 - \operatorname{ad}_{u_2} w_1 - \operatorname{ad}_{u_1} w_2$$

$$= - \operatorname{ad}_{u_1 + u_2} w_2 - \operatorname{ad}_{u_2} w_1,$$

$$u_3' - \dot{w}_3 = - \operatorname{ad}_{u_3} w_3 - \operatorname{ad}_{u_3}(w_1 + w_2) - \operatorname{ad}_{u_1 + u_2} w_3$$

$$= - \operatorname{ad}_{u_1 + u_2 + u_3} w_3 - \operatorname{ad}_{u_3}(w_1 + w_2).$$

$$(9.4.11)$$

The sequence of EP variational relations for compositions of group actions follows a clear pattern. Note the changes in sign for right action in (9.4.11), relative to equations (9.4.8) for left action [HoHuSt2023]. ★

Answer. Recalling the definitions,

$$u_1 = \dot{g}_1 g_1^{-1}, \quad u_2 =: g_{1*}(\dot{g}_2 g_2^{-1}), \quad u_3 =: g_{1*} g_{2*}(\dot{g}_3 g_3^{-1}).$$

The first relation has already been derived in equation (2.10.4):

$$u_1' = (g_1'g_1^{-1})^{\boldsymbol{\cdot}} - \mathrm{ad}_{\dot{g}_1 g_1^{-1}}\, g_1'g_1^{-1} = \dot{w}_1 - \mathrm{ad}_{u_1} w_1.$$

One derives the second formula in (9.4.11) for $u_2' - \dot{w}_2$ by direct calculation as follows:

$$
\begin{aligned}
u_2' &= (\dot{g}_2 g_2^{-1})' g_1^{-1} - \dot{g}_2 g_2^{-1} g_1^{-1} g_1' g_1^{-1} \\
&= \big[(g_2'g_2^{-1})^{\boldsymbol{\cdot}} - \mathrm{ad}_{\dot{g}_2 g_2^{-1}}\, g_2'g_2^{-1}\big] g_1^{-1} - \mathcal{L}_{w_1} u_2 \\
&= g_{1*}\big[(g_2'g_2^{-1})^{\boldsymbol{\cdot}} - \mathrm{ad}_{\dot{g}_2 g_2^{-1}}\, g_2'g_2^{-1}\big] - \mathcal{L}_{w_1} u_2 \\
&= \partial_t\big(g_{1*}(g_2'g_2^{-1})\big) + \mathcal{L}_{\dot{g}_1 g_1^{-1}}\big(g_{1*}(g_2'g_2^{-1})\big) \quad (9.4.12) \\
&\quad + \Big(\mathcal{L}_{g_{1*}\dot{g}_2 g_2^{-1}}\Big) g_{1*} g_2' g_2^{-1} - \mathcal{L}_{w_1} u_2 \\
&= \dot{w}_2 - \mathrm{ad}_{u_1} w_2 - \mathrm{ad}_{u_2} w_2 - \mathrm{ad}_{u_2} w_1 \\
&= \dot{w}_2 - \mathrm{ad}_{u_1+u_2} w_2 - \mathrm{ad}_{u_2} w_1. \qquad \blacktriangle
\end{aligned}
$$

Remark 9.4.1 After proving the first two formulas in (9.4.12) for right action and comparing with (9.4.7) for left action, the following patterns emerge for higher iterates:

$$\Omega_m' - \dot{w}_m = \mathrm{ad}_{\sum_{k=1}^m \Omega_k} w_m + \mathrm{ad}_{\Omega_m} \sum_{k=1}^{m-1} w_k \quad \text{for left action,}$$

$$u_m' - \dot{w}_m = -\,\mathrm{ad}_{\sum_{k=1}^m u_k} w_m - \mathrm{ad}_{u_m} \sum_{k=1}^{m-1} w_k \quad \text{for right action.}$$

$$(9.4.13)$$

This is how EP variations compose under iteration of semidirect-product Lie algebra action [BrGaHoRa2011, Ho2008, HoHuSt2023]. $\square$

Exercise. Show by direct calculation that HP for CoM variations $\delta u_1, \delta(g_{1*}u_2), \delta((g_1 g_2)_* u_3)$ of the frame-shifted velocities $u_1, g_{1*}u_2; (g_1g_2)_*u_3$ in the last line of the system (9.4.10) yields

$$0 = \delta S_{red} = \int_{t_1}^{t_2} \left\langle \frac{\delta\ell}{\delta u_1}, (\partial_t - \mathrm{ad}_{u_1})w_1 \right\rangle$$

$$+ \left\langle \frac{\delta\ell}{\delta(g_{1*}u_2)}, (\partial_t - \mathrm{ad}_{u_1+u_2})w_2 - \pounds_{w_1}u_2 \right\rangle$$

$$+ \left\langle \frac{\delta\ell}{\delta((g_1g_2)_*u_3)}(\partial_t - \mathrm{ad}_{u_1+u_2+u_3})w_3 \right.$$

$$\left. + \pounds_{(w_1+w_2)}u_3 \right\rangle. \qquad (9.4.14)$$

★

Exercise. Upon defining the momentum variables

$$\frac{\delta\ell}{\delta u_1} = m_1, \quad \frac{\delta\ell}{\delta(g_{1*}u_2)} = m_2, \quad \frac{\delta\ell}{\delta((g_1g_2)_*u_3)} = m_3,$$

show that integrating by parts in (9.4.14) and collecting the coefficients of variations w_1, w_2 and w_3 leads to the following three EP equations upon separately setting to zero each coefficient of variations:

$$\partial_t \begin{pmatrix} m_1 \\ m_2 \\ m_3 \end{pmatrix} = - \begin{bmatrix} \mathrm{ad}_\square^* m_1 & \mathrm{ad}_\square^* m_2 & \mathrm{ad}_\square^* m_3 \\ \mathrm{ad}_\square^* m_2 & \mathrm{ad}_\square^* m_2 & \mathrm{ad}_\square^* m_3 \\ \mathrm{ad}_\square^* m_3 & \mathrm{ad}_\square^* m_3 & \mathrm{ad}_\square^* m_3 \end{bmatrix} \begin{pmatrix} \frac{\delta h}{\delta m_1} = u_1 \\ \frac{\delta h}{\delta m_2} = u_2 \\ \frac{\delta h}{\delta m_3} = u_3 \end{pmatrix}$$

$$(9.4.15)$$

$$= - \begin{pmatrix} \mathrm{ad}_{u_1}^* m_1 + \mathrm{ad}_{u_2}^* m_2 + \mathrm{ad}_{u_3}^* m_3 \\ \mathrm{ad}_{u_1+u_2}^* m_2 + \mathrm{ad}_{u_3}^* m_3 \\ \mathrm{ad}_{u_1+u_2+u_3}^* m_3 \end{pmatrix}.$$

$$(9.4.16)$$

The matrix operator in square brackets in (9.4.15) defines an LP bracket $\{f, h\} = \langle \mu, [df, dh] \rangle$ on the dual of the following *nested* semidirect-product Lie algebra:

$$\mathfrak{s} = \mathfrak{g}_1 \, \circledS \big(\mathfrak{g}_2 \, \circledS \, \mathfrak{g}_3 \big).$$

Physically, the LP bracket in (9.4.15) refers to three types of fluid flow. It thus mimics L. F. Richardson's famous "whorls within whorls" reference in characterising fluid dynamics:

> Big whorls have little whorls, that feed on their velocity, and little whorls have lesser whorls, and so on to viscosity.
>
> —L. F. Richardson (1922)

This is called "Richardson's triple." In the composition of three non-commutative right group actions $G_1 \times G_2 \times G_3$ with velocities

$$TG_1/G_1 \times (TG_2/G_2)/G_1 \times ((TG_3/G_3)/G_2)/G_1),$$

the flow of G_1 can represent the big whorls, the flow of G_2 can represent the little whorls carried in the frame of motion of the big whorls and G_3 can represent the lesser whorls carried along successively by the two other whorls. ★

Exercise. Show that the linear transformation of variables to

$$\mu_1 = m_1 - m_2, \quad \mu_2 = m_2 - m_3, \quad \mu_3 = m_3,$$

diagonalises the *entangled Poisson matrix* in (9.4.15) and leads to equivalent equations with an *untangled Poisson matrix:*

$$
\partial_t \begin{pmatrix} \mu_1 \\ \mu_2 \\ \mu_3 \end{pmatrix} = - \begin{bmatrix} \mathrm{ad}^*_\square \mu_1 & 0 & 0 \\ 0 & \mathrm{ad}^*_\square \mu_2 & 0 \\ 0 & 0 & \mathrm{ad}^*_\square \mu_3 \end{bmatrix}
$$

$$
\times \begin{pmatrix} \frac{\delta h}{\delta \mu_1} = u_1 \\ \frac{\delta h}{\delta \mu_2} = u_1 + u_2 \\ \frac{\delta h}{\delta \mu_3} = u_1 + u_2 + u_3 \end{pmatrix} \tag{9.4.17}
$$

$$
= - \begin{pmatrix} \mathrm{ad}^*_{u_1} \mu_1 \\ \mathrm{ad}^*_{(u_1+u_2)} \mu_2 \\ \mathrm{ad}^*_{(u_1+u_2+u_3)} \mu_3 \end{pmatrix}.
$$

The untangled diagonal matrix operator in equation (9.4.17) defines an LP bracket $\{f, h\} = \langle \mu, [df, dh] \rangle$ on the dual of the following direct product Lie algebra:

$$
\mathfrak{s}_{diag} = \mathfrak{g}_1 \otimes \mathfrak{g}_2 \otimes \mathfrak{g}_3.
$$

This untangling property upon applying momentum shifts is typical in Lagrangian reduction by stages based on the composition of Lie group actions.   ★

Exercise. Consider the following HP on the product space $(TG_1 \times TG_2 \times V)$ with left actions of Lie groups G_1 and G_2 on a vector space V, and verify its symmetry reduction by stages for the following nested sum of left semidirect-product Lie algebra actions:

$$
\mathfrak{g}_1 \circledS (\mathfrak{g}_2 \oplus V) \oplus (\mathfrak{g}_2 \circledS V). \tag{9.4.18}
$$

Symmetry reduction by stages of HP in this nested semidirect-product dynamics may be written for a *fixed* element $\Theta_0 \in V$ as

$$S = \int_{t_1}^{t_2} L(g_1, \dot{g}_1; g_2, \dot{g}_2; \Theta_0)\, dt$$

$$= \int_{t_1}^{t_2} L(g_1^{-1}\dot{g}_1 \,;\, \mathrm{Ad}_{g_1^{-1}}(g_2^{-1}\dot{g}_2) \,;\, g_1^{-1}g_2^{-1}\Theta_0)\, dt$$

$$=: \int_{t_1}^{t_2} L(\Omega_1; \Omega_2; \Theta)\, dt, \qquad (9.4.19)$$

where the sequential Ad actions define two velocities and a curve in V parameterised by time t as

$$\Omega_1 := g_1^{-1}\dot{g}_1, \quad \Omega_2 := \mathrm{Ad}_{g_1^{-1}}(g_2^{-1}\dot{g}_2), \quad \Theta := g_1^{-1}g_2^{-1}\Theta_0,$$
$$(9.4.20)$$

for the nested symmetries of $(TG_1 \times TG_2 \times V)$ under the respective left actions of $G_1 \times G_2$ parametrised by time t on a fixed element $\Theta_0 \in V$.

Derive the following evolutionary equation for Θ from its definition:

$$\frac{d}{dt}\Theta = -\mathcal{L}_{\Omega_1+\Omega_2}\Theta. \qquad (9.4.21)$$

★

Answer. By a direct calculation, one finds the auxiliary evolutionary equation for Θ as

$$\frac{d}{dt}\Theta = \frac{d}{dt}\left(g_1^{-1}g_2^{-1}\Theta_0\right) = -\left(g_1^{-1}\dot{g}_1\right)\Theta - \mathrm{Ad}_{g_1^{-1}}(g_2^{-1}\dot{g}_2)\Theta$$

$$= -\mathcal{L}_{\Omega_1}\Theta - \mathcal{L}_{\Omega_2}\Theta = -\mathcal{L}_{\Omega_1+\Omega_2}\Theta.$$
$$(9.4.22)$$

▲

Exercise. Verify the line of reasoning in deriving the Lagrangian shown above. Introduce variations with the sequential Ad actions of $G_1 \times G_2$:

$$\omega_1 := g_1^{-1}g'_1, \quad \omega_2 := \mathrm{Ad}_{g_1^{-1}}(g_2^{-1}g'_2), \quad \Theta' := -\mathcal{L}_{\omega_1+\omega_2}\Theta.$$
$$(9.4.23)$$

Then, use the following equations for iterated left actions:

$$\Omega'_1 - \dot{\omega}_1 = \mathrm{ad}_{\Omega_1}\omega_1,$$
$$\Omega'_2 - \dot{\omega}_2 = \mathrm{ad}_{\Omega_1+\Omega_2}\omega_2 + \mathrm{ad}_{\Omega_2}\omega_1, \qquad (9.4.24)$$
$$\Theta' = -\mathcal{L}_{\omega_1+\omega_2}\Theta,$$

to derive the equations of motion, which follow from HP with the Lagrangian in the last line of (9.4.19).   ★

Answer. The first two equations for variations in the set (9.4.24) have already been determined in previous exercises via their definitions in equations (9.4.5) and (9.4.6). Consequently, one may calculate the corresponding EP equations as follows:

$$0 = \delta$$

$$S = \delta \int_{t_1}^{t_2} L(\Omega_1; \Omega_2; \Theta)\, dt$$

$$= \int_{t_1}^{t_2} \left\langle \frac{\delta L}{\delta \Omega_1}, \delta\Omega_1 \right\rangle + \left\langle \frac{\delta L}{\delta \Omega_2}, \delta\Omega_2 \right\rangle + \left\langle \frac{\delta L}{\delta \Theta}, \delta\Theta \right\rangle$$

$$= \int_{t_1}^{t_2} \left\langle \frac{\delta L}{\delta \Omega_1}, \dot{\omega}_1 + \mathrm{ad}_{\Omega_1}\omega_1 \right\rangle + \left\langle \frac{\delta L}{\delta \Theta}, -\mathcal{L}_{\omega_1+\omega_2}\Theta \right\rangle$$

$$+ \left\langle \frac{\delta L}{\delta \Omega_2}, \dot{\omega}_2 + \mathrm{ad}_{\Omega_1+\Omega_2}\omega_2 + \mathrm{ad}_{\Omega_2}\omega_1 \right\rangle dt$$

$$= \int_{t_1}^{t_2} \left\langle -\frac{d}{dt}\frac{\delta L}{\delta \Omega_1} + \mathrm{ad}^*_{\Omega_1}\frac{\delta L}{\delta \Omega_1} \right.$$

$$+ \mathrm{ad}^*_{\Omega_2}\frac{\delta L}{\delta \Omega_2} + \frac{\delta L}{\delta \Theta} \diamond \Theta, \omega_1 \Big\rangle$$

$$+ \left\langle -\frac{d}{dt}\frac{\delta L}{\delta \Omega_2} + \mathrm{ad}^*_{\Omega_1+\Omega_2}\frac{\delta L}{\delta \Omega_2} + \frac{\delta L}{\delta \Theta} \diamond \Theta, \omega_2 \right\rangle dt.$$

$$(9.4.25)$$

▲

Exercise. Use the symmetry-reduced Legendre transform to obtain the Hamiltonian formulation of the dynamics for nested left semidirect-product action. ★

Answer. The symmetry-reduced Legendre transform for this case is

$$H(\Pi_1, \Pi_2, \Theta) = \langle \Pi_1, \Omega_1 \rangle + \langle \Pi_2, \Omega_2 \rangle - L(\Omega_1, \Omega_2, \Theta).$$

It yields the following variational derivatives of the Hamiltonian:

$$\frac{\delta H}{\delta \Pi_1} = \Omega_1, \quad \frac{\delta H}{\delta \Pi_2} = \Omega_2, \quad \frac{\delta H}{\delta \Theta} = -\frac{\delta L}{\delta \Theta}.$$

Rearranging the equations into Hamiltonian form yields

$$\partial_t \begin{pmatrix} \Pi_1 \\ \Pi_2 \\ \Theta \end{pmatrix} = \begin{bmatrix} \mathrm{ad}^*_\square \Pi_1 & \mathrm{ad}^*_\square \Pi_2 & -\square \diamond \Theta \\ \mathrm{ad}^*_\square \Pi_2 & \mathrm{ad}^*_\square \Pi_2 & -\square \diamond \Theta \\ -\mathcal{L}_\square \Theta & -\mathcal{L}_\square \Theta & 0 \end{bmatrix} \begin{pmatrix} \frac{\delta H}{\delta \Pi_1} = \Omega_1 \\ \frac{\delta H}{\delta \Pi_2} = \Omega_2 \\ \frac{\delta H}{\delta \Theta} = -\frac{\delta L}{\delta \Theta} \end{pmatrix}$$

$$(9.4.26)$$

$$= \begin{pmatrix} \mathrm{ad}^*_{\Omega_1}\Pi_1 + \mathrm{ad}^*_{\Omega_2}\Pi_2 - \frac{\delta H}{\delta \Theta} \diamond \Theta \\ \mathrm{ad}^*_{\Omega_1+\Omega_2}\Pi_2 - \frac{\delta H}{\delta \Theta} \diamond \Theta \\ -\mathcal{L}_{\Omega_1+\Omega_2}\Theta \end{pmatrix}. \qquad (9.4.27)$$

The matrix operator in square brackets in (9.4.26) defines an LP bracket $\{f, h\} = \langle \mu, [df, dh] \rangle$ on the dual of the following *nested* semidirect-product Lie algebra:

$$\mathfrak{g}_1 \, \circledS \, (\mathfrak{g}_2 \, \oplus \, V) \oplus (\mathfrak{g}_2 \, \circledS \, V), \qquad (9.4.28)$$

with dual coordinates $\Pi_1 \in \mathfrak{g}_1^*$, $\Pi_2 \in \mathfrak{g}_2^*$ and $\Theta \in V$.   ▲

Exercise. Write the LP bracket dual to $\mathfrak{g}_1 \, \circledS \, (\mathfrak{g}_2 \, \oplus \, V) \oplus (\mathfrak{g}_2 \, \circledS \, V)$ for invariance of the Lagrangian in (9.4.25) under *right* Lie group action.   ★

Answer. Compared to left invariance $g^{-1}\dot{g}$, the right-invariant case $\dot{g}g^{-1}$ simply reverses the signs in adjoint and coadjoint actions. Consequently, one may read off the correct LP operator for right action by reversing the signs of the ad^* operations.

In particular, the LP bracket for right action in this case becomes

$$\partial_t \begin{pmatrix} m_1 \\ m_2 \\ \rho \end{pmatrix} = - \begin{bmatrix} \mathrm{ad}^*_\square m_1 & \mathrm{ad}^*_\square m_2 & \square \diamond \rho \\ \mathrm{ad}^*_\square m_2 & \mathrm{ad}^*_\square m_2 & \square \diamond \rho \\ \mathcal{L}_\square \rho & \mathcal{L}_\square \rho & 0 \end{bmatrix} \begin{pmatrix} \frac{\delta H}{\delta m_1} = u_1 \\ \frac{\delta H}{\delta m_2} = u_2 \\ \frac{\delta H}{\delta \rho} = -\frac{\delta L}{\delta \rho} \end{pmatrix},$$
$$(9.4.29)$$

where one regards m_1 and m_2 as two fluid momenta whose corresponding velocities are u_1 and u_2, both transporting the other variable, ρ, which for fluid dynamics would be the mass density, or charge density in the fluid plasma application.[1]   ▲

---

[1] In the case of stochastic transport, the velocities u_1 and u_2 in (9.4.29) would become stochastic processes. For a discussion of stochastic fluid transport, see Ref. [Ho2015].

9.5   Plasma physics applications of Lie–Poisson brackets

The result in equation (9.4.29) applies to fluid dynamics in any dimension. In the fluid context, the Poisson bracket (9.4.29) can be compared with the LP bracket in (9.4.15). The apparently slight difference between these two LP brackets turns out to matter significantly, both geometrically and physically. In a fluids interpretation, the LP bracket in (9.4.15) refers to three types of fluid flow, each carried along by the previous group action. In contrast, the LP bracket in (9.4.29) can involve physical quantities which may be carried by two different vector fields, which in turn influence each other.

Remarkably, this LP bracket for right action applies to a reduced model of Alfvén wave turbulence equations for quasi-neutral plasma flow with the magnetic field in the plane of flow. In addition, the two vector fields in this case both interact with the charge density which is carried along by the two different types of transport. For more information and explanation of the physical meaning of these variables, see Refs. [HaMe1985, HaHoMo1985].

Exercise. Find the functionals whose variational derivatives are in the kernel of the LP operator in equation (9.4.29).

These functionals are conserved for any Hamiltonian written in terms of these variables.

Hint: One efficient way to find them would be to transform variables to make the LP operator as diagonal as possible.   ★

Answer. The change of variables in the Poisson matrix of (9.4.29) from (m_1, m_2, ρ) to (μ, m_2, ρ), with

$\mu = m_1 - m_2$, yields

$$\partial_t \begin{pmatrix} \mu \\ m_2 \\ \rho \end{pmatrix} = - \begin{bmatrix} \mathrm{ad}^*_\square \mu & 0 & 0 \\ 0 & \mathrm{ad}^*_\square m_2 & \square \diamond \rho \\ 0 & \mathcal{L}_\square \rho & 0 \end{bmatrix} \begin{pmatrix} \frac{\delta C}{\delta \mu} \\ \frac{\delta C}{\delta m_2} \\ \frac{\delta C}{\delta \rho} \end{pmatrix}. \quad (9.5.1)$$

To prove this statement, note that the Jacobian matrix for this transformation is given by

$$J = \begin{bmatrix} 1 & -1 & 0 \\ 0 & 1 & 0 \\ 0 & 0 & 1 \end{bmatrix}.$$

Multiplying the Poisson matrix in (9.4.29) by Jacobian J from the left and by its transpose J^T from the right produces the transformed Poisson matrix in (9.5.1). This linear change of variables preserves the eigenvalues of the Poisson matrix. In particular, such linear transformations preserve the matrix null eigenvectors.

Consequently, one may check that the variational derivatives of the following functions C_1, C_2, C_3 are Casimirs:

$$C_1 = F_1(\mu), \quad C_2 = m_2 F_2(\rho) \quad \text{and} \quad C_3 = F_3(\rho),$$
$$(9.5.2)$$

where F_1, F_2 and F_3 are arbitrary differentiable functions of their arguments. That is, their variational derivatives are null eigenvectors of the LP bracket in (9.5.1) as well as the transformed LP bracket in (9.5.1).   ▲

Exercise. Write the LP bracket dual to the following *twice nested* semidirect-product Lie algebra:

$$\mathfrak{s} = \mathfrak{g}_1 \,\circledS\, \Big(V_1 \oplus \big(\mathfrak{g}_2 \,\circledS\, (V_2 \oplus (\mathfrak{g}_3 \,\circledS\, V_3)) \big) \Big) \quad (9.5.3)$$

for invariance of a Lagrangian under twice nested *right* Lie group actions on the product space $(TG_1 \times V_1) \times (TG_2 \times V_2) \times (TG_3 \times V_3)$.   ★

Answer. Legendre-transforming such a Lagrangian invariant, under twice nested right Lie group actions, leads to EP equations whose twice nested LP Hamiltonian formulation may be displayed in the following matrix form with reduced Hamiltonian $h(m_k, a_k)$: $\Pi_k(\mathfrak{X}_k^* \times V_k^*) \to \mathbb{R}$ with $m_k := \delta\ell/\delta u_k$:

$$
\partial_t \begin{pmatrix} m_1 \\ a_1 \\ m_2 \\ a_2 \\ m_3 \\ a_3 \end{pmatrix} = - \begin{pmatrix} \mathrm{ad}_\square^* m_1 & \square \diamond a_1 & \square \diamond m_2 & \square \diamond a_2 & \square \diamond m_3 & \square \diamond a_3 \\ \mathcal{L}_\square a_1 & 0 & 0 & 0 & 0 & 0 \\ \mathcal{L}_\square m_2 & 0 & \mathrm{ad}_\square^* m_2 & \square \diamond a_2 & \square \diamond m_3 & \square \diamond a_3 \\ \mathcal{L}_\square a_2 & 0 & \mathcal{L}_\square a_2 & 0 & 0 & 0 \\ \mathcal{L}_\square m_3 & 0 & \mathcal{L}_\square m_3 & 0 & \mathrm{ad}_\square^* m_3 & \square \diamond a_3 \\ \mathcal{L}_\square a_3 & 0 & \mathcal{L}_\square a_3 & 0 & \mathcal{L}_\square a_3 & 0 \end{pmatrix} \times \begin{pmatrix} \frac{\delta h}{\delta m_1} = u_1 \\ \frac{\delta h}{\delta a_1} = -\frac{\delta\ell}{\delta a_1} \\ \frac{\delta h}{\delta m_2} = u_2 \\ \frac{\delta h}{\delta a_2} = -\frac{\delta\ell}{\delta a_2} \\ \frac{\delta h}{\delta m_3} = u_3 \\ \frac{\delta h}{\delta a_3} = -\frac{\delta\ell}{\delta a_3} \end{pmatrix} .
$$

Again, the nested LP bracket may be "untangled" via momentum shifts into block-diagonal form.

The pattern of Lie algebraic self-similarity for extension to further nested Lie group actions on additional degrees of freedom should now be clear.   ▲

Part II

Geometric Mechanics on Manifolds

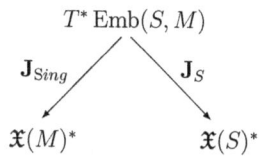

$$T^* \operatorname{Emb}(S, M)$$

$$\mathbf{J}_{Sing} \qquad\qquad \mathbf{J}_S$$

$$\mathfrak{X}(M)^* \qquad\qquad \mathfrak{X}(S)^*$$

The dual pair of momentum maps for continuum mechanics [HoMa2005].

10

GEOMETRIC STRUCTURE OF CLASSICAL MECHANICS

Contents

What is this lecture about? This lecture introduces basic vocabulary and notation for discussing the geometric mechanics of flows on manifolds.

10.1 Manifolds

Configuration space: coordinates $q \in M$, where M is a smooth manifold.

The composition $\phi_\beta \circ \phi_\alpha^{-1}$ is a smooth change of variables.

For later, smooth coordinate transformations: $q \to Q$ with $dQ = \frac{\partial Q}{\partial q} dq$.

Definition 10.1.1 *A smooth manifold M is a set of points together with a finite (or perhaps countable) set of subsets $U_\alpha \subset M$ and one-to-one mappings $\phi_\alpha \colon U_\alpha \to \mathbb{R}^n$ such that:*

1. *$\bigcup_\alpha U_\alpha = M$.*

2. *For every nonempty intersection $U_\alpha \cap U_\beta$, the set $\phi_\alpha (U_\alpha \cap U_\beta)$ is an open subset of $\mathbb{R}^n$, and the one-to-one mapping $\phi_\beta \circ \phi_\alpha^{-1}$ is a smooth function on $\phi_\alpha (U_\alpha \cap U_\beta)$.*

Remark 10.1.1 The sets U_α in the definition are called *coordinate charts*. The mappings ϕ_α are called *coordinate functions* or *local coordinates*. A collection of charts satisfying conditions (1) and (2) is called an *atlas*. Condition (2) allows the definition of manifold to be made independently of the choice of atlas. A set of charts satisfying (1) and (2) can always be extended to a maximal set; so, in practice, conditions (1) and (2) define the manifold. □

Example 10.1.1 Manifolds often arise as intersections of zero-level sets,
$$M = \{x \,|\, f_i(x) = 0, \; i = 1, \ldots, k\},$$
for a given set of functions $f_i \colon \mathbb{R}^n \to \mathbb{R}$, $i = 1, \ldots, k$. If the gradients ∇f_i are linearly independent, or, more generally, if the rank of $\{\nabla f(x)\}$ is a constant (r) for all x, then M is a smooth manifold of dimension $n - r$. The proof uses the implicit function theorem to show that an $(n-r)$-dimensional coordinate chart may be defined in a neighbourhood of each point on M. In this situation, the set M is called a *submanifold* of $\mathbb{R}^n$ (see Lee [Le2003]).

Definition 10.1.2 *If $r = k$, then the map $\{f_i\}$ is called a **submersion**.*

Exercise. Prove that the zero sets of all submersions are submanifolds (see Lee [Le2003]). ★

Definition 10.1.3 (Tangent space to level sets) *Let*

$$M = \{x \mid f_i(x) = 0, \ i = 1, \ldots, k\}$$

be a manifold in $\mathbb{R}^n$. (Note that the zero set is a manifold because the map has a constant rank.)

   *The **tangent space** at each $x \in M$, is defined by*

$$T_x M = \left\{ v \in \mathbb{R}^n \ \middle| \ \frac{\partial f_i}{\partial x^a}(x) v^a = 0, \ i = 1, \ldots, k \right\}.$$

*Note: we use the **summation convention**. That is, repeated indices are summed over their range.*

Remark 10.1.2 The tangent space is a linear vector space.   □

Example 10.1.2 (Tangent space to the sphere in $\mathbb{R}^3$) The sphere S^2 is the set of points $(x, y, z) \in \mathbb{R}^3$ solving $x^2 + y^2 + z^2 = 1$. The tangent space to the unit sphere at such a point (x, y, z) is the plane containing vectors (u, v, w) satisfying $xu + yv + zw = 0$.

Definition 10.1.4 (Tangent bundle) *The **tangent bundle** of a manifold M, denoted by TM, is the smooth manifold whose underlying set is the disjoint union of the tangent spaces to M at the points $x \in M$; that is,*

$$TM = \bigcup_{x \in M} T_x M.$$

Thus, a point of TM is a vector v which is tangent to M at some point $x \in M$.

Example 10.1.3 (Tangent bundle TS^2 of S^2) The tangent bundle TS^2 of $S^2 \in \mathbb{R}^3$ is the union of the tangent spaces of S^2:

$$TS^2 = \{(x, y, z; u, v, w) \in \mathbb{R}^6 \mid x^2 + y^2 + z^2 = 1 \text{ and } xu + yv + zw = 0\}.$$

Remark 10.1.3 (Dimension of tangent bundle TS^2) Defining TS^2 requires two independent conditions in $\mathbb{R}^6$, so $\dim TS^2 = 4$.   □

Exercise. Define the sphere S^{n-1} in $\mathbb{R}^n$. What is the dimension of its tangent space TS^{n-1}? ★

Example 10.1.4 (The two stereographic projections of $S^2 \to \mathbb{R}^2$)

The unit sphere

$$S^2 = \{(x, y, z) : x^2 + y^2 + z^2 = 1\}$$

is a smooth two-dimensional manifold realised as the level set of a submersion in $\mathbb{R}^3$. Let

$$U_N = S^2 \backslash \{0, 0, 1\}, \quad \text{and} \quad U_S = S^2 \backslash \{0, 0, -1\}$$

be the subsets obtained by deleting the north and south poles of S^2, respectively. Let

$$\chi_N : U_N \to (\xi_N, \eta_N) \in \mathbb{R}^2, \quad \text{and} \quad \chi_S : U_S \to (\xi_S, \eta_S) \in \mathbb{R}^2$$

be stereographic projections from the north and south poles onto the equatorial plane $z = 0$. See Figure 10.1.

Thus, one may place two different coordinate patches in S^2 intersecting everywhere except at the points along the z-axis at $z = 1$ (north pole) and $z = -1$ (south pole).

In the equatorial plane $z = 0$, one may define two sets of (right-handed) coordinates,

$$\phi_\alpha : U_\alpha \to \mathbb{R}^2 \backslash \{0\}, \quad \alpha = N, S,$$

obtained by the following two stereographic projections from the north and south poles:

1. (valid everywhere except $z = 1$)

$$\phi_N(x, y, z) = (\xi_N, \eta_N) = \left(\frac{x}{1 - z}, \frac{y}{1 - z} \right),$$

2. (valid everywhere except $z = -1$)

$$\phi_S(x, y, z) = (\xi_S, \eta_S) = \left(\frac{x}{1 + z}, \frac{-y}{1 + z} \right).$$

(The two complex planes are identified differently with the plane $z = 0$. An orientation reversal is necessary to maintain consistent coordinates on the sphere.)

One may check directly that, on the overlap $U_N \cap U_S$, the map

$$\phi_N \circ \phi_S^{-1} \colon \mathbb{R}^2 \backslash \{0\} \to \mathbb{R}^2 \backslash \{0\}$$

is a smooth diffeomorphism, given by the inversion

$$\phi_N \circ \phi_S^{-1}(x, y) = \left(\frac{x}{x^2 + y^2}, \frac{y}{x^2 + y^2} \right).$$

Exercise. Construct the mapping from $(\xi_N, \eta_N) \to (\xi_S, \eta_S)$ and verify that it is a diffeomorphism in $\mathbb{R}^2 \backslash \{0\}$. Hint: $(1 + z)(1 - z) = 1 - z^2 = x^2 + y^2$. ★

Answer.

$$(\xi_S, -\eta_S) = \frac{1 - z}{1 + z}(\xi_N, \eta_N) = \frac{1}{\xi_N^2 + \eta_N^2}(\xi_N, \eta_N).$$

The map $(\xi_N, \eta_N) \to (\xi_S, \eta_S)$ is smooth and invertible except at $(\xi_N, \eta_N) = (0, 0)$. ▲

Example 10.1.5 If we start with two identical circles in the xz-plane, of radius r and centred at $x = \pm 2r$, and rotate them round the z-axis in $\mathbb{R}^3$, we get a torus, written T^2. The torus T^2 is a manifold.

Exercise. If we begin with a figure eight in the xz-plane, along the x-axis and centred at the origin, and spin it round the z-axis in $\mathbb{R}^3$, we get a "pinched surface" that looks like a sphere that has been "pinched" so that the north and south poles touch. Is this a manifold? Prove it. ★

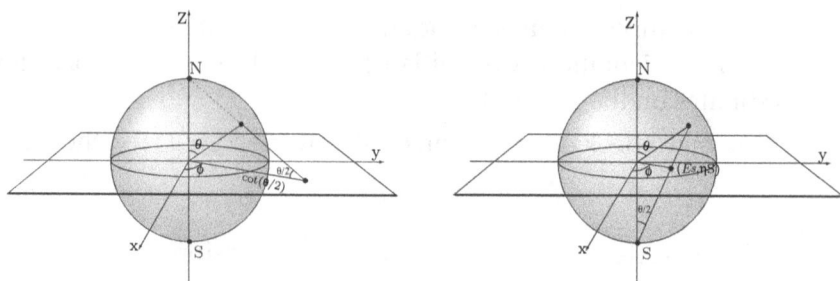

Figure 10.1. In the stereographic projection of the Riemann sphere onto the complex plane from the south pole, complex numbers lying outside (resp., inside) the unit circle are projected from points in the lower (resp., upper) hemisphere.

Answer. The origin has a neighbourhood diffeomorphic to a double cone. This is not diffeomorphic to $\mathbb{R}^2$. A proof of this is that, if the origin of the cone is removed, two components remain; while if the origin of $\mathbb{R}^2$ is removed, only one component remains.   ▲

Remark 10.1.4 The sphere $S^2 \subset \mathbb{R}^3$ in Figure 10.1 will appear in several examples as a reduced space in which motion occurs after applying a symmetry. Reduction by symmetry is associated with a classical topic in celestial mechanics known as normal form theory. Reduction may be "singular," in which case it leads to "pointed" spaces that are smooth manifolds except at one or more points. For example, different resonances of coupled oscillators correspond to the following reduced spaces: 1:1 resonance – sphere; 1:2 resonance – pinched sphere with one cone point; 1:3 resonance – pinched sphere with one cusp point; 2:3 resonance – pinched sphere with one cone point and one cusp point.   □

10.2   Motion: Tangent vectors and flows

Envisioning our later considerations of dynamical systems, we shall consider motion along curves $c(t)$ parametrised by time t on a smooth manifold M. Suppose these curves are trajectories of a flow

ϕ_t of a vector field. We anticipate that this means $\phi_t(c(0)) = c(t)$ and $\phi_t \circ \phi_s = \phi_{t+s}$ (flow property). The flow will be tangent to M along the curve. To deal with such flows, we will need the concept of *tangent vectors*.

Recall from Definition 10.1.4 that the tangent bundle of M is

$$TM = \bigcup_{x \in M} T_x M.$$

We now add a bit more to that definition. The tangent bundle is an example of a more general structure than a manifold.

Definition 10.2.1 (Bundle) *A **bundle** consists of a manifold B, another manifold M called the "base space" and a projection between them, $\Pi: B \to M$. Locally, in small enough regions of x, the inverse images of the projection Π exist. These are called the fibres of the bundle. Thus, subsets of the bundle B locally have the structure of a Cartesian product. An example is (B, M, Π), consisting of $(\mathbb{R}^2, \mathbb{R}^1, \Pi: \mathbb{R}^2 \to \mathbb{R}^1)$. In this case, $\Pi: (x, y) \in \mathbb{R}^2 \to x \in \mathbb{R}^1$. Likewise, the tangent bundle consists of M, TM and a map $\tau_M: TM \to M$.*

Let $x = (x^1, \ldots, x^n)$ be local coordinates on M, and let $v = (v^1, \ldots, v^n)$ be components of a tangent vector,

$$T_x M = \left\{ v \in \mathbb{R}^n \mid \frac{\partial f_i}{\partial x} \cdot v = 0, i = 1, \ldots, m \right\}$$

for

$$M = \left\{ x \in \mathbb{R}^n \mid f_i(x) = 0, i = 1, \ldots, m \right\}.$$

These $2n$ numbers (x, v) give local coordinates on TM, where $\dim TM = 2 \dim M$. The *tangent bundle projection* is a map $\tau_M: TM \to M$ which takes a tangent vector v to a point $x \in M$ where the tangent vector v is attached (that is, $v \in T_x M$). The inverse of this projection $\tau_M^{-1}(x)$ is called the *fibre* over x in the tangent bundle.

10.3   Summary of vector fields, integral curves and flows

Definition 10.3.1 *A **vector field** on a manifold M is a map $X \colon M \to TM$ that assigns a vector $X(x)$ at each point $x \in M$. This implies that $\tau_M \circ X = \mathrm{Id}$.*

Definition 10.3.2 *An **integral curve** of X with initial conditions x_0 at $t = 0$ is a differentiable map $c \colon \,]a, b[\to M$, where $]a, b[$ is an open interval containing 0, such that $c(0) = 0$ and $c'(t) = X\,(c(t))$ for all $t \in \,]a, b[$.*

Remark 10.3.1 A standard result from the theory of ordinary differential equations states that X being Lipschitz implies its integral curves are unique and C^1 (see Coddington and Levinson [CoLe1984]). The integral curves $c(t)$ are differentiable for smooth X. □

Definition 10.3.3 *The **flow** of X is the collection of maps $\phi_t \colon M \to M$, where $t \to \phi_t(x)$ is the integral curve of X with initial condition x.*

Remark 10.3.2

1. Existence and uniqueness results for solutions of $c'(t) = X(c(t))$ guarantee that flow ϕ of X is smooth in (x, t), for smooth X.

2. Uniqueness implies the flow property

$$\phi_{t+s} = \phi_t \circ \phi_s, \tag{FP}$$

   for the initial condition $\phi_0 = \mathrm{Id}$.

3. The flow property (FP) generalises to the nonlinear case of the familiar linear situation where M is a vector space, $X(x) = Ax$ is a linear vector field for a bounded linear operator A and $\phi_t(x) = e^{At}x$. □

10.4   Differentials of functions and the cotangent bundle

We are now ready to define differentials of smooth functions and the cotangent bundle.

Let $f: M \to \mathbb{R}$ be a smooth function. We differentiate f at $x \in M$ to obtain $T_x f: T_x M \to T_{f(x)}\mathbb{R}$. As is standard, we identify $T_{f(x)}\mathbb{R}$ with $\mathbb{R}$ itself, thereby obtaining a linear map $df(x): T_x M \to \mathbb{R}$. The result $df(x)$ is an element of the cotangent space $T_x^* M$, the dual space of the tangent space $T_x M$. The natural pairing between the elements of the tangent space and the cotangent space is denoted as $\langle \cdot, \cdot \rangle: T_x^* M \times T_x M \mapsto \mathbb{R}$.

In coordinates, the linear map $df(x): T_x M \to \mathbb{R}$ may be written as the directional derivative

$$\langle df(x), v \rangle = df(x) \cdot v = \frac{\partial f}{\partial x^i} \cdot v^i,$$

for all $v \in T_x M$. (Reminder: the summation convention is intended over repeated indices.) Hence, elements $df(x) \in T_x^* M$ are dual to vectors $v \in T_x M$ with respect to the pairing $\langle \cdot, \cdot \rangle$.

Definition 10.4.1 *The symbol df denotes the **differential** of the function f.*

Definition 10.4.2 *The dual space of the tangent bundle TM is the **cotangent bundle** $T^* M$. That is,*

$$(T_x M)^* = T_x^* M \quad and \quad T^* M = \bigcup_x T_x^* M.$$

Thus, replacing $v \in T_x M$ with $df \in T_x^* M$, for all $x \in M$ and for all smooth functions $f: M \to \mathbb{R}$, yields the *cotangent bundle* $T^* M$.

10.4.1   Differential bases

When the basis of vector fields is Cartesian and denoted as $\frac{\partial}{\partial x^i}$ for $i = 1, \ldots, n$, its dual basis may be denoted as dx^i. In this notation, the differential of a function at a point $x \in M$ is expressed as

$$df(x) = \frac{\partial f}{\partial x^i} dx^i.$$

The corresponding pairing $\langle \cdot, \cdot \rangle$ of bases is written in Cartesian notation as

$$\left\langle dx^j, \frac{\partial}{\partial x^i} \right\rangle = \delta_i^j.$$

Here, δ_i^j is the Kronecker delta, which equals unity for $i = j$ and vanishes otherwise. That is, defining T^*M requires a pairing, $\langle \cdot, \cdot \rangle : T^*M \times TM \to \mathbb{R}$.

Although different pairings can exist, e.g., for curvilinear coordinates, Riemannian manifolds, etc., for simplicity of notation in this text, we will usually apply the Cartesian pairing, as above.

11

INTRODUCTION TO VECTOR FIELDS

Contents

> **What is this lecture about?** This lecture defines vector fields as tangents to the curves obtained from actions of smooth invertible maps on smooth manifolds; it then explains how this definition determines the properties of the vector fields.

Definition 11.0.1 (Vector fields) *A **vector field** X on a manifold M is a map $X : M \rightarrow TM$ that assigns a vector $X(q)$ at every point $q \in M$. The real vector space of vector fields on a manifold M is denoted by $\mathfrak{X}(M)$.*

Definition 11.0.2 *A **time-dependent vector field** is a map*

$$X : M \times \mathbb{R} \rightarrow TM$$

such that $X(q, t) \in T_q M$ for each $q \in M$ and $t \in \mathbb{R}$.

Definition 11.0.3 (Integral curves) *An **integral curve** of vector field $X(q)$ with initial condition q_0 is a differentiable map $q :]t_1, t_2[\rightarrow M$ such* *that the open interval* $]t_1, t_2[$ *contains the initial time* $t = 0$, *at which* $q(0) = q_0$ *and the tangent vector coincides with the vector field*

$$\dot{q} = X(q(t))$$

for all $t \in]t_1, t_2[$.

Remark 11.0.1 In what follows, we shall always assume that we are dealing with vector fields that satisfy the conditions required for their integral curves to exist and be unique.   □

Definition 11.0.4 (Vector field basis) *The components of a vector field* $\dot{q}$ *are defined by its directional derivatives in the chosen coordinate basis.*

Example 11.0.1 (Vector field basis) An example of a vector basis for the components of a vector field is given by

$$\dot{q} = \dot{q}^a \frac{\partial}{\partial q^a} \quad \text{(Vector basis).} \qquad (11.0.1)$$

That is, in the vector basis defined by the coordinate derivatives $\partial/\partial q^a$, the vector field $\dot{q}$ has components $\dot{q}^a$, $a = 1, \ldots, K$.

11.1   Motions, pull-backs, push-forwards and commutators

A *motion* is defined as a smooth curve $q(t) \in M$ parameterised by $t \in \mathbb{R}$ that solves the *motion equation*, which is a system of differential equations,

$$\dot{q}(t) = \frac{dq}{dt} = f(q) \in TM, \qquad (11.1.1)$$

or, in components,

$$\dot{q}^i(t) = \frac{dq^i}{dt} = f^i(q) \quad i = 1, 2, \ldots, n. \qquad (11.1.2)$$

The map $f : q \in M \to f(q) \in T_q M$ is a *vector field*.

According to standard theorems about differential equations that are not proven in this course, the solution, or integral curve, $q(t)$ exists, provided f is sufficiently smooth, which will always be assumed to hold.

Vector fields can also be defined as *differential operators* that act on functions, as

$$\frac{d}{dt}G(q) = \dot{q}^i(t)\frac{\partial G}{\partial q^i} = f^i(q)\frac{\partial G}{\partial q^i}$$

$$i = 1, 2, \ldots, n, \quad \text{(sum on repeated indices)} \qquad (11.1.3)$$

for any smooth function $G(q) : M \to \mathbb{R}$.

To indicate the dependence of the solution of its initial condition $q(0) = q_0$, we write the motion as a smooth transformation:

$$q(t) = \phi_t(q_0).$$

Because the vector field f is independent of time t, for any fixed value of t, we may regard ϕ_t as a mapping from M into itself that satisfies the *composition law*

$$\phi_t \circ \phi_s = \phi_{t+s}$$

and

$$\phi_0 = \text{Id}.$$

Setting $s = -t$ shows that ϕ_t has a smooth inverse. A smooth mapping that has a smooth inverse is called a *diffeomorphism*. Geometric mechanics deals with diffeomorphisms.

The smooth mapping $\phi_t : \mathbb{R} \times M \to M$ that determines the solution $\phi_t \circ q_0 = q(t) \in M$ of the motion equation (11.1.1) with initial condition $q(0) = q_0$ is called the *flow* of the vector field Q.

A point $q^* \in M$ at which $f(q^*) = 0$ is called a *fixed point* of the flow ϕ_t, or an *equilibrium*.

Vice versa, the vector field f is called the *infinitesimal transformation* of the mapping ϕ_t since

$$\frac{d}{dt}\Big|_{t=0} (\phi_t \circ q_0) = f(q).$$

That is, $f(q)$ is the *linearisation* of the flow map ϕ_t at the point $q \in M$.

More generally, the *directional derivative* of the function h along the vector field f is given by the action of a differential operator, as

$$\frac{d}{dt}\bigg|_{t=0} h \circ \phi_t = \left[\frac{\partial h}{\partial \phi_t} \frac{d}{dt} (\phi_t \circ q_0) \right]_{t=0} = \frac{\partial h}{\partial q^i} \dot{q}^i = \frac{\partial h}{\partial q^i} f^i(q) =: Qh.$$

Under a smooth change of variables $q = c(r)$, the vector field Q in the expression Qh transforms as

$$Q = f^i(q) \frac{\partial}{\partial q^i} \quad \mapsto \quad R = g^j(r) \frac{\partial}{\partial r^j}, \tag{11.1.4}$$

with

$$g^j(r) \frac{\partial c^i}{\partial r^j} = f^i(c(r)) \quad \text{or} \quad g = c_r^{-1} f \circ c, \tag{11.1.5}$$

where c_r is the *Jacobian matrix* of the transformation. That is, since $h(q)$ is a function of q,

$$(Qh) \circ c = R(h \circ c).$$

We express the transformation between the vector fields as $R = c^*Q$, and write this relation as

$$(Qh) \circ c =: c^*Q(h \circ c). \tag{11.1.6}$$

The expression c^*Q is called the *pull-back* of the vector field Q by the map c. Two vector fields are equivalent under a map c if one is the pull-back of the other, and fixed points are mapped into fixed points.

The inverse of the pull-back is called the *push-forward*. Namely, the push-forward by the map c is the pull-back by the inverse map c^{-1}.

The *commutator*

$$QR - RQ =: [Q, R]$$

of two vector fields Q and R defines another vector field. Indeed, if

$$Q = f^i(q) \frac{\partial}{\partial q^i} \quad \text{and} \quad R = g^j(q) \frac{\partial}{\partial q^j},$$

then

$$[Q, R] = \left(f^i(q) \frac{\partial g^j(q)}{\partial q^i} - g^i(q) \frac{\partial f^j(q)}{\partial q^i} \right) \frac{\partial}{\partial q^j}$$

because the second-order derivative terms cancel. By the pull-back relation (11.1.6), we have

$$c^* [Q, R] = [c^*Q, c^*R] \qquad (11.1.7)$$

under a change of variables defined by a smooth map, c. This means the definition of the vector field commutator is independent of the choice of coordinates. As we shall see in Corollary 11.1.1, the *tangent* to the relation $c_t^* [Q, R] = [c_t^*Q, c_t^*R]$ at the identity $t = 0$ is the *Jacobi condition* for the vector fields to form an algebra.

Pullbacks of vector fields lead to Lie derivative expressions.

Definition 11.1.1 (Lie derivative of a vector field) *The **Lie deriva-tive** of a vector field $Y \in \mathfrak{X}$ by another vector field $X \in \mathfrak{X}$ is defined by linearising the flow ϕ_t of X around the identity $t = 0$:*

$$\pounds_X Y = \frac{d}{dt} \Big|_{t=0} \phi_t^* Y \quad maps \quad \pounds_X \in \mathfrak{X} \mapsto \mathfrak{X}.$$

Theorem 11.1.1 *The Lie derivative $\pounds_X Y$ of a vector field Y by a vector field X satisfies*

$$\pounds_X Y = \frac{d}{dt} \Big|_{t=0} \phi_t^* Y = [X, Y], \qquad (11.1.8)$$

where $[X, Y] = XY - YX$ is the commutator of the vector fields X and Y.

Proof. Denote the vector fields in components as

$$X = X^i(q) \frac{\partial}{\partial q^i} = \frac{d}{dt} \Big|_{t=0} \phi_t^* \quad \text{and} \quad Y = Y^j(q) \frac{\partial}{\partial q^j}.$$

Then, by the pull-back relation (11.1.6), a direct computation yields, on using the matrix identity $dM^{-1} = -M^{-1}dM\,M^{-1}$,

$$
\pounds_X Y = \frac{d}{dt}\bigg|_{t=0} \phi_t^* Y = \frac{d}{dt}\bigg|_{t=0} \left(Y^j(\phi_t q) \frac{\partial}{\partial(\phi_t q)^j} \right)
$$

$$
= \frac{d}{dt}\bigg|_{t=0} \left(Y^j(\phi_t q) \left[\frac{\partial(\phi_t q)}{\partial q}^{-1} \right]_j^k \frac{\partial}{\partial q^k} \right)
$$

$$
= \left(X^j \frac{\partial Y^k}{\partial q^j} - Y^j \frac{\partial X^k}{\partial q^j} \right) \frac{\partial}{\partial q^k}
$$

$$
= [X, Y].
$$
■

Corollary 11.1.1 *The Lie derivative of the relation (11.1.7) for the pull-back of the commutator $c_t^*[Y, Z] = [c_t^* Y, c_t^* Z]$ yields the Jacobi condition for the vector fields to form an algebra.*

Proof. By the product rule and the definition of the Lie bracket (11.1.8), we have

$$
\frac{d}{dt}\bigg|_{t=0} \phi_t^*[Y, Z] = [X, [Y, Z]] = [[X, Y], Z] + [Y, [X, Z]]
$$

$$
= \frac{d}{dt}\bigg|_{t=0} [\phi_t^* Y, \phi_t^* Z].
$$

This is the *Jacobi identity* for vector fields.   ■

11.2   Lie algebras of vector fields

Definition 11.2.1 (The ad-operation) *For $A \in \mathfrak{g}$, we define the operator ad_A to be the operator $\mathrm{ad} : \mathfrak{g} \times \mathfrak{g} \to \mathfrak{g}$ that maps $B \in \mathfrak{g}$ to $[A, B]$. We write $\mathrm{ad}_A B = [A, B]$.*

Definition 11.2.2 *A **representation** of a Lie algebra $\mathfrak{g}$ on a vector space V is a mapping ρ from $\mathfrak{g}$ to the linear transformations of V such that, for $A, B \in \mathfrak{g}$ and any constant scalar c,*

1. $\rho(A + cB) = \rho(A) + c\rho(B)$,

2. $\rho([A, B]) = \rho(A)\rho(B) - \rho(B)\rho(A)$.

If the map ρ is one-to-one, the representation is said to **faithful**.

Exercise. For a Lie algebra $\mathfrak{g}$, show that the map $A \to$ (ad A) is a representation of the Lie algebra $\mathfrak{g}$, with $\mathfrak{g}$ itself the vector space of the representation. This is called the *adjoint representation*.   ★

Example 11.2.1 (Vector field representations of Lie algebras)
The Jacobi–Lie bracket of the vector fields ξ and η may be represented in coordinate charts as

$$\eta = \frac{dx}{ds}\bigg|_{s=0} = v(x), \quad \text{and} \quad \xi = \frac{dx}{dt}\bigg|_{t=0} = u(x).$$

The Jacobi–Lie bracket of these two vector fields yields a third vector field:

$$\begin{aligned}
\xi\eta - \eta\xi &= \frac{d\eta}{dt}\bigg|_{t=0} - \frac{d\xi}{ds}\bigg|_{s=0} \\
&= \frac{dv}{dx}\frac{dx}{dt}\bigg|_{t=0} - \frac{du}{dx}\frac{dx}{ds}\bigg|_{s=0} = \frac{dv}{dx}\cdot u - \frac{du}{dx}\cdot v \\
&= u\cdot\nabla v - v\cdot\nabla u.
\end{aligned}$$

Thus, the Jacobi–Lie bracket of vector fields at the tangent space of the identity T_eG is closed and may be represented in coordinate charts by the Lie bracket (commutator of vector fields)

$$[\xi, \eta] := \xi\eta - \eta\xi = u\cdot\nabla v - v\cdot\nabla u =: [u, v].$$

This example also proves the following.

Proposition 11.2.1 *Let $\mathfrak{X}(\mathbb{R}^n)$ be the set of vector fields defined on $\mathbb{R}^n$. A Lie algebra $\mathfrak{g}$ may be represented on coordinate charts by the vector fields*

$X_\xi = X_\xi^i \frac{\partial}{\partial x^i} \in \mathfrak{X}(\mathbb{R}^n)$ *for each element* $\xi \in \mathfrak{g}$. *This vector field represen-*
tation satisfies

$$X_{[\xi,\eta]} = [X_\xi, X_\eta],$$

where $[\xi, \eta] \in \mathfrak{g}$ *is the Lie algebra product and* $[X_\xi, X_\eta]$ *is the vector field*
commutator.

Exercise. (Integral curves of vector fields on the real
line) Calculate the integral curves and identify the
group action generated by the two vector fields on the
real line, $v_1 = \partial_x$ and $v_2 = x\partial_x$. Find the matrix Lie group
isomorphic to the Lie group generated by the integral
curves of the vector fields v_1 and v_2.   ★

Answer. The commutator relation for v_1 and v_2 is

$$[v_1, v_2] = [\partial_x, x\partial_x] = \partial_x = v_1.$$

The integral curves for v_1 and v_2 are computed from

$$\frac{dx}{d\epsilon_i} = X_i(x) \quad i = 1, 2, \quad \text{with} \quad X_1 = 1 \quad \text{and} \quad X_2 = x$$

by integrating to find translations in ϵ_1,

$$\int d\epsilon_1 = \epsilon_1 = \int \frac{dx}{1} = x_{\epsilon_1} - x_0$$

$$\implies \quad x_{\epsilon_1} = g_{\epsilon_1} x_0 = x_0 + \epsilon_1,$$

and scaling transformations in ϵ_2,

$$\int d\epsilon_2 = \epsilon_2 = \int \frac{dx}{x} = \log(x_{\epsilon_2}/x_0)$$

$$\implies \quad x_{\epsilon_2} = g_{\epsilon_2} x_0 = x_0 e^{\epsilon_2}.$$

Combining these transformations produces the affine group action

$$g_{\epsilon_1} g_{\epsilon_2} x_0 = e^{\epsilon_2} x_0 + \epsilon_1.$$

This group action can be obtained as a left action of the upper-triangular 2×2 matrices on the column vector $[x_0, 1]^T$:

$$\begin{bmatrix} e^{\epsilon_2} & \epsilon_1 \\ 0 & 1 \end{bmatrix} \begin{bmatrix} x_0 \\ 1 \end{bmatrix} = \begin{bmatrix} e^{\epsilon_2} x_0 + \epsilon_1 \\ 1 \end{bmatrix}.$$

It can also be written as a right action of lower-triangular 2×2 matrices on the row vector $[x_0, 1]$:

$$\begin{bmatrix} x_0 & 1 \end{bmatrix} \begin{bmatrix} e^{\epsilon_2} & 0 \\ \epsilon_1 & 1 \end{bmatrix} = \begin{bmatrix} e^{\epsilon_2} x_0 + \epsilon_1 & 1 \end{bmatrix}.$$

▲

Exercise.   (Integral curves of vector fields on the real line (*Continued*)) Calculate the integral curves and identify the group action generated by the three vector fields on the real line, $v_1 = \partial_x$, $v_2 = x \partial_x$ and $v_3 = -x^2 \partial_x$. Find the matrix Lie group isomorphic to the Lie group generated by the integral curves of the vector fields v_1, v_2 and v_3.   ★

Answer. The commutator relation for v_1, v_2 and v_3 is given by

$$[v_i, v_j] = c_{ij}^k v_k = $$

$[\cdot, \cdot]$	v_1	v_2	v_3
v_1	0	v_1	$-2v_2$
v_2	$-v_1$	0	$-v_3$
v_3	$2v_2$	v_3	0

. (11.2.1)

The transformations in ϵ_3 are given by

$$\int d\epsilon_3 = \epsilon_3 = \int \frac{dx}{-x^2} = \frac{1}{x_{\epsilon_3}} - \frac{1}{x_0}$$

$$\implies x_{\epsilon_3} = g_{\epsilon_3} x_0 = \frac{x_0}{\epsilon_3 x_0 + 1}.$$

Combining these transformations produces the projective group action

$$g_{\epsilon_1} g_{\epsilon_2} g_{\epsilon_3} x_0 = \frac{e^{\epsilon_2} x_0}{1 + \epsilon_3 x_0} + \epsilon_1 = \frac{\left(e^{\epsilon_2} + \epsilon_1 \epsilon_3\right) x_0 + \epsilon_1}{\epsilon_3 x_0 + 1},$$

which can be written as a right action of 2×2 matrices on the row vector $[x_0, 1]$:

$$\begin{bmatrix} x_0 & 1 \end{bmatrix} \begin{bmatrix} e^{\epsilon_2} + \epsilon_1 \epsilon_3 & \epsilon_3 \\ \epsilon_1 & 1 \end{bmatrix} = \begin{bmatrix} \left(e^{\epsilon_2} + \epsilon_1 \epsilon_3\right) x_0 + \epsilon_1 & \epsilon_3 x_0 + 1 \end{bmatrix}$$

$$\simeq \begin{bmatrix} \frac{\left(e^{\epsilon_2} + \epsilon_1 \epsilon_3\right) x_0 + \epsilon_1}{\epsilon_3 x_0 + 1} & 1 \end{bmatrix}. \qquad \blacktriangle$$

12

DERIVATIVES OF DIFFERENTIABLE MAPS: THE TANGENT LIFT

Contents

What is this lecture about? This lecture introduces the tangent and cotangent lifts of differentiable maps between manifolds and discusses some of their properties obtained from the calculus chain rule.

12.1 Derivatives of differentiable maps between manifolds

We next define derivatives of differentiable maps between manifolds (tangent lifts).

We expect that a smooth map $f: U \to V$ from a chart $U \subset M$ to a chart $V \subset N$ will lift to a map between the tangent bundles

TM and TN so as to make sense from the viewpoint of ordinary calculus,

$$U \times \mathbb{R}^m \subset TM \longrightarrow V \times \mathbb{R}^n \subset TN$$
$$\left(q^1, \ldots, q^m; X^1, \ldots, X^m\right) \longmapsto \left(Q^1, \ldots, Q^n; Y^1, \ldots, Y^n\right).$$

Namely, the relations between the vector field components should be obtained from the differential of the map $f: U \to V$. Perhaps not unexpectedly, these vector field components will be related by

$$Y^i \frac{\partial}{\partial Q^i} = X^j \frac{\partial}{\partial q^j}, \quad \text{so } Y^i = \frac{\partial Q^i}{\partial q^j} X^j,$$

in which the quantity called the *tangent lift* of the function f, denoted

$$Tf = \frac{\partial Q}{\partial q},$$

arises from the chain rule and is equal to the Jacobian for the transformation

$$Tf: TM \mapsto TN.$$

The dual of the tangent lift is the cotangent lift, discussed further in 16.0.1. Basically, the *cotangent lift* of the function f,

$$T^* f = \frac{\partial q}{\partial Q},$$

arises from

$$\beta_i dQ^i = \alpha_j dq^j, \quad \text{so } \beta_i = \alpha_j \frac{\partial q^j}{\partial Q^i}$$

and $T^* f: T^* N \mapsto T^* M$. Note the directions of these maps:

$$Tf: q, X \in TM \mapsto Q, Y \in TN$$
$$f: q \in M \mapsto Q \in N$$
$$T^* f: Q, \beta \in T^* N \mapsto q, \alpha \in T^* M \quad (T^* f \text{ map goes the other way}).$$

12.2 Summary remarks about derivatives on manifolds

Definition 12.2.1 (Differentiable map) *A map $f\colon M \to N$ from manifold M to manifold N is said to be **differentiable** (resp., C^k) if it is represented in local coordinates on M and N by differentiable (resp., C^k) functions.*

Definition 12.2.2 (Derivative of a differentiable map) *The **derivative** of a differentiable map*

$$f\colon M \to N$$

at a point $x \in M$ is defined to be the linear map

$$T_x f\colon T_x M \to T_x N$$

constructed as follows. For $v \in T_x M$, choose a curve $c(t)$ that maps an open interval $t \in (-\epsilon, \epsilon)$ around the point $t = 0$ to the manifold M:

$$c\colon (-\epsilon, \epsilon) \to M$$

with $c(0) = x$ and velocity vector $c'(0) := \frac{dc}{dt}\big|_{t=0} = v$.

*   Then, $T_x f \cdot v$ is the velocity vector at $t = 0$ of the curve $f \circ c\colon \mathbb{R} \to N$. That is,*

$$T_x f \cdot v = \frac{d}{dt} f(c(t))\Big|_{t=0} = \frac{\partial f}{\partial c}\frac{d}{dt}c(t)\Big|_{t=0}.$$

Definition 12.2.3 *The union $Tf = \bigcup_x T_x f$ of the derivatives $T_x f\colon T_x M \to T_x N$ over points $x \in M$ is called the **tangent lift** of the map $f\colon M \to N$.*

Remark 12.2.1 The chain-rule definition of the derivative $T_x f$ of a differentiable map at a point x depends on the function f and the vector v. Other degrees of differentiability are possible. For example, if M and N are manifolds and $f\colon M \to N$ is of class C^{k+1}, then the tangent lift (Jacobian) $T_x f\colon T_x M \to T_x N$ is C^k. $\square$

Exercise. Let $\phi_t\colon S^2 \to S^2$ rotate points on S^2 about a fixed axis through an angle $\psi(t)$. Show that ϕ_t is the flow of a certain vector field on S^2.   ★

Exercise. Let $f\colon S^2 \to \mathbb{R}$ be defined by $f(x, y, z) = z$. Compute df using spherical coordinates (θ, ϕ).   ★

Exercise. Compute the tangent lifts for the two stereographic projections of $S^2 \to \mathbb{R}^2$ in 11.1.4. That is, assuming (x, y, z) depend smoothly on t, determine:

1. How $(\dot{\xi}_N, \dot{\eta}_N)$ depend on $(\dot{x}, \dot{y}, \dot{z})$? Likewise for $(\dot{\xi}_S, \dot{\eta}_S)$.

2. How $(\dot{\xi}_N, \dot{\eta}_N)$ depend on $(\dot{\xi}_S, \dot{\eta}_S)$?

Hint: Recall $(1 + z)(1 - z) = 1 - z^2 = x^2 + y^2$ and use $x\dot{x} + y\dot{y} + z\dot{z} = 0$ when $(\dot{x}, \dot{y}, \dot{z})$ is tangent to S^2 at (x, y, z).   ★

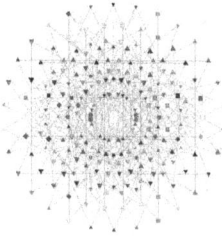

13

LIFTED ACTIONS AND THE JACOBI–LIE BRACKET ON VECTOR FIELDS

Contents

> **What is this lecture about?** This lecture discusses the properties of the tangent and cotangent lifted actions of differentiable maps on manifolds and their relations to the Jacobi–Lie bracket.

13.1 Lifted actions

Definition 13.1.1 *Let* $\Phi \colon G \times M \to M$ *be a left action, and write* $\Phi_g(x) = \Phi(g, x)$ *for* $x \in M$. *The **tangent lift action** of* G *on the tangent bundle* TM *is defined by* $gv = T_x \Phi_g(v)$ *for every* $v \in T_x M$.

Remark 13.1.1 In standard calculus notation, the expression for *tangent lift* may be written as

$$T_x \Phi \cdot v = \frac{d}{dt} \Phi(c(t)) \Big|_{t=0} = \frac{\partial \Phi}{\partial c} c'(t) \Big|_{t=0} =: D\Phi(x) \cdot v,$$

with $c(0) = x$ and $c'(0) = v$. Thus, $T_x\Phi = D\Phi(x)$ is nothing but the Jacobian on the map Φ.   □

Definition 13.1.2 *If X is a vector field on M and ϕ is a differentiable map from M to itself, then the **push-forward** of X by ϕ is the vector field ϕ_*X defined by $(\phi_*X)(\phi(x)) = T_x\phi(X(x))$. Consequently, the following diagram commutes:*

$$
\begin{array}{ccc}
& T\phi & \\
TM & \longrightarrow & TM \\
\uparrow & & \uparrow \\
X & & \phi_*X \\
& \phi & \\
M & \longrightarrow & M
\end{array}
$$

*If ϕ is a diffeomorphism, then the **pull-back** ϕ^*X is also defined: $(\phi^*X)(x) = T_{\phi(x)}\phi^{-1}(X(\phi(x)))$. Hence, one sees that push-forward by the map ϕ is the pull-back by the inverse map ϕ^{-1}.*

Definition 13.1.3 *Let $\Phi\colon G \times M \to M$ be a left action, and write $\Phi_g(m) = \Phi(g,m)$. Then, G has a left action on $X \in \mathfrak{X}(M)$ (the set of vector fields on M) by the push-forward: $gX = (\Phi_g)_* X$.*

Definition 13.1.4 *Let G act on M on the left. A vector field X on M is **invariant** with respect to this action (we often say "G-invariant" if the action is understood) if $gX = X$ for all $g \in G$; equivalently (using all of the above definitions!), $g(X(x)) = X(gx)$ for all $g \in G$ and all $x \in X$.*

Definition 13.1.5 *Consider the left action of G on itself by left multiplication, $\Phi_g(h) = L_g(h) = gh$, for $g, h \in G$. A vector field on G that is invariant with respect to this action is called **left invariant**. From Definition 13.1.4, we see that X is left invariant if and only if $g(X(h)) = X(gh)$, which, in less compact notation, means $T_hL_gX(h) = X(gh)$. The set of all such vector fields is written as $\mathfrak{X}^L(G)$.*

Proposition 13.1.1 *Given a $\xi \in T_eG$, define $X_\xi^L(g) = g\xi$ (recall: $g\xi \equiv T_eL_g\xi$). Then, X_ξ^L is the unique left-invariant vector field such that $X_\xi^L(e) = \xi$.*

Proof. To show that X_ξ^L is left invariant, we need to show that $g\big(X_\xi^L(h)\big) = X_\xi^L(gh)$ for every $g, h \in G$. This follows from the definition of X_ξ^L and the associativity property of group actions:

$$g\big(X_\xi^L(h)\big) = g(h\xi) = (gh)\xi = X_\xi^L(gh).$$

We repeat the last line in less compact notation:

$$T_h L_g\big(X_\xi^L(h)\big) = T_h L_g(h\xi) = T_e L_{gh}\xi = X_\xi^L(gh).$$

For uniqueness, suppose X is left invariant and $X(e) = \xi$. Then, for any $g \in G$, we have $X(g) = g(X(e)) = g\xi = X_\xi^L(g)$.   ∎

Remark 13.1.2 Note that the map $\xi \mapsto X_\xi^L$ is a vector space isomorphism from $T_e G$ to $\mathfrak{X}^L(G)$.   □

All of the above definitions have analogues for right actions. The definitions of *right invariant*, $\mathfrak{X}^R(G)$ and X_ξ^R use the right action of G on itself, defined by $\Phi(g, h) = R_g(h) = hg$.

> **Exercise.** There is a left action of G on itself defined by $\Phi_g(h) = hg^{-1}$.   ★

We will use the map $\xi \mapsto X_\xi^L$ to relate the Lie bracket on $\mathfrak{g}$, defined as $[\xi, \eta] = \mathrm{ad}_\xi\, \eta$, with the Jacobi–Lie bracket on vector fields.

13.2   Jacobi–Lie bracket on vector fields

Definition 13.2.1 *The **Jacobi–Lie bracket** on $\mathfrak{X}(M)$ is defined in local coordinates by*

$$[X, Y]_{J\text{-}L} \equiv (DX) \cdot Y - (DY) \cdot X,$$

which, in finite dimensions, is equivalent to

$$[X, Y]_{J\text{-}L} \equiv -(X \cdot \nabla)Y + (Y \cdot \nabla)X \equiv -[X, Y].$$

Theorem 13.2.1 (Properties of the Jacobi–Lie bracket)

1. *The Jacobi–Lie bracket satisfies*

$$[X, Y]_{J\text{-}L} = \mathcal{L}_X Y \equiv \frac{d}{dt}\Big|_{t=0} \Phi_t^* Y,$$

   where Φ is the flow of X. (This relation is coordinate-free and can be used as an alternative definition.)

2. *This bracket makes $\mathfrak{X}^L(M)$ a Lie algebra with $[X, Y]_{J\text{-}L} = -[X, Y]$, where $[X, Y]$ is the Lie algebra bracket on $\mathfrak{X}(M)$.*

3. *$\phi_*[X, Y] = [\phi_* X, \phi_* Y]$ for any differentiable $\phi\colon M \to M$.*

Theorem 13.2.2 *$\mathfrak{X}^L(G)$ is a subalgebra of $\mathfrak{X}(G)$.*

Proof. Let $X, Y \in \mathfrak{X}^L(G)$. Using the last item of the previous theorem and then the G invariance of X and Y gives the push-forward relations

$$(L_g)_* [X, Y]_{J\text{-}L} = [(L_g)_* X, (L_g)_* Y]_{J\text{-}L}$$

for all $g \in G$. Hence, $[X, Y]_{J\text{-}L} \in \mathfrak{X}^L(G)$. This is the second property in Theorem 13.2.1. ∎

Theorem 13.2.3 *Set $\left[X_\xi^L, X_\eta^L\right]_{J\text{-}L}(e) = [\xi, \eta]$ for every $\xi, \eta \in \mathfrak{g}$, where the bracket on the right is the Jacobi–Lie bracket. (One can says that the Lie bracket on $\mathfrak{g}$ is the pull-back of the Jacobi–Lie bracket by the map $\xi \mapsto X_\xi^L$.)*

Proof. The proof of Theorem 13.2.3 for matrix Lie algebras is relatively easy: we have already seen that $\operatorname{ad}_A B = AB - BA$. On the other hand, since $X_A^L(C) = CA$ for all C and this is linear in C, we have $DX_B^L(I) \cdot A = AB$, so

$$[A, B] = \left[X_A^L, X_B^L\right]_{J\text{-}L}(I) = DX_B^L(I) \cdot X_A^L(I) - DX_A^L(I) \cdot X_B^L(I)$$

$$= DX_B^L(I) \cdot A - DX_A^L(I) \cdot B = AB - BA.$$

This is the third property of the Jacobi–Lie bracket listed in Theorem 13.2.1. For the general proof, see Marsden and Ratiu [MaRa1994, Proposition 9.14]. ∎

Remark 13.2.1 Theorem 13.2.3, together with Item 2 in Theorem 13.2.1, proves that the Jacobi–Lie bracket makes $\mathfrak{g}$ into a Lie algebra. □

Remark 13.2.2 By Theorem 13.2.2, the vector field $\left[X_\xi^L, X_\eta^L\right]$ is left-invariant. Since $\left[X_\xi^L, X_\eta^L\right]_{J-L}(e) = [\xi, \eta]$, it follows that

$$\left[X_\xi^L, X_\eta^L\right] = X_{[\xi,\eta]}^L.$$

□

Definition 13.2.2 *Let* $\Phi \colon G \times M \to M$ *be a left action, and let* $\xi \in \mathfrak{g}$. *Let* $g(t)$ *be a path in* G *such that* $g(0) = e$ *and* $g'(0) = \xi$. *Then, the* ***infinitesimal generator*** *of the action in the* ξ *direction is the vector field* ξ_M *on* M, *defined by*

$$\xi_M(x) = \frac{d}{dt}\bigg|_{t=0} \Phi_{g(t)}(x).$$

Remark 13.2.3 Note: this definition does not depend on the choice of $g(t)$. For example, the choice made by Marsden and Ratiu [MaRa1994] is $\exp(t\xi)$, where $\exp$ denotes the exponentiation on Lie groups (not defined here). □

Exercise. Consider the action of $SO(3)$ on the unit sphere S^2 around the origin, and let $\xi = (0, 0, 1)^\wedge$. Sketch the vector field ξ_M.

Hint: The vectors all point "eastward." ★

Theorem 13.2.4 *For any left action of* G, *the Jacobi–Lie bracket of infinitesimal generators is related to the Lie bracket on* $\mathfrak{g}$ *as follows (note the minus sign):*

$$[\xi_M, \eta_M] = -[\xi, \eta]_M.$$

For a proof, see Marsden and Ratiu [MaRa1994, Proposition 9.3.6].

Exercise. Express the statements and formulas of this lecture for the case of $SO(3)$ action on its Lie algebra $\mathfrak{so}(3)$. (Hint: look at the previous lecture.) Wherever possible, translate these formulas to $\mathbb{R}^3$ by using the $\widehat{}$ map: $\mathfrak{so}(3) \to \mathbb{R}^3$.

Write the Lie algebra for $\mathfrak{so}(3)$ using the Jacobi–Lie bracket in terms of linear vector fields on $\mathbb{R}^3$. What are the characteristic curves of these linear vector fields? ★

Definition 13.2.3 *Let X and Y be two vector fields on the same manifold M. The **Lie derivative** of Y with respect to X is $\mathcal{L}_X Y \equiv \frac{d}{dt}\Phi_t^* Y\big|_{t=0}$, where Φ is the flow of X.*

Remark 13.2.4 The Lie derivative $\mathcal{L}_X Y$ is "the derivative of Y in the direction given by X." Its definition is coordinate-independent. By contrast, $DY \cdot X$ (also written as $X[Y]$) is also "the derivative of Y in the X direction"; however, the value of $DY \cdot X$ depends on the coordinate system and, in particular, does not usually equal $\mathcal{L}_X Y$ in the chosen coordinate system. □

Theorem 13.2.5 $\mathcal{L}_X Y = [X, Y]$, *where the bracket on the right is the Jacobi–Lie bracket.*

Proof. In the following calculation, we assume that M is finite dimensional, and we work in local coordinates. Thus, we may consider everything as matrices, which allows us to use the product rule and the identities $(M^{-1})' = -M^{-1}M'M^{-1}$ and $\frac{d}{dt}(D\Phi_t(x)) = D(\frac{d}{dt}\Phi_t)(x)$:

$$\mathcal{L}_X Y(x) = \frac{d}{dt}\Phi_t^* Y(x)\Big|_{t=0}$$

$$= \frac{d}{dt}(D\Phi_t(x))^{-1} Y(\Phi_t(x))\Big|_{t=0}$$

$$= \left[\left(\frac{d}{dt} \left(D\Phi_t(x) \right)^{-1} \right) Y \left(\Phi_t(x) \right) \right.$$

$$\left. + \left(D\Phi_t(x) \right)^{-1} \frac{d}{dt} Y \left(\Phi_t(x) \right) \right]_{t=0}$$

$$= \left[- \left(D\Phi_t(x) \right)^{-1} \left(\frac{d}{dt} D\Phi_t(x) \right) \left(D\Phi_t(x) \right)^{-1} Y \left(\Phi_t(x) \right) \right.$$

$$\left. + \left(D\Phi_t(x) \right)^{-1} \frac{d}{dt} Y \left(\Phi_t(x) \right) \right]_{t=0}$$

$$= \left[- \left(\frac{d}{dt} D\Phi_t(x) \right) Y(x) + \frac{d}{dt} Y \left(\Phi_t(x) \right) \right]_{t=0}$$

$$= -D \left(\frac{d}{dt} \Phi_t(x) \Big|_{t=0} \right) Y(x) + DY(x) \left(\frac{d}{dt} \Phi_t(x) \Big|_{t=0} \right)$$

$$= -DX(x) \cdot Y(x) + DY(x) \cdot X(x)$$

$$= [X, Y]_{\text{J-L}}(x)$$

Therefore, $\mathcal{L}_X Y = [X, Y]_{\text{J-L}}$. ∎

14

LIE GROUP ACTION ON ITS TANGENT BUNDLE

Contents

What is this lecture about? This lecture discusses the left-invariant and right-invariant subalgebras of the Lie algebra of vector fields and explains the sign differences in their associated Jacobi–Lie brackets.

14.1 Definitions of actions

Definition 14.1.1 *A Lie group G acts on its tangent bundle TG by tangent lifts.*[1] *Given $X \in T_h G$, we can consider the action of G on X by either*

---

[1]Recall that Section 12.2.3 deals with tangent lifts of a differentiable manifold.

left or right translations, denoted as $T_h L_g X$ or $T_h R_g X$, respectively. These expressions may be abbreviated as

$$T_h L_g X = L_g^* X = gX \qquad and \qquad T_h R_g X = R_g^* X = Xg.$$

The left action of a Lie group G on its tangent bundle TG is illustrated in the following figure.

$$
\begin{array}{ccc}
TG & \xrightarrow{\ TL_g\ } & TG \\
\big\uparrow{\scriptstyle X} & & \big\uparrow{\scriptstyle gX} \\
G & \xrightarrow{\ L_g\ } & G
\end{array}
$$

For matrix Lie groups, this action is simply multiplication on the left or right, respectively.

14.2   Left- and right-invariant vector fields

A vector field X on G is called left invariant if for every $g \in G$, one has $L_g^* X = X$, that is, if

$$(T_h L_g) X(h) = X(gh)$$

for every $h \in G$. The commutative diagram for a left-invariant vector field is illustrated in the following figure.

$$
\begin{array}{ccc}
TG & \xrightarrow{\ TL_g\ } & TG \\
\big\uparrow{\scriptstyle X} & & \big\uparrow{\scriptstyle X} \\
G & \xrightarrow{\ L_g\ } & G
\end{array}
$$

Proposition 14.2.1 *The set $\mathfrak{X}_L(G)$ of left-invariant vector fields on the Lie group G is a subalgebra of $\mathfrak{X}(G)$, the set of all vector fields on G.*

Proof. If $X, Y \in \mathfrak{X}_L(G)$ and $g \in G$, then

$$L_g^*[X, Y] = \left[L_g^* X, L_g^* Y\right] = [X, Y].$$

Consequently, the Lie bracket $[X, Y] \in \mathfrak{X}_L(G)$. Therefore, $\mathfrak{X}_L(G)$ is a subalgebra of $\mathfrak{X}(G)$, the set of all vector fields on G. ∎

Proposition 14.2.2 *The linear maps $\mathfrak{X}_L(G)$ and $T_e G$ are isomorphic as vector spaces.*

Demonstration of proposition. For each $\xi \in T_e G$, define a vector field X_ξ on G by letting $X_\xi(g) = T_e L_g(\xi)$. Then,

$$
\begin{aligned}
X_\xi(gh) &= T_e L_{gh}(\xi) = T_e(L_g \circ L_h)(\xi) \\
&= T_h L_g(T_e L_h(\xi)) = T_h L_g(X_\xi(h)),
\end{aligned}
$$

which shows that X_ξ is left invariant. (This proposition is stated by Marsden and Ratiu [MaRa1994, Chapter 9], who refer to Abraham and Marsden [AbMa1978] for the full proof.)

14.3   Jacobi–Lie bracket of vector fields

Definition 14.3.1 (Jacobi–Lie bracket of vector fields) *Let $g(t)$ and $h(s)$ be curves in G with $g(0) = e$ and $h(0) = e$, and define vector fields at the identity of G by the tangent vectors $g'(0) = \xi$ and $h'(0) = \eta$. Compute the linearisation of the adjoint action of G on $T_e G$ as*

$$[\xi, \eta] := \frac{d}{dt}\frac{d}{ds} g(t)h(s)g(t)^{-1}\Big|_{s=0,t=0} = \frac{d}{dt} g(t)\eta g(t)^{-1}\Big|_{t=0} = \xi\eta - \eta\xi.$$

*This is the **Jacobi–Lie bracket** of the vector fields ξ and η.*

Definition 14.3.2 *The **Lie bracket** in $T_e G$ is defined by*

$$[\xi, \eta] := [X_\xi, X_\eta](e),$$

for $\xi, \eta \in T_e G$ and for $[X_\xi, X_\eta]$, the Jacobi–Lie bracket of vector fields. This makes $T_e G$ into a Lie algebra. Note that

$$[X_\xi, X_\eta] = X_{[\xi, \eta]},$$

for all $\xi, \eta \in T_e G$.

Definition 14.3.3 *The vector space $T_e G$ with this Lie algebra structure is called the **Lie algebra of** G and is denoted by $\mathfrak{g}$.*

If we let $\xi_L(g) = T_e L_g \xi$, then the Jacobi–Lie bracket of two such left-invariant vector fields, in fact, gives the Lie algebra bracket

$$[\xi_L, \eta_L](g) = [\xi, \eta]_L(g).$$

Remark 14.3.1 For the right-invariant case, the right-hand side obtains a minus sign, namely,

$$[\xi_R, \eta_R](g) = -[\xi, \eta]_R(g).$$

The relative minus sign arises because of the difference in action $(xh^{-1})g^{-1} = x(gh)^{-1}$ on the right versus $(gh)x = g(hx)$ on the left.   □

14.4   Infinitesimal generators

In mechanics, group actions often appear as symmetry transformations which arise through their infinitesimal generators, defined as follows.

Definition 14.4.1 *Suppose $\Phi: G \times M \to M$ is an action. For $\xi \in \mathfrak{g}$, $\Phi^\xi(t, x): \mathbb{R} \times M \to M$ defined by $\Phi^\xi(x) = \Phi(\exp t\xi, x) = \Phi_{\exp t\xi}(x)$ is an $\mathbb{R}$-action on M. In other words, $\Phi_{\exp t\xi} \to M$ is a flow on M. The vector field on M defined by[2]*

$$\xi_M(x) = \left.\frac{d}{dt}\right|_{t=0} \Phi_{\exp t\xi}(x)$$

*is called the **infinitesimal generator** of the action $\Phi: G \times M \to M$, corresponding to ξ.*

---

[2]Recall Definition 11.3.1 of vector fields.

The Jacobi–Lie bracket of infinitesimal generators is related to the Lie algebra bracket as follows:

$$[\xi_M, \eta_M] = -[\xi, \eta]_M.$$

See, for example, Marsden and Ratiu [MaRa1994, Chapter 9] for the proof.

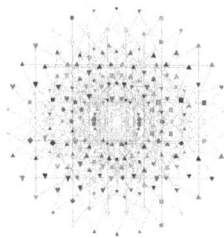

15

HAMILTON'S PRINCIPLE ON MANIFOLDS

Contents

What is this lecture about? This lecture surveys the properties of Hamilton's principle on manifolds and discusses the implications of their Lie symmetries in Noether's theorem.

15.1 Hamilton's principle of stationary action

Theorem 15.1.1 (Hamilton's principle of stationary action) *Let the smooth function* $L: TQ \to \mathbb{R}$ *be a Lagrangian on* TQ. *A* C^2 *curve*

$c: [a, b] \rightarrow Q$ *joining* $q_a = c(a)$ *to* $q_b = c(b)$ *satisfies the Euler–Lagrange equations if and only if*

$$\delta \int_a^b L(c(t), \dot{c}(t)) + \left\langle p, \dot{c} - \frac{dc}{dt} \right\rangle dt = 0.$$

Proof. The meaning of the variational derivative in the statement is the following. Consider a family of C^2 curves $c(t, s)$ for $|s| < \varepsilon$ satisfying $c_0(t) = c(t)$, $c(a, s) = q_a$ and $c(b, s) = q_b$ for all $s \in (-\varepsilon, \varepsilon)$. Then,

$$\delta \int_a^b L(c(t), \dot{c}(t))dt := \frac{d}{ds}\Big|_{s=0} \int_a^b L(c(t, s), \dot{c}(t, s))dt.$$

Differentiating under the integral sign, working in local coordinates (covering the curve $c(t)$ by a finite number of coordinate charts), integrating by parts, denoting the variation as

$$v(t) := \frac{d}{ds}\Big|_{s=0} c(t, s),$$

taking into account that $v(a) = v(b) = 0$, and applying the constraint $\dot{c} = \frac{dc}{dt}$ so that $\dot{v} = \frac{dv}{dt}$ yields

$$\int_a^b \left(\frac{\partial L}{\partial q^i} v^i + \frac{\partial L}{\partial \dot{q}^i} \dot{v}^i \right) dt = \int_a^b \left(\frac{\partial L}{\partial q^i} - \frac{d}{dt}\frac{\partial L}{\partial \dot{q}^i} \right) v^i dt + \left\langle \frac{\partial L}{\partial \dot{q}^i}, v^i \right\rangle \Big|_0^T.$$

This vanishes for any C^1 function $v(t)$ if and only if the Euler–Lagrange equations hold and the endpoint terms vanish. ∎

Remark 15.1.1 The integral appearing in this theorem,

$$\mathcal{S}(c(\cdot)) := \int_a^b L(c(t), \dot{c}(t)) + \left\langle p, \dot{c} - \frac{dc}{dt} \right\rangle dt,$$

is called the *action integral*. It is defined on C^2 curves $c: [a, b] \rightarrow Q$ with fixed endpoints, $c(a) = q_a$ and $c(b) = q_b$. □

Remark 15.1.2   (Variational derivatives of functionals vs. Lie derivatives of functions) The variational derivative of a functional $S[u]$ is defined as the linearisation

$$\lim_{\epsilon \to 0} \frac{S[u + \epsilon v] - S[u]}{\epsilon} = \frac{d}{d\epsilon}\bigg|_{\epsilon=0} S[u + \epsilon v] = \left\langle \frac{\delta S}{\delta v}, v \right\rangle.$$

Compare this to the expression for the Lie derivative of a function. If f is a real-valued function on a manifold M and X is a vector field on M, the Lie derivative of f along X is defined as the directional derivative

$$\mathcal{L}_X f = X(f) := \mathbf{d}f \cdot X.$$

If M is finite dimensional, this is

$$\mathcal{L}_X f = X[f] := \mathbf{d}f \cdot X = \frac{\partial f}{\partial x^i} X^i = \lim_{\epsilon \to 0} \frac{f(x + \epsilon X) - f(x)}{\epsilon}.$$

The similarity is suggestive; namely, the Lie derivative of a function and the variational derivative of a functional are both defined as linearisations of smooth maps in certain directions.   □

The following theorem emphasises the role of Lagrangian 1-forms and 2-forms in the variational principle. It is a direct corollary of the previous theorem.

Theorem 15.1.2 *Given a C^k Lagrangian $L: TQ \to \mathbb{R}$ for $k \geq 2$, there exists a unique C^{k-2} map $\mathcal{E}L(L): \ddot{Q} \to T^*Q$, where*

$$\ddot{Q} := \left\{ \frac{d^2 q}{dt^2}\bigg|_{t=0} \in T(TQ) \,\middle|\, q(t) \text{ is a } C^2 \text{ curve in } Q \right\}$$

is a submanifold of $T(TQ)$, and a unique C^{k-1} 1-form $\Theta_L \in \Lambda^1(TQ)$, such that for all C^2 variations $q(t, s)$ (defined on a fixed t interval) of $q(t, 0) = q_0(t) := q(t)$, we have

$$\delta \mathcal{S} := \frac{d}{ds}\bigg|_{s=0} \mathcal{S}[c(\cdot, s)] = \mathbf{D}\mathcal{S}[q(\cdot)] \cdot \delta q(\cdot) \tag{15.1.1}$$

$$= \int_a^b \langle \mathcal{E}L(L)\,(q, \dot{q}, \ddot{q})\,,\, \delta q \rangle \, dt + \underbrace{\langle \Theta_L\,(q, \dot{q})\,,\, \delta q \rangle \big|_a^b}_{\text{cf. Noether quantity}}.$$

In this equation, $\Theta_L(q, \dot{q})$ is the fibre derivative, and δq could, for example, be a Lie derivative:

$$\Theta_L(q, \dot{q}) = \frac{\partial L}{\partial \dot{q}} \quad \text{and} \quad \delta q := \frac{d}{ds}\Big|_{s=0} q(t, s).$$

15.2 Symmetries and conservation laws: Noether's theorem

In Theorem 15.1.2,

$$\delta S := \frac{d}{ds}\Big|_{s=0} S[c(\cdot, s)] = \mathbf{D}S[q(\cdot)] \cdot \delta q(\cdot) \tag{15.2.1}$$

$$= \int_a^b \mathcal{E}L(L)(q, \dot{q}, \ddot{q}) \cdot \delta q \, dt + \underbrace{\Theta_L(q, \dot{q}) \cdot \delta q\Big|_a^b}_{\text{Noether quantity}},$$

where

$$\delta q(t) = \frac{d}{ds}\Big|_{s=0} q(t, s),$$

the map $\mathcal{E}L \colon \ddot{Q} \to T^*Q$ is called the *Euler–Lagrange operator* and its expression in local coordinates is

$$\mathcal{E}L(q, \dot{q}, \ddot{q})_i = \frac{\partial L}{\partial q^i} - \frac{d}{dt}\frac{\partial L}{\partial \dot{q}^i}.$$

One understands that the formal time derivative is taken in the second summand and everything is expressed as a function of $(q, \dot{q}, \ddot{q})$.

Theorem 15.2.1 (Symmetries and conservation laws, Noether [No1918]) *If the action variation in equation (15.2.1) vanishes $\delta S = 0$ because of a symmetry transformation which does not preserve the end points and the Euler–Lagrange equations hold, then the term marked*

Noether quantity must also vanish. However, the vanishing of this term is now interpreted as a constant of motion. Namely, the term

$$A(v, w) := \langle \mathbb{F}L(v), w \rangle, \quad \text{or, in coordinates} \quad A(q, \dot{q}, \delta q) = \frac{\partial L}{\partial \dot{q}^i} \delta q^i,$$

is constant for solutions of the Euler–Lagrange equations.

This result was first reported by Noether [No1918]. In fact, that result is more general than the one given here. In particular, in the partial differential equation (PDE) setting, one must also include the transformation of the volume element in the action principle. See, for example, the work by Olver [Ol1993] for good discussions of the history, framework and applications of Noether's theorem.

Exercise. Show that conservation of energy results from Noether's theorem if, in Hamilton's principle, the variations are chosen as

$$\delta q(t) = \left. \frac{d}{ds} \right|_{s=0} q(t, s),$$

corresponding to the symmetry of the Lagrangian under reparametrizations of time along the given curve $q(t) \to q(\tau(t, s))$. ★

15.3   The canonical Lagrangian 1-form and 2-form

The 1-form Θ_L, whose existence and uniqueness is guaranteed by Theorem 15.1.2, appears as the boundary term of the derivative of the action integral when the endpoints of the curves on the configuration manifold are free. In finite dimensions, its local expression is

$$\Theta_L(q, \dot{q}) := \frac{\partial L}{\partial \dot{q}^i} dq^i \quad \left(= p_i(q, \dot{q}) dq^i \right).$$

The corresponding closed two-form $\Omega_L = d\Theta_L$ obtained by taking its exterior derivative may be expressed as

$$\Omega_L := -d\Theta_L = \frac{\partial^2 L}{\partial \dot{q}^i \partial q^j} dq^i \wedge dq^j + \frac{\partial^2 L}{\partial \dot{q}^i \partial \dot{q}^j} dq^i \wedge d\dot{q}^j$$

$$(= dp_i(q, \dot{q}) \wedge dq^i).$$

These coefficients may be written as the $2n \times 2n$ skew-symmetric matrix

$$\Omega_L = \begin{pmatrix} \mathcal{A} & \frac{\partial^2 L}{\partial \dot{q}^i \partial \dot{q}^j} \\ -\frac{\partial^2 L}{\partial \dot{q}^i \partial \dot{q}^j} & 0 \end{pmatrix}, \tag{15.3.1}$$

where $\mathcal{A}$ is the skew-symmetric $n \times n$ matrix $\left(\frac{\partial^2 L}{\partial \dot{q}^i \partial q^j}\right) - \left(\frac{\partial^2 L}{\partial \dot{q}^i \partial q^j}\right)^T$.

The non-degeneracy of Ω_L is equivalent to the invertibility of the matrix $\left(\frac{\partial^2 L}{\partial \dot{q}^i \partial \dot{q}^j}\right)$.

Definition 15.3.1 *The **Legendre transformation** $\mathbb{F}L: TQ \to T^*Q$ is a smooth map near the identity defined by*

$$\langle \mathbb{F}L(v_q), w_q \rangle := \left. \frac{d}{ds} \right|_{s=0} L(v_q + s w_q).$$

In the finite-dimensional case, the local expression of $\mathbb{F}L$ is

$$\mathbb{F}L(q^i, \dot{q}^i) = \left(q^i, \frac{\partial L}{\partial \dot{q}^i} \right) = (q^i, p_i(q, \dot{q})).$$

If the skew-symmetric matrix (15.3.1) is invertible, the Lagrangian L is said to be *regular*. In this case, by the implicit function theorem, $\mathbb{F}L$ is locally invertible. If $\mathbb{F}L$ is a diffeomorphism, L is called *hyperregular*.

Definition 15.3.2 *Given a Lagrangian L, the **action** of L is the map $A: TQ \to \mathbb{R}$ given by*

$$A(v) := \langle \mathbb{F}L(v), v \rangle, \quad \text{or, in coordinates,} \quad A(q, \dot{q}) = \frac{\partial L}{\partial \dot{q}^i} \dot{q}^i, \tag{15.3.2}$$

*and the **energy** of L is*

$$E(v) := A(v) - L(v), \quad \text{or, in coordinates,} \quad E(q, \dot{q}) = \frac{\partial L}{\partial \dot{q}^i} \dot{q}^i - L(q, \dot{q}). \tag{15.3.3}$$

15.4 Lagrangian vector fields and conservation laws

Definition 15.4.1 *A vector field Z on TQ is called a **Lagrangian vector field** if*
$$\Omega_L(v)(Z(v), w) = \langle \mathbf{d}E(v), w \rangle,$$
for all $v \in T_q Q$, $w \in T_v(TQ)$.

Proposition 15.4.1 *The energy is conserved along the flow of a Lagrangian vector field Z.*

Proof. Let $v(t) \in TQ$ be an integral curve of Z. The skew symmetry of Ω_L implies

$$\frac{d}{dt}E(v(t)) = \langle \mathbf{d}E(v(t)), \dot{v}(t) \rangle = \langle \mathbf{d}E(v(t)), Z(v(t)) \rangle$$

$$= \Omega_L(v(t))\,(Z(v(t)), Z(v(t))) = 0.$$

Thus, the energy $E(v(t))$ is constant in time t. ∎

15.5 Equivalent dynamics for hyperregular Lagrangians and Hamiltonians

Recall that a Lagrangian L is said to be *hyperregular* if its Legendre transformation $\mathbb{F}L\colon TQ \to T^*Q$ is a diffeomorphism.

The equivalence between the Lagrangian and Hamiltonian formulations for hyperregular Lagrangians and Hamiltonians is summarised as follows, referring to Marsden and Ratiu [MaRa1994]:

(a) Let L be a hyperregular Lagrangian on TQ and $H = E \circ (\mathbb{F}L)^{-1}$, where E is the energy of L and $(\mathbb{F}L)^{-1}\colon T^*Q \to TQ$ is the inverse of the Legendre transformation. Then, the Lagrangian vector field Z on TQ and the Hamiltonian vector field X_H on T^*Q are related by the identity

$$(\mathbb{F}L)^* X_H = Z.$$

Furthermore, if $c(t)$ is an integral curve of Z and $d(t)$ is an integral curve of X_H with $\mathbb{F}L(c(0)) = d(0)$, then $\mathbb{F}L(c(t)) = d(t)$, and their integral curves coincide on the manifold Q. That is, $\tau_Q(c(t)) = \pi_Q(d(t)) = \gamma(t)$, where $\tau_Q \colon TQ \to Q$ and $\pi_Q \colon T^*Q \to Q$ are the canonical bundle projections.

In particular, the pull-back of the inverse Legendre transformation $\mathbb{F}L^{-1}$ induces a one-form Θ and a closed two-form Ω on T^*Q by

$$\Theta = (\mathbb{F}L^{-1})^*\Theta_L, \quad \Omega = -d\Theta = (\mathbb{F}L^{-1})^*\Omega_L.$$

In coordinates, these are the canonical presymplectic and symplectic forms, respectively:

$$\Theta = p_i\,dq^i, \quad \Omega = -d\Theta = dp_i \wedge dq^i.$$

(b) A Hamiltonian $H \colon T^*Q \to \mathbb{R}$ is said to be *hyperregular* if the smooth map $\mathbb{F}H \colon T^*Q \to TQ$, defined by

$$\langle \mathbb{F}H(\alpha_q), \beta_q \rangle := \frac{d}{ds}\bigg|_{s=0} H(\alpha_q + s\beta_q), \quad \alpha_q, \beta_q \in T_q^*Q,$$

is a diffeomorphism. Define the *action* of H by $G := \langle \Theta, X_H \rangle$. If H is a hyperregular Hamiltonian, then the energies of L and H and the actions of L and H are related by

$$E = H \circ (\mathbb{F}H)^{-1}, \qquad A = G \circ (\mathbb{F}H)^{-1}.$$

Also, the Lagrangian $L = A - E$ is hyperregular and $\mathbb{F}L = \mathbb{F}H^{-1}$.

(c) These constructions define a bijective correspondence between hyperregular Lagrangians and Hamiltonians.

Remark 15.5.1 For thorough discussions of many additional results arising from Hamilton's principle for hyperregular Lagrangians, see, for example, Refs. [MaRa1994, Chapters 7 and 8] and [HoScSt2009]. □

16

EULER–LAGRANGE EQUATIONS ON MANIFOLDS

Contents

What is this lecture about? This lecture is about the Euler–Lagrange approach to geodesic motion and its ramifications, such as the covariant derivative.

Definition 16.0.1 (Cotangent lift) *Given two manifolds Q and S related by a diffeomorphism $f: Q \mapsto S$, the **cotangent lift** $T^*f: T^*S \mapsto T^*Q$ of f is defined by*

$$\langle T^*f(\alpha), v \rangle = \langle \alpha, Tf(v) \rangle, \qquad (16.0.1)$$

where

$$\alpha \in T_s^*S, \quad v \in T_qQ, \quad and \quad s = f(q).$$

*As explained by Marsden and Ratiu [MaRa1994, Chapter 6], cotangent lifts preserve the **action** of the Lagrangian L, which we write as*

$$\langle \mathbf{p}, \dot{\mathbf{q}} \rangle = \langle \alpha, \dot{\mathbf{s}} \rangle, \qquad (16.0.2)$$

where $\mathbf{p} = T^*f(\alpha)$ is the cotangent lift of α under the diffeomorphism f and $\dot{\mathbf{s}} = Tf(\dot{\mathbf{q}})$ is the tangent lift of $\dot{\mathbf{q}}$ under the function f, which is written in Euclidean coordinate components as $q^i \rightarrow s^i = f^i(\mathbf{q})$. Preservation of the action in (16.0.2) yields the coordinate relations:

$$\textit{(Tangent lift in coordinates)} \quad \dot{s}^j = \frac{\partial f^j}{\partial q^i} \dot{q}^i \quad \Longrightarrow$$

$$p_i = \alpha_k \frac{\partial f^k}{\partial q^i} \quad \begin{array}{l} \textit{(Cotangent lift in} \\ \textit{coordinates)} \end{array}$$

Thus, in coordinates, the cotangent lift is the inverse transpose of the tangent lift.

Remark 16.0.1 The cotangent lift of a function preserves the induced action 1-form,

$$\langle \mathbf{p}, d\mathbf{q} \rangle = \langle \alpha, d\mathbf{s} \rangle,$$

so it is a source of (pre-)symplectic transformations. □

16.1   The classic Euler–Lagrange example: Geodesic flow

An important example of a Lagrangian vector field is the geodesic spray of a Riemannian metric. A *Riemannian manifold* is a smooth manifold Q endowed with a symmetric non-degenerate covariant tensor g, which is positive definite. Thus, on each tangent space T_qQ, there is a non-degenerate definite inner product defined by pairing with $g(q)$.

If (Q, g) is a Riemannian manifold, there is a natural Lagrangian on it given by the *kinetic energy* K of the metric g, namely,

$$K(v) := \tfrac{1}{2}g(q)(v_q, v_q),$$

for $q \in Q$ and $v_q \in T_qQ$. In finite dimensions, in a local chart,

$$K(q, \dot{q}) = \tfrac{1}{2}g_{ij}(q)\dot{q}^i\dot{q}^j.$$

The fibre derivative in this case is $\mathbb{F}K(v_q) = g(q)(v_q, \cdot)$, for $v_q \in T_qQ$. In coordinates, this is

$$\mathbb{F}K(q, \dot{q}) := \left(q^i, \frac{\partial K}{\partial \dot{q}^i} \right) := (q^i, g_{ij}(q)\dot{q}^j) =: (q^i, p_i).$$

The Euler–Lagrange equations become the *geodesic equations* for the Riemannian metric g, given (for finite-dimensional Q in a local chart) by

$$\ddot{q}^i + \Gamma^i_{jk}\dot{q}^j\dot{q}^k = 0, \quad i = 1, \dots n,$$

where the three-index quantities

$$\Gamma^h_{jk} = \tfrac{1}{2}g^{hl}\left(\frac{\partial g_{jl}}{\partial q^k} + \frac{\partial g_{kl}}{\partial q^j} - \frac{\partial g_{jk}}{\partial q^l} \right), \quad \text{with } g_{ih}g^{hl} = \delta^l_i,$$

are the *Christoffel symbols* of the Levi–Civita connection on (Q, g).

Exercise. Explicitly compute the geodesic equation as an Euler–Lagrange equation for the kinetic-energy Lagrangian $K(q, \dot{q}) = \tfrac{1}{2}g_{ij}(q)\dot{q}^i\dot{q}^j$. ★

Exercise. For the kinetic-energy Lagrangian $K(q, \dot{q}) = \frac{1}{2} g_{ij}(q) \dot{q}^i \dot{q}^j$ with $i, j = 1, 2, \ldots, N$:

1. Compute the momentum p_i canonical to q^i for geodesic motion.

2. Perform the Legendre transformation to obtain the Hamiltonian for geodesic motion.

3. Write out the geodesic equations in terms of q^i and its canonical momentum p_i.

4. Check directly that Hamilton's equations are satisfied. ★

Exercise. Consider the Lagrangian

$$L_\epsilon(\mathbf{q}, \dot{\mathbf{q}}) = \tfrac{1}{2} \|\dot{\mathbf{q}}\|^2 - \tfrac{1}{2\epsilon}(1 - \|\mathbf{q}\|^2)^2$$

for a particle in $\mathbb{R}^3$. Let $\gamma_\epsilon(t)$ be the curve in $\mathbb{R}^3$ obtained by solving the Euler–Lagrange equations for L_ϵ with the initial conditions $\mathbf{q}_0 = \gamma_\epsilon(0), \dot{\mathbf{q}}_0 = \dot{\gamma}_\epsilon(0)$. Show that

$$\lim_{\epsilon \to 0} \gamma_\epsilon(t)$$

is a great circle on the two-sphere S^2, provided that $\mathbf{q}_0$ has unit length and the initial conditions satisfy $\mathbf{q}_0 \cdot \dot{\mathbf{q}}_0 = 0$. ★

Remark 16.1.1 The Lagrangian vector field associated with $K(q, \dot{q})$ is called the *geodesic spray*. Since the Legendre transformation is a diffeomorphism (in finite dimensions or in infinite dimensions if the metric is assumed to be strong), the geodesic spray is always a second-order equation. □

16.2   Covariant derivative

The variational approach to geodesics recovers the classical formulation using covariant derivatives as follows. Let $\mathfrak{X}(Q)$ denote the set of vector fields on the manifold Q. The *covariant derivative*

$$\nabla: \mathfrak{X}(Q) \times \mathfrak{X}(Q) \to \mathfrak{X}(Q) \quad (X, Y) \mapsto \nabla_X(Y),$$

of the Levi–Civita connection on (Q, g) is given in local charts by

$$\nabla_X(Y) = \Gamma^k_{ij} X^i Y^j \frac{\partial}{\partial q^k} + X^i \frac{\partial Y^k}{\partial q^i} \frac{\partial}{\partial q^k}.$$

If $c(t)$ is a curve on Q and $Y \in \mathfrak{X}(Q)$, the covariant derivative of Y along $c(t)$ is defined by

$$\frac{DY}{Dt} := \nabla_{\dot{c}} Y,$$

or, locally,

$$\left(\frac{DY}{Dt}\right)^k = \Gamma^k_{ij}(c(t))\dot{c}^i(t) Y^j(c(t)) + \frac{d}{dt} Y^k(c(t)).$$

A vector field is said to be *parallel transported* along $c(t)$ if

$$\frac{DY}{Dt} = 0.$$

Thus, $\dot{c}(t)$ is parallel transported along $c(t)$ if and only if

$$\ddot{c}^i + \Gamma^i_{jk} \dot{c}^j \dot{c}^k = 0.$$

In classical differential geometry, a *geodesic* is defined to be a curve $c(t)$ in Q whose tangent vector $\dot{c}(t)$ is parallel transported along $c(t)$. As the expression above shows, geodesics are integral curves of the Lagrangian vector field defined by the kinetic energy of g.

Remark 16.2.1 A classic problem is to determine the metric tensors $g_{ij}(q)$ for which these geodesic equations admit enough additional conservation laws to be integrable.   □

Exercise. Consider a geodesic flow on a Riemannian manifold M with metric $g_{ij}(x)$ for $x \in M$, so that $i, j = 1, 2, \ldots, n$ for $\dim M = n$. Show that if, in Hamilton's principle, the Lagrangian

$$L(x, \dot{x}) = \frac{1}{2}\dot{x}^i g_{ij}(x)\dot{x}^j$$

admits an isometry group G, then the Poisson brackets among the conservation laws of the corresponding geodesic flow are isomorphic to the Lie algebra $\mathfrak{g}$ of the isometry group G. For this calculation, it will be helpful to recall the diamond ($\diamond$) operation defined in equation (2.11.2) as

$$\langle p \diamond q, \xi \rangle_{\mathfrak{g}} := \langle p, -\mathcal{L}_\xi q \rangle_{TQ},$$

where $\delta q = -\mathcal{L}_\xi q$ with $\xi \in \mathfrak{g}$ is the infinitesimal symmetry transformation of q (i.e., minus the Lie derivative of q in the direction ξ).   ★

Answer. Hamilton's principle with Lagrangian $L(q, \dot{q})$ yields the Euler–Lagrange equations, plus an endpoint term obtained from integration by parts in time defined by the (non-degenerate) pairing

$$\left\langle \frac{\partial L}{\partial \dot{q}}, \delta q \right\rangle_{TQ} \bigg|_a^b =: \langle p, \delta q \rangle_{TQ} \bigg|_a^b.$$

The *Noether quantity* is defined to be $\langle p, \delta q \rangle_{TQ}$ for $\delta q = -\mathcal{L}_\xi q$. The corresponding *cotangent lift momentum map* is obtained in the form

$$\langle p, \delta q \rangle_{TQ} = \langle p, -\mathcal{L}_\xi q \rangle_{TQ} =: \langle p \diamond q, \xi \rangle_{\mathfrak{g}}$$
$$=: \langle J(q, p), \xi \rangle_{\mathfrak{g}} = J_\xi(q, p),$$

where $J_\xi(q, p)$ is the Noether Hamiltonian, and one defines the diamond $(\diamond)$ operation as

$$\langle p \diamond q, \xi \rangle_{\mathfrak{g}} := \langle p, -\mathcal{L}_\xi q \rangle_{TQ}. \tag{16.2.1}$$

Thus, the momentum map $J(q, p)$ of the isometry group G of the geodesic Lagrangian $L_{geo}(q, \dot{q})$ is the quantity $J(q, p) = p \diamond q$. This form holds in general for geodesic problems.

The canonical Poisson brackets $\{J_\xi(q, p), J_\eta(q, p)\}_{can}$ among the Noether Hamiltonians for the symmetries of the Lagrangian $L_{geo}(q, \dot{q})$ may be shown to be isomorphic to the Lie algebra $\mathfrak{g}$ of the isometry group G via the following calculation, which uses the product rule for the canonical Poisson bracket:

$$\begin{aligned}
\{J_\xi(q, p), J_\eta(q, p)\}_{can} &= \{J_\xi(q, p), \langle p \diamond q, \eta \rangle_{\mathfrak{g}}\}_{can} \\
&= \langle \{J_\xi, p\}_{can} \diamond q + p \diamond \{J_\xi, q\}_{can}, \eta \rangle_{\mathfrak{g}} \\
&= \langle (-\mathcal{L}_\xi^T p) \diamond q + p \diamond (\mathcal{L}_\xi q), \eta \rangle_{\mathfrak{g}} \\
&= \langle \mathcal{L}_\xi(p \diamond q), \eta \rangle_{\mathfrak{g}} = \langle \mathrm{ad}_\xi^*(p \diamond q), \eta \rangle_{\mathfrak{g}} \\
&= \langle p \diamond q, \mathrm{ad}_\xi \eta \rangle_{\mathfrak{g}} = \langle p, -\mathcal{L}_{[\xi, \eta]} q \rangle_{TQ} \\
&= J_{[\eta, \xi]}(q, p) \quad \text{(Anti-homomorphism)}.
\end{aligned}$$

$\blacktriangle$

Definition 16.2.1 *A simple mechanical system [Sm1970a, Sm1970b] is given by a Lagrangian of the form $L(v_q) = K(v_q) - V(q)$, for $v_q \in T_q Q$. The smooth function $V : Q \to \mathbb{R}$ is called the **potential energy**. The total energy of this system is given by $E = K + V$, and the Euler–Lagrange equations (which are always second order for a hyperregular Lagrangian) are*

$$\ddot{q}^i + \Gamma^i_{jk} \dot{q}^j \dot{q}^k + g^{il} \frac{\partial V}{\partial q^l} = 0, \quad i = 1, \dots n,$$

where g^{ij} are the entries of the inverse matrix of the Riemannian metric (g_{ij}).

Exercise. (Gauge invariance) Show that the Euler–Lagrange equations are unchanged under

$$L(\mathbf{q}(t), \dot{\mathbf{q}}(t)) \to L' = L + \frac{d}{dt}\gamma(\mathbf{q}(t), \dot{\mathbf{q}}(t)), \qquad (16.2.2)$$

for any function $\gamma \colon \mathbb{R}^{6N} = \{(\mathbf{q}, \dot{\mathbf{q}}) \mid \mathbf{q}, \dot{\mathbf{q}} \in \mathbb{R}^{3N}\} \to \mathbb{R}$.

★

Exercise. (Generalized coordinate theorem) Show that the Euler–Lagrange equations are *unchanged in form* under any smooth invertible mapping $f \colon \{\mathbf{q} \mapsto \mathbf{s}\}$. That is, with

$$L(\mathbf{q}(t), \dot{\mathbf{q}}(t)) = \check{L}(\mathbf{s}(t), \dot{\mathbf{s}}(t)), \qquad (16.2.3)$$

show that

$$\frac{d}{dt}\left(\frac{\partial L}{\partial \dot{\mathbf{q}}}\right) - \frac{\partial L}{\partial \mathbf{q}} = 0 \quad \Longleftrightarrow \quad \frac{d}{dt}\left(\frac{\partial \check{L}}{\partial \dot{\mathbf{s}}}\right) - \frac{\partial \check{L}}{\partial \mathbf{s}} = 0.$$

$$(16.2.4)$$

★

Exercise. How do the Euler–Lagrange equations transform under $\mathbf{q}(t) = \mathbf{r}(t) + \mathbf{s}(t)$?   ★

Exercise. (Other example Lagrangians) Write the Euler–Lagrange equations, and then apply the Legendre transformation to determine the Hamiltonian and Hamilton's canonical equations for the following Lagrangians. Determine which of them are hyperregular:

1. $L(q, \dot{q}) = \left(g_{ij}(q) \dot{q}^i \dot{q}^j \right)^{1/2}$. (Is it possible to assume that $L(q, \dot{q}) = 1$? Why?)

2. $L(\mathbf{q}, \dot{\mathbf{q}}) = -(1 - \dot{\mathbf{q}} \cdot \dot{\mathbf{q}})^{1/2}$.

3. $L(\mathbf{q}, \dot{\mathbf{q}}) = \frac{m}{2} \dot{\mathbf{q}} \cdot \dot{\mathbf{q}} + \frac{e}{c} \dot{\mathbf{q}} \cdot \mathbf{A}(\mathbf{q})$, for constants m and c and prescribed function $\mathbf{A}(\mathbf{q})$. How do the Euler–Lagrange equations for this Lagrangian differ from free motion in a moving frame with velocity $\frac{e}{mc} \mathbf{A}(\mathbf{q})$?

4. Calculate the action and energy for each of these Lagrangians. ★

16.2.1 Example: Free special relativistic particle motion

To illustrate the approach, consider the Lagrangian $L(\mathbf{q}, \dot{\mathbf{q}}) = -(1 - \dot{\mathbf{q}} \cdot \dot{\mathbf{q}})^{1/2}$:

1. **Action**

$$S = \int_0^T L(q, \dot{q}) \, dt = - \int_0^T (1 - \dot{\mathbf{q}} \cdot \dot{\mathbf{q}})^{1/2} \, dt$$

2. **Fibre derivative**

$$\mathbf{p} = \frac{\partial L}{\partial \dot{\mathbf{q}}} = \frac{\dot{\mathbf{q}}}{\sqrt{1 - \dot{\mathbf{q}} \cdot \dot{\mathbf{q}}}} =: \gamma \, \dot{\mathbf{q}} \quad \Longrightarrow \quad \dot{\mathbf{q}} = \pm \frac{\mathbf{p}}{\sqrt{1 + \mathbf{p} \cdot \mathbf{p}}}.$$

so this Lagrangian is hyperregular, after making a choice of sign convention, that $\mathbf{p} \cdot \dot{\mathbf{q}} > 0$, for example; so that $\gamma = \sqrt{1 + \mathbf{p} \cdot \mathbf{p}} = 1/\sqrt{1 - \dot{\mathbf{q}} \cdot \dot{\mathbf{q}}}$.

3. **Euler–Lagrange equations**

$$\frac{d(\gamma \, \dot{\mathbf{q}})}{dt} = 0.$$

4. **Hamiltonian and canonical equations**
   The Hamiltonian for this system is

   $$H = \mathbf{p} \cdot \dot{\mathbf{q}} - L = \sqrt{1 + |\mathbf{p}|^2} = \gamma,$$

   and its canonical equations are

   $$\frac{d\mathbf{q}}{dt} = \frac{\partial H}{\partial \mathbf{p}} = \frac{\mathbf{p}}{\sqrt{1 + |\mathbf{p}|^2}}, \quad \frac{d\mathbf{p}}{dt} = -\frac{\partial H}{\partial \mathbf{q}} = 0$$

   These equations represent uniform (force-free) motion in $\mathbb{R}^3$ of a relativistic particle with rest mass $m_0 = 1$ in units of $c = 1$.

   In these units, the relation $H = \gamma$ is written as

   $$H = \gamma \, m_0 c^2 = mc^2 = E.$$

16.2.2   Example: Charged particle in a magnetic field

Consider a particle of charge e and mass m moving in a magnetic field $\mathbf{B}$, where $\mathbf{B} = \nabla \times \mathbf{A}$ is a given magnetic field on $\mathbb{R}^3$. The Lagrangian for the motion is given by the "minimal coupling" prescription (jay-dot-ay)

$$L(\mathbf{q}, \dot{\mathbf{q}}) = \frac{m}{2} \|\dot{\mathbf{q}}\|^2 + \frac{e}{c} \mathbf{A}(\mathbf{q}) \cdot \dot{\mathbf{q}},$$

in which the constant c is the speed of light. The derivatives of this Lagrangian are

$$\frac{\partial L}{\partial \dot{\mathbf{q}}} = m\dot{\mathbf{q}} + \frac{e}{c}\mathbf{A} =: \mathbf{p} \quad \text{and} \quad \frac{\partial L}{\partial \mathbf{q}} = \frac{e}{c}\nabla \mathbf{A}^T \cdot \dot{\mathbf{q}}.$$

Hence, the Euler–Lagrange equations for this system are

$$m\ddot{\mathbf{q}} = \frac{e}{c}(\nabla \mathbf{A}^T \cdot \dot{\mathbf{q}} - \nabla \mathbf{A} \cdot \dot{\mathbf{q}}) = \frac{e}{c}\dot{\mathbf{q}} \times \mathbf{B}$$

(Newton's equations for the Lorentz force). The Lagrangian L is hyperregular because

$$\mathbf{p} = \mathbb{F}L(\mathbf{q}, \dot{\mathbf{q}}) = m\dot{\mathbf{q}} + \frac{e}{c}\mathbf{A}(\mathbf{q})$$

has the inverse

$$\dot{\mathbf{q}} = \mathbb{F}H(\mathbf{q}, \mathbf{p}) = \frac{1}{m}\left(\mathbf{p} - \frac{e}{c}\mathbf{A}(\mathbf{q})\right).$$

The corresponding Hamiltonian is given by the invertible change of variables:

$$H(\mathbf{q}, \mathbf{p}) = \mathbf{p} \cdot \dot{\mathbf{q}} - L(\mathbf{q}, \dot{\mathbf{q}}) = \frac{1}{2m}\left\|\mathbf{p} - \frac{e}{c}\mathbf{A}\right\|^2. \tag{16.2.5}$$

The Hamiltonian H is hyperregular since

$$\dot{\mathbf{q}} = \mathbb{F}H(\mathbf{q}, \mathbf{p}) = \frac{1}{m}\left(\mathbf{p} - \frac{e}{c}\mathbf{A}\right) \quad \text{has the inverse}$$

$$\mathbf{p} = \mathbb{F}L(\mathbf{q}, \dot{\mathbf{q}}) = m\dot{\mathbf{q}} + \frac{e}{c}\mathbf{A}.$$

The canonical equations for this Hamiltonian recover Newton's equations for the Lorentz force law.

16.2.3   Kaluza–Klein construction for a charged particle in a magnetic field

Although the minimal-coupling Lagrangian is not expressed as the kinetic energy of a metric, Newton's equations for the Lorentz force law may still be obtained as geodesic equations. This is accomplished by suspending them in a higher-dimensional space via the *Kaluza–Klein construction*, which proceeds as follows.

Let Q_{KK} be the manifold $\mathbb{R}^3 \times S^1$ with variables $(\mathbf{q}, \theta)$. On Q_{KK}, introduce the 1-form $A + d\theta$ (which defines a connection 1-form on the trivial circle bundle $\mathbb{R}^3 \times S^1 \to \mathbb{R}^3$), and introduce the *Kaluza–Klein Lagrangian* $L_{KK}: TQ_{KK} \simeq T\mathbb{R}^3 \times TS^1 \mapsto \mathbb{R}$ as

$$L_{KK}(\mathbf{q}, \theta, \dot{\mathbf{q}}, \dot{\theta}) = \tfrac{1}{2}m\|\dot{\mathbf{q}}\|^2 + \tfrac{1}{2}\|\langle A + d\theta, (\mathbf{q}, \dot{\mathbf{q}}, \theta, \dot{\theta})\rangle\|^2$$

$$= \tfrac{1}{2}m\|\dot{\mathbf{q}}\|^2 + \tfrac{1}{2}(\mathbf{A} \cdot \dot{\mathbf{q}} + \dot{\theta})^2.$$

The Lagrangian L_{KK} is positive definite in $(\dot{\mathbf{q}}, \dot{\theta})$; therefore, it may be regarded as the kinetic energy of a metric, the *Kaluza–Klein metric* on TQ_{KK}. (This construction fits the idea of $U(1)$ gauge symmetry for electromagnetic fields in $\mathbb{R}^3$. It can be generalised to a principal bundle with compact structure group endowed with a connection. The Kaluza–Klein Lagrangian in this generalisation leads to Wong's equations for a colour-charged particle moving in a classical Yang–Mills field.) The Legendre transformation for L_{KK} gives the momenta

$$\mathbf{p} = m\dot{\mathbf{q}} + (\mathbf{A} \cdot \dot{\mathbf{q}} + \dot{\theta})\mathbf{A} \quad \text{and} \quad \pi = \mathbf{A} \cdot \dot{\mathbf{q}} + \dot{\theta}. \qquad (16.2.6)$$

Since L_{KK} does not depend on θ, the Euler–Lagrange equation

$$\frac{d}{dt}\frac{\partial L_{KK}}{\partial \dot{\theta}} = \frac{\partial L_{KK}}{\partial \theta} = 0,$$

shows that $\pi = \partial L_{KK}/\partial \dot{\theta}$ is conserved. The *charge* is now defined by $e := c\pi$. The Hamiltonian H_{KK} associated with L_{KK} by the Legendre transformation (16.2.6) is

$$\begin{aligned}
H_{KK}(\mathbf{q}, \theta, \mathbf{p}, \pi) &= \mathbf{p} \cdot \dot{\mathbf{q}} + \pi\dot{\theta} - L_{KK}(\mathbf{q}, \dot{\mathbf{q}}, \theta, \dot{\theta}) \\
&= \mathbf{p} \cdot \tfrac{1}{m}(\mathbf{p} - \pi\mathbf{A}) + \pi(\pi - \mathbf{A} \cdot \dot{\mathbf{q}}) - \tfrac{1}{2}m\|\dot{\mathbf{q}}\|^2 - \tfrac{1}{2}\pi^2 \\
&= \mathbf{p} \cdot \tfrac{1}{m}(\mathbf{p} - \pi\mathbf{A}) + \tfrac{1}{2}\pi^2 - \pi\mathbf{A} \cdot \tfrac{1}{m}(\mathbf{p} - \pi\mathbf{A}) - \tfrac{1}{2m}\|\mathbf{p} - \pi\mathbf{A}\|^2 \\
&= \tfrac{1}{2m}\|\mathbf{p} - \pi\mathbf{A}\|^2 + \tfrac{1}{2}\pi^2. \qquad (16.2.7)
\end{aligned}$$

On the constant level set $\pi = e/c$, the Kaluza–Klein Hamiltonian H_{KK} is a function of only the variables $(\mathbf{q}, \mathbf{p})$ and is equal to the Hamiltonian (16.2.5) for charged particle motion under the Lorentz force up to an additive constant. This example provides an easy but fundamental illustration of the geometry of (Lagrangian) reduction by symmetry. The canonical equations for the Kaluza–Klein Hamiltonian H_{KK} now reproduce Newton's equations for the Lorentz force law.

Exercise. (Spherical pendulum) Spherical pendulum dynamics is equivalent to a particle rolling on the interior of a spherical surface under gravity. Write down the

Lagrangian and the equations of motion for a spherical pendulum with S^2 as its configuration space. Show explicitly that the Lagrangian is hyperregular. Use the Legendre transformation to convert the equations to Hamiltonian form. Find the conservation law corresponding to angular momentum about the axis of gravity by "bare hands" methods. ★

Exercise. (Euler–Lagrange equations for differentially rotating frames) The Lagrangian for a free particle of unit mass relative to a moving frame is obtained by setting

$$L(\dot{\mathbf{q}}, \mathbf{q}, t) = \tfrac{1}{2} \|\dot{\mathbf{q}}\|^2 + \dot{\mathbf{q}} \cdot \mathbf{R}(\mathbf{q}, t)$$

for a function $\mathbf{R}(\mathbf{q}, t)$ which prescribes the space and time dependence of the moving-frame velocity. For example, a frame rotating with time-dependent frequency $\Omega(t)$ about the vertical axis $\hat{\mathbf{z}}$ is obtained by choosing $\mathbf{R}(\mathbf{q}, t) = \mathbf{q} \times \Omega(t)\hat{\mathbf{z}}$. Calculate $\Theta_L(q, \dot{q})$, the Euler–Lagrange operator $\mathcal{E}L(L)(q, \dot{q}, \ddot{q})$, the Hamiltonian and its corresponding canonical equations. ★

16.2.4 The free particle in $\mathbb{H}^2$: Part #1

Exercise. In Appendix I of V. I. Arnold's book [Ar2013], we read:

> EXAMPLE. We consider the upper half-plane $y > 0$ of the plane of complex

numbers $z = x + iy$ with the metric

$$ds^2 = \frac{dx^2 + dy^2}{y^2}.$$

It is easy to compute that the geodesics of this two-dimensional Riemannian manifold are circles and straight lines perpendicular to the x-axis. Linear fractional transformations with real coefficients

$$z \to \frac{az + b}{cz + d} \qquad (16.2.8)$$

are isometric transformations of our manifold ($\mathbb{H}^2$), which is called the *Lobachevsky plane*. These isometric transformations of $\mathbb{H}^2$ have deep significance in physics. They correspond to the most general Lorentz transformation of space-time.

Consider a free particle of mass m moving on the Lobachevsky half-plane $\mathbb{H}^2$. Its Lagrangian is the kinetic energy corresponding to the Lobachevsky metric. Namely,

$$L = \frac{m}{2} \left(\frac{\dot{x}^2 + \dot{y}^2}{y^2} \right). \qquad (16.2.9)$$

1. a. Write the fibre derivatives of the Lagrangian (16.2.9), and

    b. Compute its Euler–Lagrange equations. These equations represent geodesic motion on $\mathbb{H}^2$.

    c. Evaluate the Christoffel symbols.   ★

Answer.

a. Fibre derivatives:

$$\frac{\partial L}{\partial \dot{x}} = \frac{m\dot{x}}{y^2} =: p_x \quad \text{and} \quad \frac{\partial L}{\partial \dot{y}} = \frac{m\dot{y}}{y^2} =: p_y.$$

b. The Euler–Lagrange equations $\frac{d}{dt}\frac{\partial L}{\partial \dot{x}} = \frac{\partial L}{\partial x}$ and $\frac{d}{dt}\frac{\partial L}{\partial \dot{y}} = \frac{\partial L}{\partial y}$ yield, respectively,

$$\frac{d}{dt}\left(\frac{\dot{x}}{y^2}\right) = 0 \quad \text{and} \quad \frac{d}{dt}\left(\frac{\dot{y}}{y^2}\right) = -\frac{\dot{x}^2 + \dot{y}^2}{y^3}.$$

$$(16.2.10)$$

c. Expanding these equations yield the Christoffel symbols for the geodesic motion:

$$\ddot{x} - \frac{2}{y}\dot{x}\dot{y} = 0, \quad \ddot{y} + \frac{1}{y}\dot{x}^2 - \frac{1}{y}\dot{y}^2 = 0.$$

Hence,

$$\Gamma^1_{12} = -\frac{2}{y} = \Gamma^1_{21}, \quad \Gamma^2_{11} = \frac{1}{y}, \quad \Gamma^2_{22} = -\frac{1}{y}. \quad \blacktriangle$$

Exercise. (*Continued*) Hint: The Lagrangian in (16.2.9) is invariant under the group of linear fractional transformations with real coefficients. These have an $SL(2, \mathbb{R})$ matrix representation:

$$\begin{bmatrix} a & b \\ c & d \end{bmatrix}\begin{bmatrix} z \\ 1 \end{bmatrix} = \frac{az + b}{cz + d}. \qquad (16.2.11)$$

2. Show that the quantities

$$u = \frac{\dot{x}}{y} \quad \text{and} \quad v = \frac{\dot{y}}{y} \qquad (16.2.12)$$

are invariant under a subgroup of these symmetry transformations.

3. Specify this subgroup in terms of the representation (16.2.11). ★

Answer. The quantities (16.2.12) are invariant under a subgroup of translations and scalings:

$$T_\tau : (x, y) \mapsto (x + \tau, y) \quad \text{Flow of } X_T = \partial_x,$$
$$(\delta x, \delta y) = (1, 0),$$
$$[X_T, X_S] = X_T.$$
$$S_\sigma : (x, y) \mapsto (e^\sigma x, e^\sigma y) \quad \text{Flow of } X_S = x\partial_x + y\partial_y,$$
$$(\delta x, \delta y) = (x, y).$$

These transformations are translations T along the x-axis and scalings S centred at $(x, y) = (0, 0)$. They are represented by the elements of (16.2.11) as

$$T = \begin{bmatrix} 1 & b \\ 0 & 1 \end{bmatrix} \quad \text{and} \quad S = \begin{bmatrix} a & 0 \\ 0 & 1 \end{bmatrix}.$$

That is, the transformations T and S are isometries of the metric $ds^2 = (dx^2 + dy^2)/y^2$ on $\mathbb{H}^2$, with $T : a = 1 = d, c = 0, b \neq 0$ and $S : a \neq 0, b = 0 = c, d = 1$.   ▲

Exercise. (*Continued*)

4. a. Use the invariant quantities (u, v) in (16.2.12) as new variables in Hamilton's principle.
    Hint: The transformed Lagrangian is

    $$\ell(u, v) = \frac{m}{2}(u^2 + v^2).$$

    b. Find the corresponding conserved Noether quantities.   ★

Answer.

a. The translations T along the x-axis and scalings S centred at $(x, y) = (0, 0)$ leave invariant the quantities

$$u = \frac{\dot{x}}{y} \quad \text{and} \quad v = \frac{\dot{y}}{y},$$

in terms of which the Lagrangian L in (16.2.9) reduces to

$$\ell(u, v) = \frac{m}{2}(u^2 + v^2).$$

The reduced Hamilton's principle in the variables u and v yields

$$0 = \delta S = \delta \int_a^b \ell(u, v)\, dt = \int_a^b m(u\delta u + v\delta v)\, dt$$

$$= m \int_a^b \frac{u}{y}(\delta \dot{x} - u\delta y) + \frac{v}{y}(\delta \dot{y} - v\delta y)\, dt$$

$$= -m \int_a^b \left(\frac{d}{dt}\frac{u}{y}\right) \delta x$$

$$+ \left(\frac{d}{dt}\frac{v}{y} + \frac{u^2 + v^2}{y}\right) \delta y\, dt$$

$$+ m \left[\frac{u}{y}\delta x + \frac{v}{y}\delta y\right]_a^b.$$

Thus, Hamilton's principle recovers equations (16.2.10) in the variables u and v.

b. Applying Noether's theorem to the endpoint term in these variables yields the conservation of

$$C_T = \frac{u}{y}, \quad \text{for } (\delta x, \delta y) = (1, 0) \text{ translations,}$$

and

$$C_S = \frac{ux + vy}{y} \quad \text{for } (\delta x, \delta y) = (x, y) \text{ scaling.} \quad \blacktriangle$$

Exercise. (*Continued*)

5. Transform the Euler–Lagrange equations from x and y to the variables u and v that are invariant under the symmetries of the Lagrangian.

Then:

a. Show that the resulting system conserves the kinetic energy expressed in these variables.

b. Discuss its integral curves and critical points in the uv-plane.

c. Show that the u and v equations can be integrated explicitly in terms of sech and tanh.

Hint: In the uv variables, the Euler–Lagrange equations for the Lagrangian (16.2.9) are expressed as

$$\frac{d}{dt}\frac{u}{y} = 0 \quad \text{and} \quad \frac{d}{dt}\frac{v}{y} + \frac{u^2 + v^2}{y} = 0.$$

Expanding these equations using $u = \dot{x}/y$ and $v = \dot{y}/y$ yields

$$\dot{u} = uv, \quad \dot{v} = -u^2. \tag{16.2.13}$$

★

Answer.

a. Equations (16.2.13) imply conservation of the kinetic energy

$$\ell(u, v) = \frac{m}{2}(u^2 + v^2) = E.$$

b. The integral curves of the system of equations (16.2.13) in the uv-plane are either critical points along the axis $u = 0$ or heteroclinic connections between these points that are semi-circles around the origin on the level sets of the energy E.

The critical points at $u = \dot{x}/y = 0$ are relative equilibria of the system corresponding to vertical motion on the xy-plane. Those corresponding to "upward motion" ($\dot{y} > 0$) are unstable, and the ones corresponding to "downward motion" ($\dot{y} < 0$) are stable.

c. The trial solutions $u = \tanh$ and $v = \text{sech}$ quickly converge to the exact solutions of the uv system. ▲

Exercise. (*Continued*)

6.   a. Legendre-transform the Lagrangian (16.2.9) to the Hamiltonian side, and obtain the canonical equations.

   b. Then, derive Poisson brackets for the variables $\{u, v\}$.

   Hint: $\{y p_x, y p_y\} = y p_x$.   ★

Answer.

a. The equations of motion on the Hamiltonian formulation are defined by introducing the momenta

$$p_x = \frac{\partial L}{\partial \dot{x}} = \frac{m\dot{x}}{y^2}, \quad p_y = \frac{\partial L}{\partial \dot{y}} = \frac{m\dot{y}}{y^2}$$

and the Hamiltonian

$$H = \frac{y^2}{2m} \left(p_x^2 + p_y^2 \right).$$

One obtains

$$\dot{x} = \frac{y^2 p_x}{m} \quad \dot{p}_x = 0,$$

$$\dot{y} = \frac{y^2 p_y}{m} \quad \dot{p}_y = \frac{-y}{m} \left(p_x^2 + p_y^2 \right).$$

(16.2.14)

By defining

$$u = y p_x / m, \quad v = y p_y / m,$$

the Hamiltonian can be written as

$$H = h(u, v) = \frac{1}{2} \left(u^2 + v^2 \right),$$

and the equations of motion (16.2.14) become, using $\{y p_x, y p_y\} = y p_x$,

$$\dot{u} = uv, \quad \dot{v} = -u^2.$$

(16.2.15)

b. These equations are Hamiltonian with respect to the Lie–Poisson bracket

$$\{u, v\} = u,$$

and the reduced Hamiltonian $h(u, v)$ in terms of the invariant variables. Namely,

$$\dot{u} = \{u, h\} = uv, \quad \dot{v} = \{v, h\} = -u^2. \qquad \blacktriangle$$

16.2.5   The free particle in $\mathbb{H}^2$: Part #2

Exercise. Consider the following pair of differential equations for $(u, v) \in \mathbb{R}^2$:

$$\dot{u} = uv, \quad \dot{v} = -u^2.$$

(16.2.16)

These equations have discrete symmetries under combined reflection and time reversal, $(u, t) \to (-u, -t)$ and $(v, t) \to (-v, -t)$. (This is called PT symmetry in the (u, v) plane.)

1. Find 2×2 real matrices L and B for which the system (16.2.16) may be written as a commutator, namely, as

$$\frac{dL}{dt} = [L, B].$$

Hint: A basis for 2×2 real matrices is given by

$$\sigma_1 = \begin{bmatrix} 0 & 1 \\ 1 & 0 \end{bmatrix}, \quad \sigma_2 = \begin{bmatrix} 0 & 1 \\ -1 & 0 \end{bmatrix}, \quad \sigma_3 = \begin{bmatrix} 1 & 0 \\ 0 & -1 \end{bmatrix}.$$

Explain what the commutator relation means and determine a constant of the motion from it.   ★

Answer. We introduce two linear 2×2 matrices, one symmetric ($L^T = L$) and the other skew-symmetric ($B^T = -B$), as required for the commutator $[L, B]$ to be symmetric:

$$L = \begin{bmatrix} -v & u \\ u & v \end{bmatrix} = u \begin{bmatrix} 0 & 1 \\ 1 & 0 \end{bmatrix} - v \begin{bmatrix} 1 & 0 \\ 0 & -1 \end{bmatrix} = u\sigma_1 - v\sigma_3,$$

$$B = \frac{1}{2} \begin{bmatrix} 0 & u \\ -u & 0 \end{bmatrix} = \frac{u}{2} \begin{bmatrix} 0 & 1 \\ -1 & 0 \end{bmatrix} = \frac{u}{2}\sigma_2.$$

Both matrices must be linear homogeneous, so that the commutator $[L, B]$ and time derivative $\frac{dL}{dt}$ can match powers using (16.2.16). The $\mathfrak{sl}(2, \mathbb{R})$ σ-matrices satisfy

$$[\sigma_1, \sigma_2] = 2\sigma_3, \quad [\sigma_2, \sigma_3] = 2\sigma_1, \quad \text{and} \quad [\sigma_3, \sigma_1] = -2\sigma_2.$$

Thus, we find the commutator relation

$$\frac{dL}{dt} = u^2 \sigma_3 + uv\, \sigma_1 = [L, B] = \left[u\sigma_1 - v\sigma_3,\ \frac{u}{2}\sigma_2\right].$$

What the commutator relation means: isospectrality.
The commutator relation implies that the flow generates
a similarity transformation of the 2×2 symmetric matrix
$L(0)$. The traceless matrix $L(t)$ has one independent
eigenvalue, and system (16.2.16) has only one conserved
quantity. The conserved quantity is the determinant
$\det L(t) = \det L(0)$. However, this conservation law
introduces no constraints on u. ▲

Exercise. (*Continued*)

2. Write system (16.2.16) as a double matrix commu-
   tator, $\frac{dL}{dt} = [L, [L, N]]$. In particular, find N explicitly and explain what this means for the solutions.

   Hint: Compute $\frac{d}{dt} \operatorname{tr}(LN)$. ★

Answer. Substituting $N := \begin{bmatrix} a & 0 \\ 0 & b \end{bmatrix}$ into $\frac{dL}{dt} = [L, [L, N]]$
yields $b - a =$, so for example, we may set

$$N := \begin{bmatrix} 1 & 0 \\ 0 & 2 \end{bmatrix}.$$

What this means for the solutions: dissipative flow.
The evolution by the double-bracket relation $\frac{dL}{dt} = [L, [L, N]]$ is a *dissipative flow* that decreases the quantity $\operatorname{tr}(LN)$ according to

$$\frac{d}{dt} \operatorname{tr}(LN) = -\operatorname{tr}([L, N]^T[L, N]),$$

until L becomes diagonal, and hence $[L, N] \to 0$ because N is diagonal. Thus, the dynamics (16.2.16) becomes asymptotically steady as L tends to a diagonal matrix. This means system (16.2.16) must asymptotically approach a stable equilibrium that is consistent with its initial conditions and conservation laws. For the current case, substituting the explicit forms of L and N yields

$$\frac{d}{dt} \operatorname{tr}(LN) = \frac{1}{2}\dot{v} = -\frac{1}{2}u^2 = -\operatorname{tr}([L, N]^T[L, N])$$

$$= -\|[L, N]\|^2,$$

which holds by (16.2.16) and thus verifies the previous calculation. In the current case, it will turn out that $\lim_{t\to\infty} u(t) = 0$, which will verify $[L, N] \to 0$, as the off-diagonal parts of L will vanish asymptotically.   ▲

Exercise. (*Continued*)

3. Find explicit solutions and discuss their motion and asymptotic behaviour:

   a. in time;

   b. in the (u, v) phase plane.

   Hint: Keep the *tanh* function in mind.   ★

Answer. Keeping the tanh function in mind and recalling that

$$\frac{d\tanh(ct)}{dt} = c\operatorname{sech}^2(ct)$$

$$\frac{d\operatorname{sech}(ct)}{dt} = -c\operatorname{sech}(ct)\tanh(ct),$$

we find, for $u(0) = c$ and $v(0) = 0$,

$$v(t) = -c\tanh(ct) \quad \text{and} \quad u(t) = c\operatorname{sech}(ct),$$

and of course, we check that

$$2h = u^2 + v^2 = c^2(\tanh^2(ct) + \operatorname{sech}^2(ct)) = c^2.$$

Motion and asymptotic behaviour.

a. **In time:** We have $\lim_{t\to\infty}(u(t), v(t)) = (0, -c)$. Consequently, the quantity $u(t)$ falls exponentially with time, from $u(0)$ towards the line of fixed points at $u = 0$, while $u(t)$ goes to a constant equal to $-u(0)$.

b. **In the (u, v) phase plane:** Since h is conserved, the motion is along a family of semi-circles, each parameterised by its radius $c = \sqrt{2h}$, as

$$u^2 + v^2 = c^2 \quad \text{for } u > 0 \quad \text{and} \quad u < 0,$$

lying in the upper and lower (u, v) half-planes. These semi-circular motions are mirror images, reflected across the vertical line of fixed points at $u = 0$ in the (u, v) plane. The equations of motion are PT-symmetric, so the fixed points along $u = 0$ in the (u, v) plane are stable for $v < 0$ and unstable for $v > 0$.

Thus, the two families of semi-circular motion both connect the line of fixed points at $u = 0$ to itself. One family of semi-circles lies in the upper half (u, v) plane, and the other lies symmetrically placed to complete the circles in the lower half (u, v) plane. The flows along each reflection-symmetric pair of semi-circles pass in the same (negative) v direction from $v = c$ to $v = -c$. ▲

Exercise. (*Continued*)

4. Explain why the solution behaviour found in the previous part is consistent with the behaviour predicted by the double-bracket relation. ★

Answer. This analysis is consistent with the conclusion from the double-bracket relation $\frac{dL}{dt} = [L, [L, N]]$ that the dynamics of L-matrix

$$
L = \begin{bmatrix} -v & u \\ u & v \end{bmatrix}
$$

asymptotically becomes steady. In fact, since $\lim_{t \to \infty} u = 0$ and $\lim_{t \to \infty} v(t) = -c$, the L-matrix asymptotically diagonalises, and hence $[L, N] \to 0$ because N is diagonal. ▲

The following exercise illustrates the differences between Hamiltonian and dissipative evolution, written in matrix commutator form.

Exercise. Commutator form of the 3D Volterra system.

Consider the dynamical system in $(x_1, x_2, x_3) \in \mathbb{R}^3$:

$$
\begin{bmatrix} \dot{x}_1 \\ \dot{x}_2 \\ \dot{x}_3 \end{bmatrix} = \begin{bmatrix} x_1 x_2 \\ x_2 x_3 - x_1 x_2 \\ -x_2 x_3 \end{bmatrix} = x_2 \begin{bmatrix} x_1 \\ x_3 - x_1 \\ -x_3 \end{bmatrix} \quad (16.2.17)
$$

This is a 3D version of the Volterra (1931) model of competition among species, which for more species is given by

$$
\dot{x}_n = x_n(x_{n+1} - x_{n-1}), \quad n = 1, 2, \ldots, N,
$$

with $x_0 = 0 = x_{N+1}$.

1. Find two conservation laws for the system (16.2.17). ★

Answer. The flow of the vector field $(\dot{x}_1, \dot{x}_2, \dot{x}_3) \in T\mathbb{R}^3$ preserves the sum $H = x_1 + x_2 + x_3$ and the product $C = x_1 x_3$. ▲

Exercise. (*Continued*)

2. Verify that this system may be written in commutator form as

$$\frac{dL}{dt} = [L, B]$$

for the 4×4 matrices

$$L := \begin{bmatrix} x_1 & 0 & \sqrt{x_1 x_2} & 0 \\ 0 & x_1 + x_2 & 0 & \sqrt{x_2 x_3} \\ \sqrt{x_1 x_2} & 0 & x_2 + x_3 & 0 \\ 0 & \sqrt{x_2 x_3} & 0 & x_3 \end{bmatrix}$$

$$B := \frac{1}{2}\begin{bmatrix} 0 & 0 & -\sqrt{x_1 x_2} & 0 \\ 0 & 0 & 0 & -\sqrt{x_2 x_3} \\ \sqrt{x_1 x_2} & 0 & 0 & 0 \\ 0 & \sqrt{x_2 x_3} & 0 & 0 \end{bmatrix}.$$

3. Explain how the two conservation laws for the sum $H = x_1 + x_2 + x_3$ and the product $C = x_1 x_3$ found earlier are related to the matrices L and B. Explain what this means for the system in (16.2.17). ★

Answer.

$$\operatorname{tr} L = 2H = 2(x_1 + x_2 + x_3) \quad \text{and} \quad \det L = C^2 = (x_1 x_3)^2.$$

Preservation of the trace and determinant of the 4×4 matrix L means that two of its four eigenvalues are preserved. ▲

Exercise. (*Continued*)

4. Give the geometrical interpretation of the formula $\frac{dL}{dt} = [L, B]$ with 4×4 matrices L and B.   ★

Answer. The commutator form has little effect on the spectrum of the 4×4 symmetric matrix $L(t)$ because the system has only two conserved quantities.   ▲

Exercise. (*Continued*)

5. Write system (16.2.17) as a double matrix commutator. In particular, find the diagonal matrix N for which $B = [L, N]$.   ★

Answer.
$$\frac{dL}{dt} = [L, B] = [L, [L, N]].$$

$$N := \begin{bmatrix} 1 & 0 & 0 & 0 \\ 0 & 1 & 0 & 0 \\ 0 & 0 & \frac{1}{2} & 0 \\ 0 & 0 & 0 & \frac{1}{2} \end{bmatrix}.$$

▲

Exercise. (*Continued*)

6. Give the geometrical interpretation of the formula $\frac{dL}{dt} = [L, [L, N]]$.   ★

Answer. The evolution of the matrix $L(t)$ preserves its trace and determinant as it tends towards an equilibrium $x_2 \to 0$, and matrix L diagonalises to commute with diagonal N. The eigenvalues in the equilibrium diagonal form of L are doubly degenerate.   ▲

17

MOMENTUM MAPS

Contents

> **What is this lecture about?** This lecture describes the definitions
> and properties of cotangent-lift momentum maps and symplec-
> tic momentum maps.

17.1 The main idea

Symmetries are often associated with conserved quantities. For
example, the flow of any $SO(3)$-invariant Hamiltonian vector field
on $T^*\mathbb{R}^3$ conserves angular momentum, $\mathbf{q} \times \mathbf{p}$. More generally, given
a Hamiltonian H on a phase space P and a group action of G on
P that conserves H, there is often an associated "momentum map"

$J: P \to \mathfrak{g}^*$ that is conserved by the flow of the Hamiltonian vector field.

Note: all group actions in this section will be left actions unless otherwise specified.

Let G be a Lie group, $\mathfrak{g}$ be its Lie algebra and $\mathfrak{g}^*$ be its dual. Suppose that G acts symplectically on a symplectic manifold P with symplectic form denoted by Ω. Denote the infinitesimal generator associated with the Lie algebra element ξ by ξ_P, and let the Hamiltonian vector field associated with a function $f: P \to \mathbb{R}$ be denoted X_f, so that $df = \iota_{X_f}\Omega$, or, in equivalent expanded notation, $df = X_f \lrcorner \Omega$.

17.2   Definition, history and overview

A *momentum map* $J: P \to \mathfrak{g}^*$ is defined by the condition relating the infinitesimal generator ξ_P of a symmetry to the vector field of its corresponding conservation law: $\langle J, \xi \rangle$,

$$\xi_P = X_{\langle J, \xi \rangle}$$

for all $\xi \in \mathfrak{g}$. Here, $\langle J, \xi \rangle: P \to \mathbb{R}$ is defined by the natural pointwise pairing.

A momentum map is said to be *equivariant* when it is equivariant with respect to the given action on P and the coadjoint action on $\mathfrak{g}^*$. That is,

$$J(g \cdot p) = \mathrm{Ad}^*_{g^{-1}} J(p)$$

for every $g \in G$, $p \in P$, where $g \cdot p$ denotes the action of g on the point p and where Ad denotes the adjoint action.

According to Weinstein [We1983b], Lie [Li1890] already knew the following:

1. An action of a Lie group G with Lie algebra $\mathfrak{g}$ on a symplectic manifold P should be accompanied by such an equivariant momentum map $J: P \to \mathfrak{g}^*$.

2. The orbits of this action are themselves symplectic manifolds.

The links with mechanics were developed in the works by Lagrange, Poisson, Jacobi and, later, Noether. In particular, Noether showed that a momentum map for the action of a group G that is a symmetry of the Hamiltonian for a given system is a *conservation law* for that system.

In modern form, the momentum map and its equivariance were rediscovered by Kostant [Ko1966] and Souriau [So1970] in the general symplectic case and by Smale [Sm1970a, Sm1970b] for the case of the lifted action from a manifold Q to its cotangent bundle $P = T^*Q$. In this case, the equivariant momentum map is given explicitly by

$$\langle J(\alpha_q), \xi \rangle = \langle \alpha_q, \xi_Q(q) \rangle,$$

where $\alpha_q \in T^*Q$, $\xi \in \mathfrak{g}$ and where the angular brackets denote the natural pairing on the appropriate spaces. See Marsden and Ratiu [MaRa1994] and Ortega and Ratiu [OrRa2004] for additional history and description of the momentum map and its properties. See the textbooks in Refs. [Ho2011b, Ho2011c, HoScSt2009] and the survey in Ref. [Ho2011a] for further details.

17.3 Hamiltonian systems on Poisson manifolds

Definition 17.3.1 *A **Poisson bracket** on a manifold P is a skew-symmetric bilinear operation on*

$$\mathcal{F}(P) := C^\infty (P, \mathbb{R})$$

satisfying the Jacobi identity and the Leibniz identity:

$$\{FG, H\} = F\{G, H\} + \{F, H\}G$$

*The pair $(P, \{\cdot, \cdot\})$ is called a **Poisson manifold**.*

Remark 17.3.1 The Leibniz identity is sometimes not included in the definition. Note that bilinearity, skew symmetry and the Jacobi identity are the axioms of a Lie algebra. In what follows, a Poisson bracket is a binary operation that makes $\mathcal{F}(P)$ into a Lie algebra and also satisfies the Leibniz identity. □

Exercise. Show that the *classical Poisson bracket*, defined in cotangent-lifted coordinates

$$\left(q^1, \ldots, q^N, p_1, \ldots, p_N\right)$$

on an $2N$-dimensional cotangent bundle T^*Q by

$$\{F, G\} = \sum_{i=1}^{N} \left(\frac{\partial F}{\partial q^i} \frac{\partial G}{\partial p_i} - \frac{\partial F}{\partial p_i} \frac{\partial G}{\partial q^i} \right),$$

satisfies the axioms of a Poisson bracket. Show also that the definition of this bracket is independent of the choice of local coordinates $\left(q^1, \ldots, q^N\right)$. ★

Definition 17.3.2 *A **Poisson map** between two Poisson manifolds is a map*

$$\varphi \colon \left(P_1, \{\cdot, \cdot\}_1\right) \to \left(P_2, \{\cdot, \cdot\}_2\right)$$

that preserves the brackets, meaning

$$\{F \circ \varphi, G \circ \varphi\}_1 = \{F, G\}_2 \circ \varphi, \quad \text{for all } F, G \in \mathcal{F}\left(P_2\right).$$

Definition 17.3.3 *An action Φ of G on a Poisson manifold $(P, \{,\})$ is **canonical** if Φ_g is a Poisson map for every g, that is,*

$$\{F \circ \Phi_g, K \circ \Phi_g\} = \{F, K\} \circ \Phi_g$$

for every $F, K \in \mathcal{F}(P)$.

Definition 17.3.4 *Let $(P, \{\cdot, \cdot\})$ be a Poisson manifold, and let $H \colon P \to \mathbb{R}$ be differentiable. The **Hamiltonian vector field** for H is the vector field X_H defined by*

$$X_H(F) = \{F, H\}, \quad \text{for any } F \in \mathcal{F}(P).$$

Remark 17.3.2 X_H is well defined because of the Leibniz identity and the correspondence between vector fields and derivations (see Lee [Le2003]). □

Remark 17.3.3 $X_H(F) = \mathcal{L}_{X_H} F = \dot{F}$, the Lie derivative of F along the flow of X_H. The equations

$$\dot{F} = \{F, H\},$$

called "Hamilton's equations," have already appeared in Theorem 3.2.1, and they are an equivalent definition of X_H. □

Exercise. Show that Hamilton's equations for the classical Poisson bracket are the canonical Hamilton's equations

$$\dot{q}^i = \frac{\partial H}{\partial p_i}, \quad \dot{p}_i = -\frac{\partial H}{\partial q^i}.$$

★

17.4   Infinitesimal invariance under Hamiltonian vector fields

Let G act smoothly on P, and let $\xi \in \mathfrak{g}$. Recall (from Lecture 14) that the infinitesimal generator ξ_P is the vector field on P defined by

$$\xi_P(x) = \frac{d}{dt} g(t) x \bigg|_{t=0},$$

for some path $g(t)$ in G such that $g(0) = e$ and $g'(0) = \xi$.

Remark 17.4.1 For matrix groups, we can take $g(t) = \exp(t\xi)$. This works in general for the exponential map of an arbitrary Lie group. For matrix groups,

$$\xi_P(\mathbf{x}) = \frac{d}{dt} \exp(t\xi)\mathbf{x} \bigg|_{t=0} = \xi\mathbf{x} \quad \text{(matrix multiplication)}. \quad □$$

Definition 17.4.1 *If $H: P \to \mathbb{R}$ is G-invariant, meaning that $H(gx) = H(x)$ for all $g \in G$ and $x \in P$, then $\mathcal{L}_{\xi_P} H = 0$ for all $\xi \in \mathfrak{g}$. This property is called* **infinitesimal invariance.**

Example 17.4.1 (The momentum map for the rotation group) Consider the cotangent bundle of ordinary Euclidean space $\mathbb{R}^3$. This is the Poisson (symplectic) manifold with coordinates $(\mathbf{q}, \mathbf{p}) \in T^*\mathbb{R}^3 \simeq \mathbb{R}^6$, equipped with the canonical Poisson bracket. An element g of the rotation group $SO(3)$ acts on $T^*\mathbb{R}^3$ according to

$$g(\mathbf{q}, \mathbf{p}) = (g\mathbf{q}, g\mathbf{p}).$$

Set $g(t) = \exp(tA)$, so that $\frac{d}{dt}\big|_{t=0} g(t) = A$ and the corresponding Hamiltonian vector field is

$$X_A = (\dot{\mathbf{q}}, \dot{\mathbf{p}}) = (A\mathbf{q}, A\mathbf{p}),$$

where $A \in \mathfrak{so}(3)$ is a skew-symmetric matrix. The corresponding Hamiltonian equations read

$$\dot{\mathbf{q}} = A\mathbf{q} = \frac{\partial J_A}{\partial \mathbf{p}}, \quad \dot{\mathbf{p}} = A\mathbf{p} = -\frac{\partial J_A}{\partial \mathbf{q}}.$$

Hence,

$$J_A(\mathbf{q}, \mathbf{p}) = -A\mathbf{p} \cdot \mathbf{q} = a_i \epsilon_{ijk} p_k q_j = \mathbf{a} \cdot \mathbf{q} \times \mathbf{p}.$$

for a vector $\mathbf{a} \in \mathbb{R}^3$ with components a_i, $i = 1, 2, 3$. So, the momentum map for the rotation group is the angular momentum $J = \mathbf{q} \times \mathbf{p}$.

Example 17.4.2 Consider angular momentum $\mathbf{J} = \mathbf{q} \times \mathbf{p}$, defined on $P = T^*\mathbb{R}^3$. For every $\xi \in \mathbb{R}^3$, define

$$\mathbf{J}_\xi(\mathbf{q}, \mathbf{p}) := \xi \cdot (\mathbf{q} \times \mathbf{p}) = \mathbf{p} \cdot (\xi \times \mathbf{q})$$

From Section 17.3 and Example 17.4.1,

$$X_{J_\xi}(\mathbf{q}, \mathbf{p}) = \left(\frac{\partial J_\xi}{\partial \mathbf{p}}, -\frac{\partial J_\xi}{\partial \mathbf{q}}\right) = (\xi \times \mathbf{q}, \xi \times \mathbf{p}) = \widehat{\xi}_P(\mathbf{q}, \mathbf{p}),$$

where the last line is the infinitesimal generator corresponding to $\widehat{\xi} \in \mathfrak{so}(3)$. Now, suppose $H: P \to \mathbb{R}$ is $SO(3)$-invariant. From Example 17.4.1, we have $\mathcal{L}_{\widehat{\xi}} H = 0$. It follows that

$$\mathcal{L}_{X_H} J_\xi = \{J_\xi, H\} = -\{H, J_\xi\} = -\mathcal{L}_{X_{J_\xi}} H = -\mathcal{L}_{\xi_P} H = 0.$$

Since this holds for all ξ, we have shown that J is conserved by the Hamiltonian flow. This is the Hamiltonian version of Noether's theorem.

17.5   Defining momentum maps

In order to generalise this example, we recast it using the hat map $\widehat{\ }: \mathbb{R}^3 \to \mathfrak{so}(3)$, the associated map $\widetilde{\ }: (\mathbb{R}^3)^* \to \mathfrak{so}(3)^*$ and the standard identification $(\mathbb{R}^3)^* \cong \mathbb{R}^3$ via the Euclidean dot product. We consider J as a function from P to $\mathfrak{so}(3)^*$ given by $J(\mathbf{q}, \mathbf{p}) = (\mathbf{q} \times \mathbf{p})\widetilde{\ }$. For any $\xi = \widehat{\mathbf{v}}$, we define $J_\xi(\mathbf{q}, \mathbf{p}) = \langle (\mathbf{q} \times \mathbf{p})\widetilde{\ }, \widehat{\mathbf{v}} \rangle = (\mathbf{q} \times \mathbf{p}) \cdot \mathbf{v}$. As before, we find that $X_{J_\xi} = \xi_P$ for every ξ, and J is conserved by the Hamiltonian flow. We take the first property, $X_{J_\xi} = \xi_P$, as the general definition of a momentum map. The conservation of J follows using the same Poisson bracket calculation as in the example; the result is Noether's theorem.

Definition 17.5.1 *A* ***momentum map*** *for a canonical action of Lie group G on Poisson manifold P is a map $J: P \to \mathfrak{g}^*$ such that, for every $\xi \in \mathfrak{g}$, the map $J_\xi: P \to \mathbb{R}$ defined by $J_\xi(p) = \langle J(p), \xi \rangle$ satisfies*

$$X_{J_\xi} = \xi_P.$$

Theorem 17.5.1 (Noether's theorem) *Let G act canonically on $(P, \{\cdot, \cdot\})$ with momentum map J. If H is G-invariant, then J is conserved by the flow of X_H.*

Proof. For every $\xi \in \mathfrak{g}$,

$$\mathcal{L}_{X_H} J_\xi = \{J_\xi, H\} = -\{H, J_\xi\} = -\mathcal{L}_{X_{J_\xi}} H = -\mathcal{L}_{\xi_P} H = 0.$$

■

Exercise. Show that momentum maps are unique up to a choice of a constant element of $\mathfrak{g}^*$ on every connected component of M. ★

Exercise. Show that the S^1 action on the torus $T^2 :=$ $S^1 \times S^1$ given by $\alpha(\theta, \phi) = (\alpha + \theta, \phi)$ is canonical with respect to the classical bracket (with θ, ϕ in place of q, p), but it does not have a momentum map. ★

Exercise. Show that the Petzval invariant for Fermat's principle in axisymmetric, translation-invariant media is a momentum map, $T^*\mathbb{R}^2 \mapsto sp(2, \mathbb{R})^*$ taking $(\mathbf{q}, \mathbf{p}) \mapsto$ (X, Y, Z). What is its corresponding symmetry? What is its Hamiltonian vector field? ★

Theorem 17.5.2 (Also due to Noether) *Let G act on Q and by cotangent lifts on T^*Q. Then, $J: T^*Q \to \mathfrak{g}^*$ defined by, for every $\xi \in \mathfrak{g}$,*

$$J_\xi(\alpha_q) = \langle \alpha_q, \xi_Q(q) \rangle, \text{ for every } \alpha_q \in T_q^*Q,$$

is a momentum map (the "standard one") for the G action with respect to the classical Poisson bracket.

Proof. We need to show that $X_{J_\xi} = \xi_{T^*Q}$, for every $\xi \in \mathfrak{g}$. From the definition of Hamiltonian vector fields, this is equivalent to showing that $\xi_{T^*Q}[F] = \{F, J_\xi\}$ for every $F \in \mathcal{F}(T^*Q)$. We verify this for finite-dimensional Q by using cotangent-lifted local coordinates:

$$\frac{\partial J_\xi}{\partial p}(q, p) = \xi_Q(q)$$

$$\frac{\partial J_\xi}{\partial q^i}(q, p) = \left\langle p, \frac{\partial}{\partial q^i}(\xi_Q(q)) \right\rangle$$

$$= \left\langle p, \frac{\partial}{\partial q^i} \left(\frac{\partial}{\partial t} \Phi_{(\exp(t\xi))}(q) \Big|_{t=0} \right) \right\rangle$$

$$= \left\langle p, \frac{\partial}{\partial t} \left(\frac{\partial}{\partial q^i} \Phi_{(\exp(t\xi))}(q) \right) \Big|_{t=0} \right\rangle$$

$$= \frac{\partial}{\partial t} \left\langle p, T\Phi_{(\exp(t\xi))} \frac{\partial}{\partial q^i}(q) \right\rangle \Big|_{t=0}$$

$$= \frac{\partial}{\partial t} \left\langle T^*\Phi_{(\exp(t\xi))}p, \frac{\partial}{\partial q^i}(q) \right\rangle \Big|_{t=0}$$

$$= \left\langle -\xi_{T^*Q}(q, p), \frac{\partial}{\partial q^i}(q) \right\rangle$$

$$\frac{\partial J_\xi}{\partial q}(q, p) = -\xi_{T^*Q}(q, p).$$

So, for every $F \in \mathcal{F}(T^*Q)$,

$$\xi_{T^*Q}[F] = \frac{\partial}{\partial t} F(\exp(t\xi)q, \exp(t\xi)p) \Big|_{t=0}$$

$$= \frac{\partial F}{\partial q} \xi_Q(q) + \frac{\partial F}{\partial p} \xi_{T^*Q}(q, p)$$

$$= \frac{\partial F}{\partial q} \frac{\partial J_\xi}{\partial p} - \frac{\partial F}{\partial p} \frac{\partial J_\xi}{\partial q} = \{F, J_\xi\},$$

which completes the proof. ∎

Example 17.5.1 Let $G \subset M_n(\mathbb{R})$ be a matrix group, with cotangent-lifted action on $(q, p) \in T^*\mathbb{R}^n$. For every $g \subset M_n(\mathbb{R})$, $q \mapsto gq$.

The cotangent-lifted action is $(q, p) \mapsto (gq, g^{-T}p)$. Thus, writing $g = \exp(t\xi)$, the linearisation of this group action yields the vector field

$$X_\xi = (\xi q, -\xi^T p).$$

The corresponding Hamiltonian equations read

$$\xi q = \frac{\partial J_\xi}{\partial p}, \quad -\xi^T p = -\frac{\partial J_\xi}{\partial q}.$$

This yields the momentum map $J(q, p)$ given by

$$J_\xi(q, p) = \langle J(q, p), \xi \rangle = p^T \xi_Q(q) = p^T \xi q.$$

In coordinates, $p^T \xi q = p_i \xi^i_j q^j$, so $J(q, p) = q^i p_j$.

Exercise. Calculate the momentum map of the cotangent-lifted action of the group of translations of $\mathbb{R}^3$. ★

Answer. The element $\mathbf{x} \in \mathbb{R}^3$ acts on $\mathbf{q} \in \mathbb{R}^3$ by addition of vectors,

$$\mathbf{x} \cdot (\mathbf{q}) = \mathbf{q} + \mathbf{x}.$$

The infinitesimal generator is $\lim_{\mathbf{x} \to 0} \frac{d}{d\mathbf{x}}(\mathbf{q} + \mathbf{x}) = \text{Id}$. Thus, $\xi_\mathbf{q} = \text{Id}$ and

$$\langle J_k, \xi \rangle = \langle (\mathbf{q}, \mathbf{p}), \xi_\mathbf{q} \rangle = \langle \mathbf{p}, \text{Id} \rangle = p_i \delta^i_k = p_k$$

This is also Hamiltonian with $J_\xi = \mathbf{p}$, so that $\{\mathbf{p}, J_\xi\} = 0$ and $\{\mathbf{q}, J_\xi\} = \text{Id}$. ▲

Example 17.5.2 Let G act on itself by left multiplication and by cotangent lifts on T^*G. We first note that the infinitesimal action on G is

$$\xi_G(g) = \frac{d}{dt} \exp(t\xi)g \Big|_{t=0} = TR_g \xi.$$

Let J_L be the momentum map for this action. Let $\alpha_g \in T_g^*G$. For every $\xi \in \mathfrak{g}$, we have

$$\langle J_L(\alpha_g), \xi \rangle = \langle \alpha_g, \xi_G(g) \rangle = \langle \alpha_g, TR_g\xi \rangle = \langle TR_g^*\alpha_g, \xi \rangle,$$

so $J_L(\alpha_g) = TR_g^*\alpha_g$. Alternatively, writing $\alpha_g = T^*L_{g^{-1}}\mu$ for some $\mu \in \mathfrak{g}^*$, we have

$$J_L\left(T^*L_{g^{-1}}\mu\right) = TR_g^*T^*L_{g^{-1}}\mu = Ad_{g^{-1}}^*\mu.$$

Exercise. Show that the momentum map for the right multiplication action $R_g(h) = hg$ is $J_R(\alpha_g) = TL_g^*\alpha_g$.

★

For matrix groups, the tangent lift of the left (or right) multiplication action is again matrix multiplication. Indeed, to compute $TR_G(A)$ for any $A \in T_QSO(3)$, let $B(t)$ be a path in $SO(3)$ such that $B(0) = Q$ and $B'(0) = A$. Then,

$$TR_G(A) = \frac{d}{dt}B(t)G\bigg|_{t=0} = AG.$$

Similarly, $TL_G(A) = GA$. To compute the cotangent lift similarly, we need to be able to consider the elements of T^*G as matrices. This can be done using any non-degenerate bilinear form on each tangent space T_QG. We will use the Frobenius pairing, defined by

$$\langle\langle A, B \rangle\rangle := -\tfrac{1}{2}\mathrm{tr}\left(A^TB\right) = -\tfrac{1}{2}\mathrm{tr}\left(AB^T\right).$$

(The equivalence of the two formulas follows from the properties $\mathrm{tr}(CD) = \mathrm{tr}(DC)$ and $\mathrm{tr}(C^T) = \mathrm{tr}(C)$.)

Exercise. Check that this pairing, restricted to $\mathfrak{so}(3)$, corresponds to the Euclidean inner product via the hat map.

★

Example 17.5.3 Consider the previous example for a *matrix* group G. For any $Q \in G$, the pairing given above allows us to consider any element $P \in T_Q^* G$ as a matrix. The natural pairing of $T_Q^* G$ with $T_Q G$ now has the formula

$$< P, A >= -\tfrac{1}{2}\text{tr}(P^T A), \quad \text{for all } A \in T_Q G.$$

We compute the cotangent lifts of the left and right multiplication actions:

$$\langle T^* L_Q(P), A \rangle = \langle P, T L_Q(A) \rangle = \langle P, QA \rangle$$
$$= -\tfrac{1}{2}\text{tr}(P^T QA) = -\tfrac{1}{2}\text{tr}((Q^T P)^T A) = \langle Q^T P, A \rangle$$
$$\langle T^* R_Q(P), A \rangle = \langle P, T R_Q(A) \rangle =< P, AQ >$$
$$= -\tfrac{1}{2}\text{tr}(P(AQ)^T) = -\tfrac{1}{2}\text{tr}(PQ^T A^T) = \langle PQ^T, A \rangle.$$

In summary,

$$T^* L_Q(P) = Q^T P \quad \text{and} \quad T^* R_Q(P) = PQ^T.$$

We thus compute the momentum maps as

$$J_L(Q, P) = T^* R_Q P = PQ^T,$$
$$J_R(Q, P) = T^* L_Q P = Q^T P.$$

In the special case of $G = SO(3)$, these matrices PQ^T and $Q^T P$ are skew symmetric since they are elements of $\mathfrak{so}(3)$. Therefore,

$$J_L(Q, P) = T^* R_Q P = \tfrac{1}{2}(PQ^T - QP^T),$$
$$J_R(Q, P) = T^* L_Q P = \tfrac{1}{2}(Q^T P - P^T Q).$$

Exercise. Show that the cotangent-lifted action on $SO(n)$ can be expressed as a matrix multiplication:

$$Q \cdot P = Q^T P. \qquad \bigstar$$

17.6   Equivariance

Definition 17.6.1 (Equivariance of momentum maps) *A momentum map is said to be **equivariant** when it is equivariant with respect to the given action on P and the coadjoint action on $\mathfrak{g}^*$. That is,*

$$J(g \cdot p) = \mathrm{Ad}^*_{g^{-1}} J(p)$$

for every $g \in G$, $p \in P$, where $g \cdot p$ denotes the action of g on the point p and where Ad denotes the adjoint action.

Proposition 17.6.1 *All cotangent-lifted actions are Ad^*-equivariant.*

Proposition 17.6.2 *Every Ad^*-equivariant momentum map $J: P \to \mathfrak{g}^*$ is a Poisson map, with respect to the "+" Lie–Poisson bracket on $\mathfrak{g}^*$.*

Exercise. Prove the previous two propositions.   ★

Exercise. Show that the momentum map derived from the cotangent theorem 17.5.2 is equivariant.   ★

Example 17.6.1 (Momentum map for symplectic representations) Let (V, Ω) be a symplectic vector space, and let G be a Lie group acting linearly and symplectically on V. This action admits an equivariant momentum map $\mathbf{J}: V \to \mathfrak{g}$ given by

$$J^\xi(v) = \langle \mathbf{J}(v), \xi \rangle = \tfrac{1}{2}\Omega(\xi \cdot v, v),$$

where $\xi \cdot v$ denotes the Lie algebra representation of the element $\xi \in \mathfrak{g}$ acting on the vector $v \in V$. To verify this, note that the infinitesimal generator $\xi_V(v) = \xi \cdot v$, by the definition of the Lie algebra representation induced by the given Lie group representation and that $\Omega(\xi \cdot u, v) = -\Omega(u, \xi \cdot v)$ for all $u, v \in V$. Therefore,

$$\mathbf{d}J^{\xi}(u)(v) = \tfrac{1}{2}\Omega(\xi \cdot u, v) + \tfrac{1}{2}\Omega(\xi \cdot v, u) = \Omega(\xi \cdot u, v).$$

The equivariance of $\mathbf{J}$ follows from the obvious relation $g^{-1}\cdot\xi\cdot g\cdot v = (\mathrm{Ad}_{g^{-1}}\xi) \cdot v$ for any $g \in G, \xi \in \mathfrak{g}$, and $v \in V$.

Example 17.6.2 (Cayley–Klein parameters and the Hopf fibration) Consider the natural action of $SU(2)$ on $\mathbb{C}^2$. Since this action is through isometries of the Hermitian metric, it is automatically symplectic and therefore has a momentum map $\mathbf{J}: \mathbb{C}^2 \to \mathfrak{su}(2)^*$ given in Example 17.6.1, that is,

$$\langle \mathbf{J}(z, w), \xi \rangle = \tfrac{1}{2}\Omega(\xi \cdot (z, w), (z, w)),$$

where $z, w \in \mathbb{C}$ and $\xi \in \mathfrak{su}(2)$. Now, the symplectic form on $\mathbb{C}^2$ is given by minus the imaginary part of the Hermitian inner product. That is, $\mathbb{C}^n$ has a Hermitian inner product given by

$$\mathbf{z} \cdot \mathbf{w} := \sum_{j=1}^{n} z_j \overline{w}_j,$$

where $\mathbf{z} = (z_1, \ldots, z_n)$, and $\mathbf{w} = (w_1, \ldots, w_n) \in \mathbb{C}^n$.

The symplectic form is thus given by $\Omega(\mathbf{z}, \mathbf{w}) := -\operatorname{Im}(\mathbf{z}\cdot\mathbf{w})$, and it is identical to the one given before on $\mathbb{R}^{2n}$ by identifying $\mathbf{z} = \mathbf{u} + i\mathbf{v} \in \mathbb{C}^n$ with $(\mathbf{u}, \mathbf{v}) \in \mathbb{R}^{2n}$ and $\mathbf{w} = \mathbf{u}' + i\mathbf{v}' \in \mathbb{C}^n$ with $(\mathbf{u}', \mathbf{v}') \in \mathbb{R}^{2n}$.

The Lie algebra $\mathfrak{su}(2)$ of $SU(2)$ consists of 2×2 skew Hermitian matrices of trace zero. This Lie algebra is isomorphic to $\mathfrak{so}(3)$ and, therefore, to $(\mathbb{R}^3, \times)$ by the isomorphism given by the **tilde map**,

$$\mathbf{x} = (x^1, x^2, x^3) \in \mathbb{R}^3 \quad \mapsto$$

$$\tilde{\mathbf{x}} := \tfrac{1}{2}\begin{bmatrix} -ix^3 & -ix^1 - x^2 \\ -ix^1 + x^2 & ix^3 \end{bmatrix} \in \mathfrak{su}(2).$$

Thus, we have $[\tilde{\mathbf{x}}, \tilde{\mathbf{y}}] = (\mathbf{x} \times \mathbf{y})\tilde{}$ for any $\mathbf{x}, \mathbf{y} \in \mathbb{R}^3$. Other useful relations are $\det(2\tilde{\mathbf{x}}) = \|\mathbf{x}\|^2$ and $\operatorname{trace}(\tilde{\mathbf{x}}\tilde{\mathbf{y}}) = -\frac{1}{2}\mathbf{x} \cdot \mathbf{y}$.

Identify $\mathfrak{su}(2)^*$ with $\mathbb{R}^3$ by the **check map** $(\check{\cdot})\colon \mu \in \mathfrak{su}(2)^* \mapsto \check{\mu} \in \mathbb{R}^3$, defined by

$$\check{\mu} \cdot \mathbf{x} := -2\langle \mu, \tilde{\mathbf{x}} \rangle$$

for any $\mathbf{x} \in \mathbb{R}^3$. With these notations, the momentum map $\check{\mathbf{J}}\colon \mathbb{C}^2 \to \mathbb{R}^3$ can be explicitly computed in coordinates: for any $\mathbf{x} \in \mathbb{R}^3$, we have

$$\check{\mathbf{J}}(z, w) \cdot \mathbf{x} = -2\langle \mathbf{J}(z, w), \tilde{\mathbf{x}} \rangle$$

$$= \tfrac{1}{2} \operatorname{Im} \left(\begin{bmatrix} -ix^3 & -ix^1 - x^2 \\ -ix^1 + x^2 & ix^3 \end{bmatrix} \begin{bmatrix} z \\ w \end{bmatrix} \cdot \begin{bmatrix} z \\ w \end{bmatrix} \right)$$

$$= -\tfrac{1}{2}(2\operatorname{Re}(w\bar{z}), 2\operatorname{Im}(w\bar{z}), |z|^2 - |w|^2) \cdot \mathbf{x}.$$

Therefore,

$$\check{\mathbf{J}}(z, w) = -\tfrac{1}{2}(2w\bar{z}, |z|^2 - |w|^2) \in \mathbb{R}^3.$$

Thus, $\check{\mathbf{J}}$ is a Poisson map from $\mathbb{C}^2$, endowed with the canonical symplectic structure, to $\mathbb{R}^3$, endowed with the $+$ Lie–Poisson structure.

Therefore, $-\check{\mathbf{J}}\colon \mathbb{C}^2 \to \mathbb{R}^3$ is a canonical map if $\mathbb{R}^3$ has the $-$ Lie–Poisson bracket relative to which the free rigid-body equations are Hamiltonian.

Pulling back the Hamiltonian $H(\boldsymbol{\Pi}) = \boldsymbol{\Pi} \cdot \mathbb{I}^{-1}\boldsymbol{\Pi}/2$ to $\mathbb{C}^2$ gives a Hamiltonian function (called collective) on $\mathbb{C}^2$.

The classical Hamilton equations for this function are therefore projected by $-\check{\mathbf{J}}$ to the rigid-body equations $\dot{\boldsymbol{\Pi}} = \boldsymbol{\Pi} \times \mathbb{I}^{-1}\boldsymbol{\Pi}$.

In this context, the variables (z, w) are called the **Cayley–Klein parameters**.

Exercise. Show that $-\check{\mathbf{J}}|_{S^3}\colon S^3 \to S^2$ is the *Hopf fibration*. In other words, the momentum map of the $SU(2)$-action on $\mathbb{C}^2$, the Cayley–Klein parameters and the family of Hopf fibrations on concentric three-spheres in $\mathbb{C}^2$ are all the same map. ★

Exercise. (Optical traveling wave pulses) The equation for the evolution of the complex amplitude of a polarised optical travelling wave pulse in a material medium is given as

$$\dot{z}_i = \frac{1}{\sqrt{-1}} \frac{\partial H}{\partial z_i^*}$$

with the Hamiltonian $H \colon \mathbb{C}^2 \to \mathbb{R}$ defined by

$$H = z_i^* \chi_{ij}^{(1)} z_j + 3 z_i^* z_j^* \chi_{ijkl}^{(3)} z_k z_l,$$

and the constant complex tensor coefficients $\chi_{ij}^{(1)}$ and $\chi_{ijkl}^{(1)}$ have the proper Hermitian and permutation symmetries for H to be real. Define the Stokes vectors by the isomorphism

$$\mathbf{u} = (u^1, u^2, u^3) \in \mathbb{R}^3 \mapsto$$

$$\tilde{\mathbf{u}} := \frac{1}{2} \begin{bmatrix} -iu^3 & -iu^1 - u^2 \\ -iu^1 + u^2 & iu^3 \end{bmatrix} \in \mathfrak{su}(2).$$

1. Prove that this isomorphism is an equivariant momentum map.

2. Deduce the equations of motion for the Stokes vectors of this optical travelling wave, and write it as a Lie–Poisson Hamiltonian system.

3. Determine how this system is related to the equations for an $SO(3)$ rigid body. ★

Exercise. The formula determining the momentum map for the cotangent-lifted action of a Lie group G on a smooth manifold Q may be expressed in terms of the pairing $\langle \cdot, \cdot \rangle \colon \mathfrak{g}^* \times \mathfrak{g} \mapsto \mathbb{R}$ as

$$\langle J, \xi \rangle = \langle p, \pounds_\xi q \rangle,$$

where $(q, p) \in T_q^* Q$ and $\pounds_\xi q$ is the infinitesimal generator of the action of the Lie algebra element ξ on the coordinate q.

Define appropriate pairings and determine the momentum maps explicitly for the following actions:

1. $\pounds_\xi q = \xi \times q$ for $\mathbb{R}^3 \times \mathbb{R}^3 \mapsto \mathbb{R}^3$.

2. $\pounds_\xi q = \mathrm{ad}_\xi\, q$ for ad-action $\mathrm{ad} \colon \mathfrak{g} \times \mathfrak{g} \mapsto \mathfrak{g}$ in a Lie algebra $\mathfrak{g}$.

3. AqA^{-1} for $A \in GL(3, R)$ acting on $q \in GL(3, R)$ by matrix conjugation.

4. Aq for left action of $A \in SO(3)$ on $q \in SO(3)$.

5. AqA^T for $A \in GL(3, R)$ acting on $q \in \mathrm{Sym}(3)$, that is $q = q^T$.   ★

Answer.

1. $p \cdot \xi \times q = q \times p \cdot \xi \Rightarrow J = q \times p$. (The pairing is scalar product of vectors.)

2. $\langle p, \text{ad}_\xi \, q \rangle = -\langle \text{ad}_q^* \, p, \xi \rangle \Rightarrow J = \text{ad}_q^* \, p$ for the pairing $\langle \cdot, \cdot \rangle \colon \mathfrak{g}^* \times \mathfrak{g} \mapsto \mathbb{R}$.

3. Compute $T_e(AqA^{-1}) = \xi q - q\xi = [\xi, q]$ for $\xi = A'(0) \in gl(3, R)$ acting on $q \in GL(3, R)$ by the matrix Lie bracket $[\cdot, \cdot]$. For the matrix pairing $\langle A, B \rangle = \text{tr}(A^T B)$, we have

$$\text{tr}\big(p^T[\xi, q]\big) = \text{tr}\big((pq^T - q^T p)^T \xi\big) \Rightarrow J = pq^T - q^T p.$$

4. Compute $T_e(Aq) = \xi q$ for $\xi = A'(0) \in \mathfrak{so}(3)$ acting on $q \in SO(3)$ by left matrix multiplication. For the matrix pairing $\langle A, B \rangle = \text{tr}(A^T B)$, we have

$$\text{tr}\big(p^T \xi q\big) = \text{tr}\big((pq^T)^T \xi\big) \Rightarrow J = \tfrac{1}{2}\big(pq^T - q^T p\big),$$

where we have used the antisymmetry of the matrix $\xi \in \mathfrak{so}(3)$.

5. Compute $T_e\big(AqA^T\big) = \xi q + q\xi^T$ for $\xi = A'(0) \in gl(3, R)$ acting on $q \in \text{Sym}(3)$. For the matrix pairing $\langle A, B \rangle = \text{tr}(A^T B)$, we have

$$\text{tr}\big(p^T \big(\xi q + q\xi^T\big)\big) = \text{tr}\big(q\big(p^T + p\big)\xi\big)$$
$$= \text{tr}\big((2qp)^T \xi\big) \Rightarrow J = 2qp,$$

where we have used the symmetry of the matrix $\xi q + q\xi^T$ to choose $p = p^T$. (The momentum canonical to the symmetric matrix $q = q^T$ should be symmetric to have the correct number of components!)

▲

18

HAMILTONIAN VECTOR FIELDS AND DIFFERENTIAL FORMS

Contents

What is this lecture about? This lecture discusses the coordinate-free approach to geometric mechanics in the Lagrangian and Hamiltonian formulations.

18.1   Hamilton's principle for Euler–Lagrange and Hamilton equations

The Euler–Lagrange equations for the Lagrangian $L(q, \dot{q}) : TQ \to \mathbb{R}$ follow from **Hamilton's principle** as

$$
\begin{aligned}
0 = \delta S = \delta \int_0^T & L(q, v_q) + \left\langle p, \frac{dq}{dt} - v_q \right\rangle dt \\
= \int_0^T & \left\langle \frac{\partial L}{\partial q} - \frac{dp}{dt}, \delta q \right\rangle + \left\langle \frac{\partial L}{\partial v_q} - p, \delta v_q \right\rangle \\
& + \left\langle \delta p, \frac{dq}{dt} - v_q \right\rangle + \underbrace{\langle p, \delta q \rangle \Big|_0^T}_{\text{Noether term}}.
\end{aligned}
\tag{18.1.1}
$$

Upon assuming that δq vanishes at the endpoints in time, collecting terms in (18.1.1) yields the *Euler–Lagrange equations* (2.2.4).

To pass to the Hamiltonian side in canonical position–momentum (q, p) coordinates, we Legendre-transform to write the Lagrangian in (18.6.4) in phase-space form as

$$
\begin{aligned}
0 = \delta S = \delta \int_0^T & \langle p, v_q \rangle - H(p, q) + \left\langle p, \frac{dq}{dt} - v_q \right\rangle dt \\
= \delta \int_0^T & \left\langle p, \frac{dq}{dt} \right\rangle - H(p, q) \, dt \\
= \int_0^T & \left\langle \delta p, \frac{dq}{dt} - \frac{\partial H}{\partial p} \right\rangle - \left\langle \frac{dp}{dt} + \frac{\partial H}{\partial q}, \delta q \right\rangle + \underbrace{\langle p, \delta q \rangle \Big|_0^T}_{\text{Noether term}}.
\end{aligned}
$$

$$\tag{18.1.2}$$

Variations of the Legendre-transformed phase-space Lagrangian δq and δp produce *Hamilton's canonical equations* and the *same* Noether term as before. Hamilton's equations lead to a vector field X_H on phase space, and we shall see that X_H generates symplectic transformations, i.e., transformations that preserve the area in phase space. The Noether term leads to Noether's theorem on the Hamiltonian side.

18.2 The differential

The *differential*, or *exterior derivative*, of a function F on phase space with coordinates $(q, p) \in T^*Q$ is written

$$dF = F_q dq + F_p dp,$$

in which subscripts denote partial derivatives. For the Hamiltonian itself, the exterior derivative yields

$$dH = H_q dq + H_p dp = -\frac{dp}{dt} dq + \frac{dq}{dt} dp,$$

upon inserting the canonical equations

$$\frac{dq}{dt} = H_p \quad \text{and} \quad \frac{dp}{dt} = -H_q.$$

The canonical equations also provide the definition of a Hamiltonian vector field (HVF)

$$\frac{d}{dt} = \frac{dq}{dt} \partial_q + \frac{dp}{dt} \partial_p = H_p \partial_q - H_q \partial_p =: \{\, \cdot\, , H\} := X_H.$$

This means that the action of the HVF X_H on a phase-space function F yields its time derivative

$$\frac{d}{dt} F(q, p) = X_H F(q, p) = \{F, H\}.$$

The HVF X_H may also act as a time derivative on differential forms defined on phase space. For example, X_H acts on the time-dependent 1-form $p\, dq(t)$ along the solutions of Hamilton's equations as

$$X_H (p\, dq) = \frac{d}{dt} (p\, dq) = \dot{p}\, dq + p\, d\dot{q}$$

$$= \dot{p}\, dq - \dot{q}\, dp + d(p\dot{q}) \qquad (18.2.1)$$

$$= -H_q dq - H_p dp + d(p\dot{q})$$

$$= d(-H + p\dot{q}) =: dL(q, \dot{q}),$$

upon substituting Hamilton's canonical equations and applying the Legendre transformation.

Exercise. Show that the HVF X_H commutes with its differential, or exterior derivative. Thus,

$$d(X_H F) = X_H(dF).$$ ★

The exterior derivative of the 1-form pdq yields the canonical, or symplectic 2-form[1]

$$d(pdq) = dp \wedge dq = -dq \wedge dp. \qquad (18.2.2)$$

Here, we have used the chain rule for the exterior derivative and its property that $d^2 = 0$.

Exercise. Show that the property $d^2 = 0$ for the differential amounts to equality of cross derivatives for continuous functions.   ★

Equation (18.2.2) introduces the wedge product $\wedge$, which combines two 1-forms (the line elements dq and dp) into a 2-form (the oriented surface element $dp \wedge dq = -dq \wedge dp$). As a result, the 2-form $w = dq \wedge dp$ representing area in phase space is *conserved* along the Hamiltonian flows generated by the HVFs:

$$X_H(dq \wedge dp) = \frac{d}{dt}(dq \wedge dp) = 0.$$

Proof. The calculation in (18.2.1) implies that, for $dp \wedge dq = -dq \wedge dp$,

$$d\big(X_H(p\,dq)\big) = \frac{d}{dt}(dp \wedge dq) = X_H(dp \wedge dq) = d^2 L(q, \dot{q}) = 0.$$

---

[1]The properties of differential forms are summarised in Section 18.3 and developed for PDEs in Lecture [19].

Consequently, one finds the following.   ■

Theorem 18.2.1 (Poincaré's theorem) *Hamiltonian flows preserve the area in phase space.*

Definition 18.2.1 (Symplectic two-form) *The   phase-space   area* $\omega = dq \wedge dp$ *is called the symplectic 2-form.*

Definition 18.2.2 (Symplectic flows) *Flows that preserve area in phase space are said to be **symplectic**.*

Remark 18.2.1 (Poincaré's theorem) Hamiltonian flows are symplectic.   □

18.3   Short review of exterior calculus, symplectic forms and Poincaré's theorem in higher dimensions

Exterior calculus on symplectic manifolds is the geometric language of Hamiltonian mechanics. As an introduction and motivation for more detailed studies, we begin with an overview.

In differential geometry, the operation of *contraction* (denoted equivalently as either $\iota_X \alpha$ or $X \lrcorner \alpha$) introduces a pairing between vector fields $X \in \mathfrak{X}(M)$ and differential forms $\alpha \in \Lambda(M)$.[2]

Contraction is also called *substitution*, or *insertion*, of a vector field into a differential form. For example, there are the dual relations

$$\partial_q \lrcorner dq = 1 = \partial_p \lrcorner dp, \quad \text{and} \quad \partial_q \lrcorner dp = 0 = \partial_p \lrcorner dq.$$

---

[2]The two notations ι for insertion and $\lrcorner$ for contraction can be used interchangeably. In this text, we will choose the notation that seems most convenient for enhancing the legibility of any given presentation. In particular, we will tend to favour $\lrcorner$ in the presence of subscripts and superscripts for general bases of vector fields and k-forms.

A HVF

$$X_H = \dot{q}\frac{\partial}{\partial q} + \dot{p}\frac{\partial}{\partial p} = H_p\partial_q - H_q\partial_p = \{\cdot, H\}$$

satisfies

$$X_H \lrcorner dq = H_p \quad \text{and} \quad X_H \lrcorner dp = -H_q.$$

Definition 18.3.1 *The rule for the contraction operation, ι or $\lrcorner$, also known as substitution of a vector field into a differential form, is to sum the substitutions of X_H over the permutations of the factors in the differential form that bring the corresponding dual basis element into its leftmost position.*

For example, substitution of the HVF X_H into the symplectic form $\omega = dq \wedge dp$ yields

$$X_H \lrcorner \omega = X_H \lrcorner (dq \wedge dp) = (X_H \lrcorner dq)\, dp - (X_H \lrcorner dp)\, dq.$$

In this example, $X_H \lrcorner dq = H_p$ and $X_H \lrcorner dp = -H_q$, so

$$X_H \lrcorner \omega = H_p dp + H_q dq = dH,$$

which follows because $\partial_q \lrcorner dq = 1 = \partial_p \lrcorner dp$ and $\partial_q \lrcorner dp = 0 = \partial_p \lrcorner dq$.

This calculation proves the following.

Theorem 18.3.1 (Hamiltonian vector field) *The HVF $X_H = \{\cdot, H\}$ satisfies*

$$X_H \lrcorner \omega = dH \quad \text{with} \quad \omega = dq \wedge dp. \tag{18.3.1}$$

In fact, relation (18.3.1) may be taken as the *definition* of a HVF.

As a consequence of formula (18.3.1), the flow of X_H preserves the closed exact 2-form ω for any Hamiltonian H. This preservation may be verified by a formal calculation using (18.3.1).

Along $(dq/dt, dp/dt) = (\dot{q}, \dot{p}) = (H_p, -H_q)$, we have

$$\frac{d\omega}{dt} = d\dot{q} \wedge dp + dq \wedge d\dot{p} = dH_p \wedge dp - dq \wedge dH_q$$
$$= d(H_p\, dp + H_q\, dq) = d(X_H \lrcorner \omega) = d(dH) = 0.$$

The first step here uses the chain rule for differential forms, and the third and last steps use the property of the exterior derivative d that $d^2 = 0$ for continuous forms. The latter is due to equality of cross derivatives $H_{pq} = H_{qp}$ and antisymmetry of the wedge product: $dq \wedge dp = -dp \wedge dq$.

Consequently, the relation $d(X_H \lrcorner \omega) = d^2 H = 0$ for HVFs shows the following.

Theorem 18.3.2 (Poincaré's theorem for one degree of freedom) *The flow of a HVF is symplectic, which means it preserves the phase-space area, or two-form, $\omega = dq \wedge dp$.*

Definition 18.3.2 (Cartan's formula for the Lie derivative) *The operation of **Lie derivative** of a differential form ω by a vector field X_H is defined by*

$$\pounds_{X_H}\omega := d(X_H \lrcorner \omega) + X_H \lrcorner d\omega. \qquad (18.3.2)$$

Corollary 18.3.1 *Because $d\omega = 0$, the symplectic property $d\omega/dt = d(X_H \lrcorner \omega) = 0$ in Poincaré's theorem 18.3.2 may be rewritten using Lie derivative notation as*

$$\frac{d\omega}{dt} = \pounds_{X_H}\omega := d(X_H \lrcorner \omega) + X_H \lrcorner d\omega = 0. \qquad (18.3.3)$$

Remark 18.3.1 Relation (18.3.3) associates Hamiltonian dynamics with the symplectic flow in phase space of the HVF:

$$X_H = \{\cdot, H\} = \frac{\partial H}{\partial p}\frac{\partial}{\partial q} - \frac{\partial H}{\partial q}\frac{\partial}{\partial p}, \qquad (18.3.4)$$

which is divergence-less with respect to the symplectic form $\omega = dq \wedge dp$. That is, $d(X_H \lrcorner \omega) = 0$.

The Lie derivative operation defined in relation (18.3.3) is equivalent to the time derivative along the characteristic paths (flow) of the first-order linear partial differential operator X_H, which are obtained from its characteristic equations in (18.3.4). This is the *dynamical meaning* of the Lie derivative $\pounds_{X_H}$ in (18.3.2), for which invariance $\pounds_{X_H}\omega = 0$ gives the geometric definition of symplectic flows in phase space.   □

Theorem 18.3.3 (Poincaré's theorem for N degrees of freedom)
For a system of N particles, or N degrees of freedom, the flow of a HVF preserves each subvolume in the phase space $T^*\mathbb{R}^N$. That is, let $\omega_n \equiv dq_n \wedge dp_n$ be the symplectic form expressed in terms of the position and momentum of the nth particle. Then

$$\frac{d\omega_M}{dt} = 0, \quad for \; \omega_M = \Pi_{n=1}^{M}\omega_n, \quad \forall M \leq N.$$

The proof of the preservation of these *Poincaré invariants* ω_M with $M = 1, 2, \ldots, N$ follows the same pattern as the verification above for a single degree of freedom. Basically, this is because each factor $\omega_n = dq_n \wedge dp_n$ in the wedge product of symplectic forms is preserved by its corresponding Hamiltonian flow in the sum

$$X_H = \sum_{n=1}^{M}\left(\dot{q}_n\frac{\partial}{\partial q_n} + \dot{p}_n\frac{\partial}{\partial p_n}\right)$$

$$= \sum_{n=1}^{M}\left(H_{p_n}\partial_{q_n} - H_{q_n}\partial_{p_n}\right) = \sum_{n=1}^{M}X_{H_n} = \{\cdot, H\}.$$

That is, $\pounds_{X_{H_n}}\omega_M$ vanishes for each term in the sum

$$\pounds_{X_H}\omega_M = \sum_{n=1}^{M}\pounds_{X_{H_n}}\omega_M$$

since $\partial_{q_m} \lrcorner dq_n = \delta_{mn} = \partial_{p_m} \lrcorner dp_n$ and $\partial_{q_m} \lrcorner dp_n = 0 = \partial_{p_m} \lrcorner dq_n$.

18.4   Quick summary of coordinate-free formulas for differential forms

Theorem 18.4.1 *The pull-back ϕ_t^* of a smooth flow ϕ_t generated by the action of a smooth vector field X on a smooth manifold M commutes with the exterior derivative d, wedge product $\wedge$ and contraction $\lrcorner$.*

In other words, for k-forms α, $\beta \in \Lambda^k(M)$, for each point $m \in M$, the pull-back ϕ_t^* satisfies

$$
\begin{aligned}
d(\phi_t^* \alpha) &= \phi_t^* d\alpha, \\
\phi_t^*(\alpha \wedge \beta) &= \phi_t^* \alpha \wedge \phi_t^* \beta, \\
\phi_t^*(X \lrcorner \alpha) &= \phi_t^* X \lrcorner \phi_t^* \alpha.
\end{aligned}
$$

Thus, the exterior derivative d, wedge product $\wedge$ and contraction $\lrcorner$ are said to be *natural* under pull-back by a smooth flow generated by a smooth vector field.

In addition, the *Lie derivative* $\pounds_X \alpha$ of a k-form $\alpha \in \Lambda^k(M)$ by the vector field X tangent to the flow ϕ_t on M is defined either dynamically or geometrically (by Cartan's formula) as

$$
\pounds_X \alpha = \frac{d}{dt}\bigg|_{t=0} (\phi_t^* \alpha) = X \lrcorner d\alpha + d(X \lrcorner \alpha), \qquad (18.4.1)
$$

in which the last equality is *Cartan's geometric formula* for the Lie derivative introduced in (18.3.2).

Definition 18.4.1 (The Lie chain rule) *The tangent to the pull-back $\phi_t^* \alpha$ of a differential k-form $\alpha \in \Lambda^k$ is the pull-back of the Lie derivative of α with respect to the vector field (e.g., $X = \dot{\phi}_t \phi_t^{-1}$ for right invariance) that generates the flow ϕ_t:*

$$
\frac{d}{dt}(\phi_t^* \alpha) = \phi_t^*(\pounds_X \alpha).
$$

Likewise, for the push-forward, which is the pull-back by the inverse, we have

$$
\frac{d}{dt}((\phi_t^{-1})^* \alpha) = -(\phi_t^{-1})^*(\pounds_X \alpha).
$$

Definition 18.4.2 (Advected quantity) *An advected quantity is invariant along a flow trajectory. Hence, advected quantities satisfy*

$$\alpha_0(x_0) = \alpha_t(x_t) = (\phi_t^* \alpha_t)(x_0), \text{ or equivalently,}$$

$$\alpha_t(x_t) = (\alpha_0 \circ \phi_t^{-1})(x_t) = ((\phi_t)_* \alpha_0)(x_t).$$

The dynamics of an advected quantity is given by the Lie chain rule for the push-forward as

$$\frac{d}{dt}\alpha_t(x_t) = \frac{d}{dt}(\phi_t)_*\alpha_0 = -\pounds_X \alpha_t. \tag{18.4.2}$$

The Lie chain rule implies the same advection dynamics in terms of the pull-back:

$$0 = \frac{d}{dt}\alpha_0(x_0) = \frac{d}{dt}(\phi_t^* \alpha_t)(x_0) = \phi_t^*(\partial_t + \pounds_X)\alpha_t(x_0)$$

$$= (\partial_t + \pounds_X)\alpha_t(x_t).$$

Remark 18.4.1 These formulas will enable us later to write a coordinate-free formulation of ideal fluid mechanics with advected quantities by introducing the advection dynamics as a constraint on Hamilton's principle for fluids [HoMaRa1998a].   □

18.5   Quick summary of coordinate-free Hamiltonian mechanics

A coordinate-free definition of the Poisson bracket can be formulated in terms of the operations of the *differential* (or exterior derivative), *insertion* or *contraction*, and *Lie derivative*, denoted respectively as

$$d : \to \Lambda^{k+1}$$

$$\lrcorner \ (\text{or } \iota) : \to \Lambda^{k-1}$$

$$\pounds_X = d(\iota_X) + \iota_X d : \to \Lambda^k,$$

where k is an even number for the cotangent bundle T^*M of a manifold M. For example, in two dimensions, insertion of the vector field $X = X^j\partial_j = X^1\partial_1 + X^2\partial_2$ into the two-form $\alpha = \alpha_{jk}dx^j \wedge dx^k$ with $\alpha_{21} = -\alpha_{12}$ yields

$$X \lrcorner \alpha = \iota_X\alpha = X^j\alpha_{ji_2}dx^{i_2} = X^1\alpha_{12}dx^2 + X^2\alpha_{21}dx^1$$
$$= \alpha_{12}(X^1dx^2 - X^2dx^1).$$

Consequently, the Poisson bracket can be defined by the insertion of HVFs

$$X_F := \{\,\cdot\,, F\} \quad \text{and} \quad X_H := \{\,\cdot\,, H\,\}$$

into the closed symplectic two-form $\omega \in \Lambda^2(T^*M)$ with $d\omega = 0$ as

$$\{F, H\} = \iota_{X_H}\big(\iota_{X_F}\omega(\,\cdot\,,\,\cdot\,)\big) = \omega(X_F, X_H) = X_H \lrcorner \big(X_F \lrcorner \omega(\,\cdot\,,\,\cdot\,)\big).$$

This formula is related to the original phase-space coordinates by

$$\omega = \sum_{i=1}^{2} dq_i \wedge dp_i = -\sum_{i=1}^{2} dp_i \wedge dq_i$$

and

$$X_H = \{\,\cdot\,, H\} = \sum_{i=1}^{2} \frac{\partial H}{\partial p_i}\frac{\partial}{\partial q_i} - \frac{\partial H}{\partial q_i}\frac{\partial}{\partial p_i}.$$

Note that the calculation of $\omega(X_F, X_H)$ in these coordinates yields

$$\{F, H\} = \iota_{X_H}\big(\iota_{X_F}\omega(\,\cdot\,,\,\cdot\,)\big) = \iota_{X_H}\omega(X_F,\,\cdot\,) = \iota_{X_H}dF = \omega(X_F, X_H).$$

Consequently, the dynamics along the integral curves of X_H are determined by

$$dH = \omega(X_H,\,\cdot\,) = \iota_{X_H}\omega,$$

in which the vector field X_H is inserted (ι) into the symplectic two-form ω to create the exact 1-form dH. In the original coordinates, this is

$$dH = \iota_{X_H}(dq \wedge dp) = \iota_{\left(\frac{\partial H}{\partial p}\frac{\partial}{\partial q} - \frac{\partial H}{\partial q}\frac{\partial}{\partial p}\right)}(dq \wedge dp) = \frac{\partial H}{\partial q}dq + \frac{\partial H}{\partial p}dp.$$

One can also check that $\omega(X_F, X_H) = \{F, H\}$ directly as

$$\frac{dF}{dt} = \omega(X_F, X_H)$$

$$= \iota_{X_H}(\iota_{X_F}\omega) = \iota_{X_H}dF$$

$$= \iota_{(\frac{\partial H}{\partial p}\frac{\partial}{\partial q} - \frac{\partial H}{\partial q}\frac{\partial}{\partial p})}\left(\frac{\partial F}{\partial q}dq + \frac{\partial F}{\partial p}dp\right)$$

$$= \frac{\partial H}{\partial p}\frac{\partial F}{\partial q} - \frac{\partial H}{\partial q}\frac{\partial F}{\partial p}$$

$$= \{F, H\}.$$

Definition 18.5.1 (Cartan's geometric definition of the Lie derivative) *The coordinate-free expression*

$$\pounds_{X_F}\omega = d(\iota_{X_F}\omega) + \iota_{X_F}d\omega = d(X_F \lrcorner \omega) + X_F \lrcorner d\omega$$

*is Cartan's geometric definition of the **Lie derivative** of the symplectic two-form ω with respect to the HVF $X_F = \{\cdot, F\}$.*

Lemma 18.5.1 *Since the symplectic form ω is closed ($d\omega = 0$) and $\iota_{X_F}\omega = dF$ for a HVF X_F, we have*

$$\pounds_{X_F}\omega = d(\iota_{X_F}\omega) = d^2 F = 0.$$

This means the symplectic form ω is locally invariant under the Lie algebra actions of HVFs.

Definition 18.5.2 (Symplectic flow) *The finite transformation ϕ_ϵ generated by the left-invariant HVF $X_F = \phi_\epsilon^{-1}\phi_\epsilon'|_{\epsilon=0}$ is called a **symplectic flow**.*

Theorem 18.5.1 (Symplectic flows preserve the symplectic two-form ω) *A smooth symplectic flow*

$$\phi_\epsilon^{X_F} := \exp(\epsilon X_F)$$

generated by a (time-independent) HVF X_F given by

$$X_F = \frac{d}{d\epsilon}\phi_\epsilon^{X_F}\Big|_{\epsilon=0} = \{\,\cdot\,, F\,\},$$

with $\iota_{X_F}\omega = dF$, preserves the symplectic two-form ω under pull-back by the flow $\phi_\epsilon^{X_F} = exp(\epsilon X_F)$, defined as

$$\phi_\epsilon^{X_F*}\omega(q, p) := \omega(\phi_\epsilon^{X_F}q, \phi_\epsilon^{X_F}p).$$

Proof.

$$\frac{d}{d\epsilon}(\phi_\epsilon^{X_F*}\omega) = \phi_\epsilon^{X_F*}(\pounds_{X_F}\omega) = \phi_\epsilon^{X_F*}(d(\iota_{X_F}\omega) + \iota_{X_F}d\omega)$$

$$= \phi_\epsilon^{X_F*}d(dF) = 0,$$

since $d\omega = 0$ and $d(\iota_{X_F}\omega) = d^2F = 0$. ∎

Remark 18.5.1 In the context of pull-back by smooth flows here, the first step of the proof uses the dynamic definition of the Lie derivative:

$$\pounds_X\omega = \frac{d}{d\epsilon}(\phi_\epsilon^*\omega)\Big|_{\epsilon=0} \quad \text{with } X = \phi_\epsilon^{-1}\phi_\epsilon'|_{\epsilon=0}.$$

In its second step, the proof uses the equivalence of the dynamic and Cartan definitions of the Lie derivative with respect to vector fields. □

Exercise. Demonstrate the equivalence of the dynamic and Cartan definitions of the Lie derivative $\pounds_X$ by calculating their actions on scalar functions. How is this result related to the familiar directional derivative of a scalar function? ★

18.6   Right-invariant symmetry-reduced Hamilton principle

The right-invariant form of the Euler–Poincaré equation (2.10.2) follows from a constrained Hamilton–Pontryagin variational principle. Namely,

$$
\begin{aligned}
0 = \delta S &= \delta \int_0^T \ell(\zeta) + \left\langle \mu, \dot{g}g^{-1} - \zeta \right\rangle dt \\
&= \int_0^T \left\langle \frac{\partial \ell}{\partial \zeta} - \mu, \delta\zeta \right\rangle + \left\langle \mu, \frac{d\varkappa}{dt} - \mathrm{ad}_\zeta \varkappa \right\rangle \\
&\quad + \left\langle \delta\mu, \dot{g}g^{-1} - \zeta \right\rangle dt \\
&= \int_0^T \left\langle \frac{\partial \ell}{\partial \zeta} - \mu, \delta\zeta \right\rangle - \left\langle \frac{d\mu}{dt} + \mathrm{ad}_\zeta^* \mu, \varkappa \right\rangle \\
&\quad + \left\langle \delta\mu, \dot{g}g^{-1} - \zeta \right\rangle dt + \underbrace{\left\langle \mu, \varkappa \right\rangle \big|_0^T}_{\langle \mu, \varkappa \rangle =: J_R^\varkappa},
\end{aligned}
\tag{18.6.1}
$$

where $\zeta = \dot{g}g^{-1}$ and the quantity $\varkappa := \delta g g^{-1}$ vanishes at the endpoints in time, and we have used the right-invariant version of the relation from (2.10.4):

$$
\delta(\dot{g}g^{-1}) = \frac{d\varkappa}{dt} - \mathrm{ad}_{\dot{g}g^{-1}} \varkappa.
\tag{18.6.2}
$$

Hence, we recover the right-invariant version of Euler–Poincaré equation in (2.10.2):

$$
\frac{d}{dt} \frac{\partial \ell}{\partial \zeta} + \mathrm{ad}_\zeta^* \frac{\partial \ell}{\partial \zeta} = 0 \quad \text{with } \mu = \frac{\partial \ell}{\partial \zeta}.
\tag{18.6.3}
$$

The Legendre-transformed Lagrangian for the Euler–Poincaré equations is given by

$$
0 = \delta S = \delta \int_0^T \ell(\zeta) + \left\langle \mu, \dot{g}g^{-1} - \zeta \right\rangle dt
$$

$$= \delta \int_0^T \langle \mu \,, \zeta \rangle - h(\mu) + \langle \mu \,, \dot{g}g^{-1} - \zeta \rangle \, dt$$

$$= \delta \int_0^T \langle \mu \,, \dot{g}g^{-1} \rangle - h(\mu) \, dt$$

$$= \int_0^T \left\langle \mu \,, \frac{d\varkappa}{dt} - \mathrm{ad}_\zeta \varkappa \right\rangle + \left\langle \delta\mu \,, \dot{g}g^{-1} - \frac{\partial h}{\partial \mu} \right\rangle \, dt$$

$$= \int_0^T - \left\langle \frac{d\mu}{dt} + \mathrm{ad}_\zeta^* \mu \,, \varkappa \right\rangle$$

$$+ \left\langle \delta\mu \,, \dot{g}g^{-1} - \frac{\partial h}{\partial \mu} \right\rangle dt + \langle \mu \,, \varkappa \rangle \big|_0^T, \qquad (18.6.4)$$

where the quantities $\zeta = \dot{g}g^{-1}$ and $\varkappa := \delta g g^{-1}$ both vanish at the endpoints in time, and we have used the right-invariant version of the relation from (2.10.4):

$$\delta(g^{-1}\dot{g}) = \frac{d\varkappa}{dt} - \mathrm{ad}_\zeta \varkappa. \qquad (18.6.5)$$

Thus, we recover the right-invariant version of the Lie–Poisson Hamiltonian form of the Euler–Poincaré equation in (2.10.2):

$$\frac{d\mu}{dt} + \mathrm{ad}_{\frac{\partial h}{\partial \mu}}^* \mu = 0 \quad \text{where we have used} \quad \zeta = \frac{\partial h}{\partial \mu} \in \mathfrak{g} \simeq TM/G.$$

$$(18.6.6)$$

Exercise. State the right-invariant Noether's theorem for symmetries and conservation laws on both the Lagrangian and Hamiltonian sides. Prove your statements.

Hint: $\langle \mu \,, \varkappa \rangle =: J_R^\varkappa$ is the Hamiltonian for the right-invariant momentum map, $\mu := \partial \ell / \partial \zeta = \partial \ell / \partial (\dot{g}g^{-1})$, which maps $T^*M \to T^*M/G \simeq \mathfrak{g}^*$. ★

Exercise. Prove that equation (18.6.6) preserves $\text{Ad}^*_{g_t}\mu$ with $\zeta = \dot{g}g^{-1}$ by showing that

$$\frac{d}{dt}\left(\text{Ad}^*_{g_t}\mu\right) = \text{Ad}^*_{g_t}\left(\frac{d\mu}{dt} + \text{ad}^*_\zeta\mu\right) = 0. \qquad \bigstar$$

Exercise. What is the Poisson bracket for the dynamics of $\mu \in \mathfrak{g}^* \simeq T^*M/G$ in (18.6.6)? $\qquad \bigstar$

Answer. Write $\frac{d}{dt}f(\mu) = \left\langle \frac{\partial f}{\partial \mu}, \frac{d\mu}{dt} \right\rangle$ in Poisson bracket form as

$$\frac{d}{dt}f(\mu) = \left\langle \frac{\partial f}{\partial \mu}, \frac{d\mu}{dt} \right\rangle = -\left\langle \frac{\partial f}{\partial \mu}, \text{ad}^*_{\frac{\partial h}{\partial \mu}}\mu \right\rangle$$

$$= -\left\langle \text{ad}_{\frac{\partial h}{\partial \mu}}\frac{\partial f}{\partial \mu}, \mu \right\rangle = \left\langle \left[\frac{\partial h}{\partial \mu}, \frac{\partial f}{\partial \mu}\right], \mu \right\rangle$$

$$= -\left\langle \mu, \left[\frac{\partial f}{\partial \mu}, \frac{\partial h}{\partial \mu}\right] \right\rangle =: \{f, h\}_{LPB}(\mu).$$

$$(18.6.7)$$

$\blacktriangle$

Exercise. Prove that the Lie–Poisson bracket (LPB) is Poisson, i.e., prove that $\{f, h\}_{LPB}$ satisfies the properties of a Poisson bracket. $\qquad \bigstar$

Answer. $\{f, h\}_{LPB}$ being a linear functional of the Lie algebra $\mathfrak{g}$ makes it easy to verify the Poisson properties.

Upon defining $J^{\xi_k}(\mu) = \langle \mu, \xi_k \rangle$, $k = 1, 2, 3$, we have

$$\left\{ J^{\xi_2}, J^{\xi_3} \right\} = - \left\langle \mu, [\xi_2, \xi_3] \right\rangle$$

and

$$\left\{ J^{\xi_1}, \left\{ J^{\xi_2}, J^{\xi_3} \right\} \right\} = \left\langle \mu, [\xi_1, [\xi_2, \xi_3]] \right\rangle.$$

Consequently, the vanishing of the sum of cyclic permutations of $k = 1, 2, 3$ in the vector field commutators $[\xi_1, [\xi_2, \xi_3]]$ implies that the sum of cyclic permutations of the Lie–Poisson brackets of their Hamiltonians $\left\{ J^{\xi_1}, \left\{ J^{\xi_2}, J^{\xi_3} \right\} \right\}$ also vanishes because the Hamiltonians are linear functionals of $\mu \in \mathfrak{g}^*$.   ▲

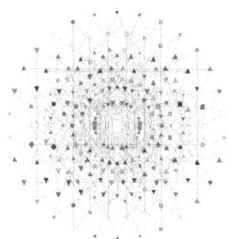

19

MORE ABOUT VECTOR FIELDS AND DIFFERENTIAL FORMS

Contents

What is this lecture about? This lecture demonstrates how to perform in coordinates the various coordinate-free calculations for differential forms discussed in the previous lecture.

Remark 19.0.1 For a survey of the basic definitions, properties and operations on differential forms, as well as useful tables of relations between differential calculus and vector calculus, see, for example, the work of Bloch [Bl2003, Chapter 2]. □

19.1   Vector fields and 1-forms

Let M be a manifold. In what follows, all maps may be assumed to be C^∞, although that's not always necessary.

A *vector field* on M is a map $X: M \to TM$ such that $X(x) \in T_x M$ for every $x \in M$. The set of all smooth vector fields on M is written $\mathfrak{X}(M)$. ("Smooth" means differentiable or C^r for some $r \leq \infty$, depending on context.)

A *(differential) 1-form* on M is a map $\theta: M \to T^*M$ such that $\theta(x) \in T_x^* M$ for every $x \in M$.

More generally, if $\pi: E \to M$ is a bundle, then a *section* of the bundle is a map $\varphi: M \to E$ such that $\pi \circ \varphi(x) = x$ for all $x \in M$. So, a vector field is a section of the tangent bundle, while a 1-form is a section of the cotangent bundle.

Vector fields can be added and also multiplied by scalar functions $k: M \to \mathbb{R}$, as follows: $(X_1 + X_2)(x) = X_1(x) + X_2(x)$, $(kX)(x) = k(x)X(x)$.

Differential forms can be added and also multiplied by scalar functions $k: M \to \mathbb{R}$, as follows: $(\alpha + \beta)(x) = \alpha(x) + \beta(x)$, $(k\theta)(x) = k(x)\theta(x)$.

We have already defined the push-forward and pull-back of a vector field. The *pull-back* of a 1-form θ on N by a map $\varphi: M \to N$ is the 1-form $\varphi^*\theta$ on M defined by

$$(\varphi^*\theta)(x) \cdot v = \theta(\varphi(x)) \cdot T\varphi(v).$$

The *push-forward* of a 1-form α on M by a diffeomorphism $\psi: M \to N$ is the pull-back of α by ψ^{-1}.

A vector field can be *contracted* with a differential form by using the pairing between tangent and cotangent vectors: $(X \lrcorner \theta)(x) = \theta(x) \cdot X(x)$. Note that $X \lrcorner \theta$ is a map from M to $\mathbb{R}$. In many books,

$i_X \theta$ is used in place of $X \lrcorner\, \theta$, and the contraction operation is often called the *interior product*.

The *differential* of $f \colon M \to \mathbb{R}$ is a 1-form df on M defined by

$$df(x) \cdot v = \frac{d}{dt} f(c(t)) \Big|_{t=0}$$

for any $x \in M$, any $v \in T_x M$ and any path $c(t)$ in M such that $c(0) = 0$ and $c'(0) = v$. The left-hand side, $df(x)\cdot v$, means the pairing between cotangent and tangent vectors, which could also be written $df(x)(v)$ or $\langle df(x), v \rangle$.

Note:

$$X \lrcorner\, df = \pounds_X f =: X[f].$$

Remark 19.1.1 df is very similar to Tf, but Tf is defined for all differentiable $f \colon M \to N$, whereas df is only defined when $N = \mathbb{R}$ (in this course, anyway). In this case, Tf is a map from TM to $T\mathbb{R}$, and $Tf(v) = df(x) \cdot v \in T_{f(x)}\mathbb{R}$ for every $v \in T_x M$. (We have identified $T_f(x)\mathbb{R}$ with $\mathbb{R}$.) □

19.1.1 In coordinates...

Let M be n-dimensional, and let $x^1, \ldots, x^n$ be differentiable local coordinates for M. This means that there's an open subset U of M and an open subset V of $\mathbb{R}^n$ such that the map $\varphi \colon U \to V$ defined by $\varphi(x) = \left(x^1(x), \ldots, x^n(x)\right)$ is a diffeomorphism. In particular, each x^i is a map from M to $\mathbb{R}$, so the differential dx^i is defined. There is also a vector field $\frac{\partial}{\partial x^i}$ for every i, which is defined by $\frac{\partial}{\partial x^i}(x) = \frac{d}{dt}\varphi^{-1}\left(\varphi(x) + t e_i\right)\big|_{t=0}$, where e_i is the ith standard basis vector.

Exercise. Verify that

$$\frac{\partial}{\partial x^i} \lrcorner\, dx^j \equiv \delta^i_j,$$

where $\equiv$ means the left-hand side is a constant function with a value of δ^i_j. ★

Remark 19.1.2 Of course, given a coordinate system $\varphi = (x^1, \ldots, x^n)$, it is usual to write $x = (x^1, \ldots, x^n)$, which means x is identified with $(x^1(x), \ldots, x^n(x)) = \varphi(x)$. $\qquad \square$

For every $x \in M$, the vectors $\frac{\partial}{\partial x^i}(x)$ form a basis for $T_x M$, so every $v \in T^x M$ can be uniquely expressed as $v = v^i \frac{\partial}{\partial x^i}(x)$. This expression defines the *tangent-lifted coordinates* $x^1, \ldots, x^n, v^1, \ldots v^n$ on TM (which are local coordinates, defined on $TU \subset TM$).

For every $x \in M$, the covectors $dx^i(x)$ form a basis for $T_x^* M$, so every $\alpha \in T^x M$ can be uniquely expressed as $\alpha = \alpha_i dx^i(x)$. This expression defines the *cotangent-lifted coordinates* $x^1, \ldots, x^n, \alpha_1, \ldots, \alpha_n$ on $T^* M$ (which are local coordinates, defined on $T^* U \subset T^* M$).

Note that the basis $\left(\frac{\partial}{\partial x^i}\right)$ is dual to the basis $(dx^1, \ldots, dx^n)$, according to the previous exercise. It follows that

$$(\alpha_i dx^i) \cdot \left(v^i \frac{\partial}{\partial x^i}\right) = \alpha_i v^i$$

(where we have used the summation convention).

In mechanics, the configuration space is often called Q, and the lifted coordinates are written as $q^1, \ldots, q^n, \dot{q}^1, \ldots, \dot{q}^n$ (on TQ) and $q^1, \ldots, q^n, p_1, \ldots, p_n$ (on $T^* Q$).

19.1.2 Why the distinction between subscripts and superscripts?

This distinction keeps track of how tensor quantities vary when coordinates are changed (see the following exercise). One benefit is that using the summation convention gives coordinate-independent answers.

Exercise. Consider two sets of local coordinates q^i and s^i on Q, related by $(s^1, \ldots, s^n) = \psi(q^1, \ldots, q^n)$. Verify that the corresponding tangent-lifted coordinates $\dot{q}^i$ and

$\dot{s}^i$ are related by

$$\dot{s}^i = \frac{\partial \psi^i}{\partial q^j} \dot{q}^j.$$

Note that the last equation can be written as $\dot{\mathbf{s}} = D\psi(q)\dot{\mathbf{q}}$, where $\dot{\mathbf{s}}$ is the column vector $(\dot{s}^1, \ldots \dot{s}^n)$, and similarly for $\dot{\mathbf{q}}$.

Perform the calculation corresponding to change of variables on the cotangent bundle side. See Definition 16.0.1.

★

19.2   The next level: TTQ, T^*T^*Q, *et cetera*

Since TQ is a manifold, we can consider vector fields on it, which are sections of $T(TQ)$. In coordinates, every vector field on TTQ has the form $X = a^i \frac{\partial}{\partial q^i} + b^i \frac{\partial}{\partial \dot{q}^i}$, where a^i and b^i are functions of q and $\dot{q}$, respectively. Note that the same symbol q^i has two interpretations: as a coordinate on TQ and as a coordinate on Q, so $\frac{\partial}{\partial q^i}$ can mean a vector field on TQ (as above) or on Q.

The tangent lift of the bundle projection $\tau \colon TQ \to Q$ is a map $T\tau \colon TTQ \to TQ$. If X is written in coordinates as above, then $T\tau \circ X = a^i \frac{\partial}{\partial q^i}$. A vector field X on TTQ is *second order* if $T\tau \circ X(v) = v$, or, in coordinates, $a^i = \dot{q}^i$. The name comes from the process of reducing second-order equations to first-order ones by introducing new variables, $\dot{q}^i = \frac{dq^i}{dt}$.

One may also consider T^*TQ, TT^*Q and T^*T^*Q. However, the subscript/superscript distinction is problematic there.

19.2.1   1-forms

The 1-forms on T^*Q are sections of T^*T^*Q. Given cotangent-lifted local coordinates

$$\left(q^1, \ldots, q^n, p_1, \ldots, p_n \right)$$

on T^*Q, the general 1-form on T^*Q has the form $a_i dq^i + b_i dp_i$, where a_i and b_i are functions of the phase space coordinates (q, p) and we sum repeated indices over their range. The *canonical* 1-*form* on T^*Q is

$$\theta = p_i dq^i,$$

also written in short form as $\langle p, dq \rangle$. Pairing $\theta(q, p)$ with an arbitrary tangent vector $v = a^j \frac{\partial}{\partial q^j} + b^j \frac{\partial}{\partial p^j} \in T_{(q,p)} T^*Q$ gives

$$\langle \theta(q, p), v \rangle = \left\langle p_i dq^i, a^j \frac{\partial}{\partial q^j} + b^j \frac{\partial}{\partial p^j} \right\rangle = p_i a^j \delta^i_j = \left\langle p_i dq^i, a^j \frac{\partial}{\partial q^j} \right\rangle.$$

19.2.2   2-forms

Recall that a 1-form on M, evaluated at a point $x \in M$, is a linear map from $T_x M$ to $\mathbb{R}$.

A 2-form on M, evaluated at a point $x \in M$, is a skew-symmetric bilinear form on $T_x M$, and the bilinear form has to vary smoothly as x changes. (Bilinear forms may be skew symmetric, symmetric or neither; *differential* forms are assumed to be skew-symmetric.)

The *pull-back* of a 2-form ω on N by a map $\varphi \colon M \to N$ is the 2-form $\varphi^* \omega$ on M, defined by

$$(\varphi^* \omega)(x)(v, w) = \theta(\varphi(x))(T\varphi(v), T\varphi(w)).$$

The *push-forward* of a 2-form ω on M by a diffeomorphism $\psi \colon M \to N$ is the pull-back of ω by ψ^{-1}.

A vector field X can be *contracted* with a 2-form ω to get a 1-form $X \lrcorner \omega$, defined by

$$(X \lrcorner \omega)(x)(v) = \omega(x)(X(x), v)$$

for any $v \in T_x M$. A shorthand for this is $(X \lrcorner \omega)(v) = \omega(X, v)$, or simply $X \lrcorner \omega = \omega(X, \cdot)$.

The *tensor product* of two 1-forms α and β is the 2-form $\alpha \otimes \beta$, defined by

$$(\alpha \otimes \beta)(v, w) = \alpha(v)\beta(w)$$

for all $v, w \in T_x^* M$.

The *wedge product* of two 1-forms α and β is the skew-symmetric 2-form $\alpha \wedge \beta$, defined by

$$(\alpha \wedge \beta)\,(v, w) = \alpha(v)\beta(w) - \alpha(w)\beta(v).$$

19.2.3   Exterior derivative

The differential df of a real-valued function is also called the exterior derivative of f. In this context, real-valued functions can be called 0-forms. The exterior derivative is a linear operation from 0-forms to 1-forms that satisfies the Leibniz identity, also known as the product rule:

$$d(fg) = f\,dg + g\,df.$$

The exterior derivative of a 1-form is an alternating 2-form, defined as follows:

$$d\left(a_i dx^i\right) = \frac{\partial a_i}{\partial x^j} dx^j \wedge dx^i.$$

An exterior derivative is a linear operation from 1-forms to 2-forms. The following identity is easily checked:

$$d(df) = 0$$

for all scalar functions f.

19.2.4 n-forms

See Marsden and Ratiu [MaRa1994], Lee [Le2003] or Abraham and Marsden [AbMa1978]. Unless otherwise specified, n-forms are assumed to be alternating. Wedge products and contractions generalise.

It is a fact that all n-forms are linear combinations of wedge products of 1-forms. Thus, we can define an exterior derivative recursively using the properties

$$d(\alpha \wedge \beta) = d\alpha \wedge \beta + (-1)^k \alpha \wedge d\beta,$$

for all k-forms α and all forms β, and

$$d \circ d = 0$$

In local coordinates, if $\alpha = \alpha_{i_1 \ldots i_k} dx^{i_1} \wedge \cdots \wedge dx^{i_k}$ (sum over all $i_1 < \cdots < i_k$), then

$$da = \frac{\partial \alpha_{i_1 \cdots i_k}}{\partial x^j} dx^j \wedge dx^{i_1} \wedge \cdots \wedge dx^{i_k}.$$

The *Lie derivative* of an n-form θ in the direction of the vector field X is defined as

$$\pounds_X \theta = \frac{d}{dt} \varphi_t^* \theta \Big|_{t=0}, \tag{19.2.1}$$

where φ_t is the flow of $X = \dot\varphi_t \varphi_t^{-1}$ for right-invariant vector fields. Formula (19.2.1) follows from the *Lie chain rule*:

$$\frac{d}{dt} \varphi_t^* \theta = \varphi_t^* \left(\pounds_{\dot\varphi_t \varphi_t^{-1}} \theta \right). \tag{19.2.2}$$

Remark 19.2.1 Pull-back commutes with the operations d, $\lrcorner$, $\wedge$ and Lie derivative. Consequently, the Lie derivative satisfies the product rule for any of these pull-back properties.   □

19.2.5   Cartan's magic formula

$$\pounds_X \alpha = d \left(X \lrcorner \alpha \right) + X \lrcorner \, d\alpha$$

This looks even more magical when written using the notation $i_X \alpha = X \lrcorner \alpha$:

$$\pounds_X = d i_X + i_X d$$

An n-form α is *closed* if $d\alpha = 0$ and *exact* if $\alpha = d\beta$ for some β. All exact forms are closed (since $d \circ d = 0$), but the converse is false. It is true that all closed forms are *locally* exact; this is the *Poincaré lemma*.

20

EULER–POINCARÉ
REDUCTION THEOREM

Contents

> **What is this lecture about?** This lecture states and proves the Euler–Poincaré (EP) reduction theorem, which proceeds by applying reduction by Lie symmetry to Hamilton's principle.

Remark 20.0.1 (Geodesic motion) As emphasised by Arnold [Ar1966], in many interesting cases, the EP equations on the dual of a Lie algebra $\mathfrak{g}^*$ correspond to *geodesic motion* on the corresponding group G. The relationship between the equations on $\mathfrak{g}^*$ and on G is the content of the basic EP theorem discussed later. Similarly, on the Hamiltonian side, the preceding paragraphs described the relation between the Hamiltonian equations on T^*G and the Lie–Poisson equations on $\mathfrak{g}^*$. The issue of geodesic motion is especially simple: if either the Lagrangian on $\mathfrak{g}$ or the Hamiltonian on $\mathfrak{g}^*$ is purely quadratic, then the corresponding motion on the group is geodesic motion.   □

20.1   We were already speaking prose (EP)

Many of our previous considerations may be recast immediately as Euler–Poincaré equations:

- rigid bodies $\simeq \big($EP $SO(n)\big)$,

- affine invariant motions $\simeq \big($EP $G(A) = GL(n, \mathbb{R})\circledS\mathbb{R}^n\big)$,

- heavy tops $\simeq \big($EP $SO(3) \times \mathbb{R}^3\big)$,

- EPDiff.

20.2   Euler–Poincaré reduction

This lecture applies reduction by symmetry to Hamilton's principle. For a G-invariant Lagrangian defined on TG, this reduction takes Hamilton's principle from TG to $TG/G \simeq \mathfrak{g}$. Stationarity of the symmetry-reduced Hamilton's principle yields the EP equations on $\mathfrak{g}^*$. The corresponding reduced Legendre transformation yields the Lie–Poisson Hamiltonian formulation of these equations.

Euler–Poincaré Reduction starts with a right (respectively, left) invariant Lagrangian $L\colon TG \to \mathbb{R}$ on the tangent bundle of a Lie group G. This means that $L(T_g R_h(v)) = L(v)$ (respectively $L(T_g L_h(v)) = L(v)$), for all $g, h \in G$ and all $v \in T_g G$. In shorter notation, the right invariance of the Lagrangian may be written as

$$L(g(t), \dot{g}(t)) = L(g(t)h, \dot{g}(t)h), \qquad (20.2.1)$$

for all $h \in G$.

Theorem 20.2.1 (Euler–Poincaré reduction) *Let G be a Lie group, $L\colon TG \to \mathbb{R}$ be a right-invariant Lagrangian and $l := L|_{\mathfrak{g}}\colon \mathfrak{g} \to \mathbb{R}$ be its* *restriction to* $\mathfrak{g}$. *For a curve* $g(t) \in G$, *let*

$$\xi(t) = \dot{g}(t) \cdot g^{-1}(t) := T_{g(t)} R_{g^{-1}(t)} \dot{g}(t) \in \mathfrak{g}.$$

Then, the following four statements are equivalent:

1. $g(t)$ *satisfies the Euler–Lagrange equations for Lagrangian* L *defined on* G.

2. *The variational principle*

$$\delta \int_a^b L(g(t), \dot{g}(t)) dt = 0$$

   holds for variations with fixed endpoints.

3. *The (right-invariant) Euler–Poincaré equations hold:*

$$\frac{d}{dt} \frac{\delta l}{\delta \xi} = - \operatorname{ad}_\xi^* \frac{\delta l}{\delta \xi}.$$

4. *The variational principle*

$$\delta \int_a^b l(\xi(t)) dt = 0$$

   holds on $\mathfrak{g}$ *using variations of the form* $\delta \xi = \dot{\eta} - [\xi, \eta]$, *where* $\eta(t)$ *is an arbitrary path in* $\mathfrak{g}$, *which vanishes at the endpoints, that is,* $\eta(a) = \eta(b) = 0$.

Proof. The proof consists of three steps.

Step I: Proof that (i) $\Longleftrightarrow$ (ii)   This is Hamilton's principle: the Euler–Lagrange equations follow from stationary action for variations δg, which vanish at the endpoints (see statement 3 above).

Step II: Proof that (ii) $\Longleftrightarrow$ (iv)   Proving equivalence of the variational principles (ii) on TG and (iv) on $\mathfrak{g}$ for a right-invariant Lagrangian requires calculation of the variations $\delta \xi$ of $\xi = \dot{g} g^{-1}$ induced by δg. To simplify the exposition, the calculation will be done first for matrix Lie groups and then generalised to arbitrary Lie groups.

Step IIA: Proof that (ii) $\Longleftrightarrow$ (iv) for a matrix Lie group   For $\xi = \dot{g}g^{-1}$, define $g_\epsilon(t)$ to be a family of curves in G such that $g_0(t) = g(t)$, and denote

$$\delta g := \frac{dg_\epsilon(t)}{d\epsilon}\bigg|_{\epsilon=0}.$$

The variation of ξ is computed in terms of δg as

$$\delta\xi = \frac{d}{d\epsilon}\bigg|_{\epsilon=0}(\dot{g}_\epsilon g_\epsilon^{-1}) = \frac{d^2g}{dtd\epsilon}\bigg|_{\epsilon=0}g^{-1} - \dot{g}g^{-1}(\delta g)g^{-1}. \qquad (20.2.2)$$

Set $\eta := g^{-1}\delta g$. That is, $\eta(t)$ is an arbitrary curve in $\mathfrak{g}$, which vanishes at the endpoints. The time derivative of η is computed as

$$\dot{\eta} = \frac{d\eta}{dt} = \frac{d}{dt}\left(\left(\frac{d}{d\epsilon}\bigg|_{\epsilon=0}g_\epsilon\right)g^{-1}\right) = \frac{d^2g}{dtd\epsilon}\bigg|_{\epsilon=0}g^{-1} - (\delta g)g^{-1}\dot{g}g^{-1}.$$
$$(20.2.3)$$

Taking the difference of (20.2.2) and (20.2.3) implies

$$\delta\xi - \dot{\eta} = -\dot{g}g^{-1}(\delta g)g^{-1} + (\delta g)g^{-1}\dot{g}g^{-1} = -\xi\eta + \eta\xi = -[\xi, \eta].$$

That is, for matrix Lie algebras,

$$\delta\xi = \dot{\eta} - [\xi, \eta],$$

where $[\xi, \eta]$ is the matrix commutator. Next, we note that the right invariance of L allows one to change variables in the Lagrangian by applying $g^{-1}(t)$ from the right, as

$$L(g(t), \dot{g}(t)) = L(e, \dot{g}(t)g^{-1}(t)) =: l(\xi(t)).$$

Combining this definition of the symmetry-reduced Lagrangian $l: \mathfrak{g} \to \mathbb{R}$ with the formula for variations $\delta\xi$ just deduced proves the equivalence of (ii) and (iv) for matrix Lie groups.

Step IIB: Proof that (ii) $\Longleftrightarrow$ (iv) for an arbitrary Lie group   The same proof extends to any Lie group G by using the following lemma.

Lemma 20.2.1 *Let* $g: U \subset \mathbb{R}^2 \to G$ *be a smooth map and denote its partial derivatives by*

$$\xi(t, \varepsilon) := T_{g(t,\varepsilon)} R_{g(t,\varepsilon)^{-1}} \frac{\partial g(t, \varepsilon)}{\partial t},$$

$$\eta(t, \varepsilon) := T_{g(t,\varepsilon)} R_{g(t,\varepsilon)^{-1}} \frac{\partial g(t, \varepsilon)}{\partial \varepsilon}. \tag{20.2.4}$$

Then,

$$\frac{\partial \xi}{\partial \varepsilon} - \frac{\partial \eta}{\partial t} = -[\xi, \eta], \tag{20.2.5}$$

where $[\xi, \eta]$ *is the Lie algebra bracket on* $\mathfrak{g}$. *Conversely, if* $U \subset \mathbb{R}^2$ *is simply connected and* $\xi, \eta: U \to \mathfrak{g}$ *are smooth functions satisfying* (20.2.5), *then there exists a smooth function* $g: U \to G$ *such that* (20.2.4) *holds.*

Proof of Lemma 20.2.1. Write $\xi = \dot{g}g^{-1}$ and $\eta = g'g^{-1}$ in natural notation, and express the partial derivatives $\dot{g} = \partial g/\partial t$ and $g' = \partial g/\partial \epsilon$ using the right translations as

$$\dot{g} = \xi \circ g \quad \text{and} \quad g' = \eta \circ g.$$

By the chain rule, these definitions have mixed partial derivatives:

$$\dot{g}' = \xi' = \nabla \xi \cdot \eta \quad \text{and} \quad g'^{\cdot} = \dot{\eta} = \nabla \eta \cdot \xi.$$

The difference between the mixed partial derivatives implies the desired formula (20.2.5):

$$\xi' - \dot{\eta} = \nabla \xi \cdot \eta - \nabla \eta \cdot \xi = -[\xi, \eta] = -\operatorname{ad}_\xi \eta.$$

(Note the minus sign in the last two terms.) ∎

Step III: Proof of equivalence (iii) $\Longleftrightarrow$ **(iv)** Let us show that the reduced variational principle produces the EP equations. We write the functional derivative of the reduced action $S_{\text{red}} = \int_a^b l(\xi) \, dt$ with

Lagrangian $l(\xi)$ in terms of the natural pairing $\langle \cdot, \cdot \rangle$ between $\mathfrak{g}^*$ and $\mathfrak{g}$ as

$$
\begin{aligned}
\delta \int_a^b l(\xi(t))dt &= \int_a^b \left\langle \frac{\delta l}{\delta \xi}, \delta \xi \right\rangle dt = \int_a^b \left\langle \frac{\delta l}{\delta \xi}, \dot{\eta} - \mathrm{ad}_\xi \, \eta \right\rangle dt \\
&= \int_a^b \left\langle \frac{\delta l}{\delta \xi}, \dot{\eta} \right\rangle dt - \int_a^b \left\langle \frac{\delta l}{\delta \xi}, \mathrm{ad}_\xi \, \eta \right\rangle dt \\
&= - \int_a^b \left\langle \frac{d}{dt} \frac{\delta l}{\delta \xi} + \mathrm{ad}_\xi^* \frac{\delta l}{\delta \xi}, \eta \right\rangle dt + \left\langle \frac{\delta l}{\delta \xi}, \eta \right\rangle \Big|_a^b \\
&= - \int_a^b \left\langle \frac{d}{dt} \frac{\delta l}{\delta \xi} + \mathrm{ad}_\xi^* \frac{\delta l}{\delta \xi}, \eta \right\rangle dt. \qquad (20.2.6)
\end{aligned}
$$

The last equality follows from integration by parts and the vanishing of the variation $\eta(t)$ at the endpoints. Thus, stationarity $\delta \int_a^b l(\xi(t))dt = 0$ for any $\eta(t)$ that vanishes at the endpoints is equivalent to

$$
\frac{d}{dt} \frac{\delta l}{\delta \xi} = - \mathrm{ad}_\xi^* \frac{\delta l}{\delta \xi}, \qquad (20.2.7)
$$

which are the EP equations arising from the right invariance of the Lagrangian in Hamilton's principle for ideal fluid dynamics. ∎

Exercise. (Noether's theorem for EP fluid equations) State and prove Noether's theorem arising from the right invariance of the Lagrangian in equation (20.2.1) under the relabelling transformation $g \to gh$ of the particle labels. ★

Answer. An infinitesimal Lie symmetry transformation δQ^A of the Lagrangian parcel labels $Q^A(\mathbf{x}, t)$ in Euclidean space $\mathbb{R}^3$, with components $A = 1, 2, 3$ obeying the Eulerian advection equation,

$$
\partial_t Q^A + \mathbf{u} \cdot \nabla Q^A = 0,
$$

can be written as $\delta Q^A = v(Q^A) = \mathbf{v}(\mathbf{x}) \cdot \nabla Q^A$ for each component $A = 1, 2, 3$ and a smooth divergence-free vector field v with Euclidean components $v = \mathbf{v} \cdot \nabla$. The corresponding momentum map is given by L^2 pairing:

$$\left\langle P_A, v(Q^A) \right\rangle_{L^2} = \int_{\mathbb{R}^3} (P_A \nabla Q^A) \cdot \mathbf{v} \, d^3 x$$

$$= \int_{\mathbb{R}^3} v \lrcorner (P_A \nabla Q^A \cdot d\mathbf{x}) \otimes d^3 x$$

$$= \int_{\mathbb{R}^3} v \lrcorner (P_A dQ^A) \otimes d^3 x$$

$$=: \left\langle P_A dQ^A, v \right\rangle_{L^2}.$$

This is the momentum map for the invariance of the Lagrangian under relabelling of the parcel labels Q^A, with $A = 1, 2, 3$ in Euclidean coordinates. ▲

Exercise. (The momentum map for an infinitesimal transformation of the velocity) What is the momentum map associated with an infinitesimal action of a diffeomorphism on the velocity vector field ξ?

Hint: Look at the second-to-last line of equation (20.2.6) in the proof of the EP equation. ★

Answer. The endpoint term in the second-to-last line of equation (20.2.6) in the proof of the EP equation provides a hint because an infinitesimal transformation of the velocity may be written as $\eta := -\mathcal{L}_v \xi$, so that

$$\left\langle \frac{\delta l}{\delta \xi}, \eta \right\rangle = \left\langle \frac{\delta l}{\delta \xi}, -\mathcal{L}_v \xi \right\rangle = \left\langle \frac{\delta l}{\delta \xi}, -\operatorname{ad}_v \xi \right\rangle$$

$$= \left\langle \frac{\delta l}{\delta \xi}, \operatorname{ad}_\xi v \right\rangle$$

when the variational vector field is defined as the infinitesimal Lie symmetry transformation generated by the vector field v for the right invariance of the fluid Lagrangian under the relabelling of fluid parcels. With this formula in mind, we may define the diamond operation ($\diamond$) as

$$\left\langle \frac{\delta l}{\delta \xi}, -\mathcal{L}_v \xi \right\rangle_{T^*G \times TG}$$

$$= \left\langle \frac{\delta l}{\delta \xi}, \mathrm{ad}_\xi v \right\rangle_{T^*G \times TG} = \left\langle \mathrm{ad}^*_\xi \frac{\delta l}{\delta \xi}, v \right\rangle_{\mathfrak{g}^* \times \mathfrak{g}}$$

$$= \left\langle \mathcal{L}_\xi \frac{\delta l}{\delta \xi}, v \right\rangle_{\mathfrak{g}^* \times \mathfrak{g}} =: \left\langle \frac{\delta l}{\delta \xi} \diamond \xi, v \right\rangle_{\mathfrak{g}^* \times \mathfrak{g}},$$

where the second pairing $\langle \cdot, \cdot \rangle_{\mathfrak{g}^* \times \mathfrak{g}}$ is between vector fields in $\mathfrak{g}$ and their L^2-dual quantities, the 1-form densities in $\mathfrak{g}^*$. ▲

Remark 20.2.1 (Left invariant) The same theorem holds for left-invariant Lagrangians on TG, except for a sign in the EP equation, cf. (20.2.7):

$$\frac{d}{dt} \frac{\delta l}{\delta \xi} = + \, \mathrm{ad}^*_\xi \frac{\delta l}{\delta \xi},$$

which arises because left-invariant variations satisfy $\delta \xi = \dot{\eta} + [\xi, \eta]$ (with the opposite sign). □

Exercise. Write out the corresponding proof of the EP reduction theorem for left-invariant Lagrangians defined on the tangent space TG of a group G. ★

20.2.1   Reconstruction

The procedure for reconstructing the solution $v(t) \in T_{g(t)}G$ of the Euler–Lagrange equations with initial conditions $g(0) = g_0$ and

$\dot{g}(0) = v_0$, starting from the solution of the Euler–Poincaré equations, is as follows. First, solve the initial-value problem for the right-invariant Euler–Poincaré equations:

$$\frac{d}{dt}\frac{\delta l}{\delta \xi} = -\mathrm{ad}^*_\xi \frac{\delta l}{\delta \xi} \quad \text{with } \xi(0) = \xi_0 := v_0 g_0^{-1}.$$

Then, from the solution for $\xi(t)$, reconstruct the curve $g(t)$ on the group by solving the "linear differential equation with time-dependent coefficients":

$$\dot{g}(t) = \xi(t)g(t) \quad \text{with } g(0) = g_0.$$

The Euler–Poincaré reduction theorem guarantees then that $v(t) = \dot{g}(t) = \xi(t) \cdot g(t)$ is a solution of the Euler–Lagrange equations with the initial condition $v_0 = \xi_0 g_0$.

Remark 20.2.2 Similar statements hold, with obvious changes for left-invariant Lagrangian systems on TG. □

20.3 Reduced Legendre transformation

As in the equivalence relation between the Lagrangian and Hamiltonian formulations discussed earlier, the relationship between symmetry-reduced Euler–Poincaré and Lie–Poisson formulations is determined by the Legendre transformation.

Definition 20.3.1 *The symmetric-reduced Legendre transformation* $\mathbb{F}l: \mathfrak{g} \to \mathfrak{g}^*$ *is defined by*

$$\mathbb{F}l(\xi) = \frac{\delta l}{\delta \xi} = \mu.$$

20.3.1 Lie–Poisson Hamiltonian formulation

Let $h(\mu) := \langle \mu, \xi \rangle - l(\xi)$. Assuming that $\mathbb{F}l$ is a diffeomorphism yields

$$\frac{\delta h}{\delta \mu} = \xi + \left\langle \mu, \frac{\delta \xi}{\delta \mu} \right\rangle - \left\langle \frac{\delta l}{\delta \xi}, \frac{\delta \xi}{\delta \mu} \right\rangle = \xi.$$

So, the Euler–Poincaré equations for l are equivalent to the Lie–Poisson equations for h:

$$\frac{d}{dt}\left(\frac{\delta l}{\delta \xi}\right) = -\mathrm{ad}^*_\xi \frac{\delta l}{\delta \xi} \iff \dot{\mu} = -\mathrm{ad}^*_{\delta h/\delta \mu}\mu.$$

The Lie–Poisson equations may be written in the Poisson bracket form:

$$\dot{f} = \{f, h\}, \tag{20.3.1}$$

where $f: \mathfrak{g}^* \to \mathbb{R}$ is an arbitrary smooth function and the bracket is the (right) Lie–Poisson bracket given by

$$\{f, h\}(\mu) = \left\langle \mu, \left[\frac{\delta f}{\delta \mu}, \frac{\delta h}{\delta \mu}\right]\right\rangle = -\left\langle \mu, \mathrm{ad}_{\delta h/\delta \mu}\frac{\delta f}{\delta \mu}\right\rangle$$

$$= -\left\langle \mathrm{ad}^*_{\delta h/\delta \mu}\mu, \frac{\delta f}{\delta \mu}\right\rangle. \tag{20.3.2}$$

In the important case when ℓ is quadratic, the Lagrangian L is the quadratic form associated with a right-invariant Riemannian metric on G. In this case, the Euler–Lagrange equations for L on G describe geodesic motion relative to this metric, and these geodesics are then equivalently described by either the EP or the Lie–Poisson equations.

Exercise. Compute the pure EP equations for geodesic motion on $SE(3)$. These equations turn out to be applicable to the motion of an ellipsoidal body through a fluid. ★

21

EPDIFF: AN EULER–POINCARÉ EQUATION ON THE DIFFEOMORPHISMS

Contents

> **What is this lecture about?** This lecture lays out the coordinate form (including tensor indices) arising when applying the Euler–Poincaré theorem for diffeomorphisms.

21.1 The n-dimensional EPDiff equation

Eulerian geodesic motion of a fluid in n dimensions is generated as an EP equation via Hamilton's principle, when the Lagrangian is given by the kinetic energy. The kinetic energy defines a norm $\|\mathbf{u}\|^2$ for the Eulerian fluid velocity, taken as $\mathbf{u}(\mathbf{x}, t)\colon \mathbb{R}^n \times \mathbb{R}^1 \to \mathbb{R}^n$. The choice of the kinetic energy as a positive functional of fluid velocity $\mathbf{u}$ is a modelling step that depends on the physics of the problem being studied. We shall choose the Lagrangian,

$$\|\mathbf{u}\|^2 = \int \mathbf{u} \cdot Q_{\mathrm{op}} \mathbf{u} \, d^n x = \int \mathbf{u} \cdot \mathbf{m} \, d^n x, \qquad (21.1.1)$$

so that the positive-definite, symmetric operator Q_{op} defines the norm $\|\mathbf{u}\|$, for appropriate (homogeneous, say, or periodic) boundary conditions. The EPDiff equation is the Euler–Poincaré equation for this Eulerian geodesic motion of a fluid. Namely,

$$\frac{d}{dt}\frac{\delta\ell}{\delta\mathbf{u}} + \mathrm{ad}^*_{\mathbf{u}}\frac{\delta\ell}{\delta\mathbf{u}} = 0, \quad \text{with } \ell[\mathbf{u}] = \tfrac{1}{2}\|\mathbf{u}\|^2. \qquad (21.1.2)$$

Here, ad^* is the dual of the vector-field ad-operation (the commutator) under the natural L^2 pairing $\langle \cdot, \cdot \rangle$ induced by the variational derivative $\delta\ell[\mathbf{u}] = \langle \delta\ell/\delta\mathbf{u}, \delta\mathbf{u} \rangle$. This pairing provides the definition of ad^*,

$$\langle \mathrm{ad}^*_{\mathbf{u}}\, \mathbf{m}, \mathbf{v} \rangle = -\langle \mathbf{m}, \mathrm{ad}_{\mathbf{u}}\, \mathbf{v} \rangle, \qquad (21.1.3)$$

where u and v are vector fields and $ad_{\mathbf{u}}\mathbf{v} = [\mathbf{u}, \mathbf{v}]$ is the commutator, that is, the *Lie bracket* given in components by (summing on repeated indices)

$$[\mathbf{u}, \mathbf{v}]^i = u^j \frac{\partial v^i}{\partial x^j} - v^j \frac{\partial u^i}{\partial x^j}, \quad \text{or} \quad [\mathbf{u}, \mathbf{v}] = \mathbf{u} \cdot \nabla \mathbf{v} - \mathbf{v} \cdot \nabla \mathbf{u}. \quad (21.1.4)$$

The notation $ad_{\mathbf{u}}\, \mathbf{v} := [\mathbf{u}, \mathbf{v}]$ formally denotes the adjoint action of the *right* Lie algebra of $\mathrm{Diff}(\mathcal{D})$ on itself, and $\mathbf{m} = \delta\ell/\delta\mathbf{u}$ is the fluid momentum, a *1-form density* whose co-vector components are also denoted as m.

If $\mathbf{u} = u^j \partial/\partial x^j$, $\mathbf{m} = m_i dx^i \otimes dV$, then the preceding formula for $\mathrm{ad}^*_{\mathbf{u}}(\mathbf{m} \otimes dV)$ has the *coordinate expression* in $\mathbb{R}^n$:

$$\left(\mathrm{ad}^*_{\mathbf{u}}\, \mathbf{m} \right)_i dx^i \otimes dV = \left(\frac{\partial}{\partial x^j}(u^j m_i) + m_j \frac{\partial u^j}{\partial x^i} \right) dx^i \otimes dV.$$

$$(21.1.5)$$

In this notation, the abstract EPDiff equation (21.1.2) may be written explicitly in Euclidean coordinates as a partial differential equation for a co-vector function,

$$\mathbf{m}(\mathbf{x}, t) \colon R^n \times R^1 \to R^n.$$

Namely,

$$\frac{\partial}{\partial t}\mathbf{m} + \underbrace{\mathbf{u}\cdot\nabla\mathbf{m}}_{\text{Convection}} + \underbrace{\nabla\mathbf{u}^T\cdot\mathbf{m}}_{\text{Stretching}} + \underbrace{\mathbf{m}(\operatorname{div}\mathbf{u})}_{\text{Expansion}} = 0,$$

$$\text{with } \mathbf{m} = \frac{\delta\ell}{\delta\mathbf{u}} = Q_{\mathrm{op}}\mathbf{u}. \tag{21.1.6}$$

To explain the terms in underbraces, we rewrite EPDiff as preservation of the 1-form density of momentum along the characteristic curves of the velocity. Namely,

$$\frac{d}{dt}\left(\mathbf{m}\cdot d\mathbf{x}\otimes dV\right) = 0 \quad \text{along } \frac{d\mathbf{x}}{dt} = \mathbf{u} = G*\mathbf{m}. \tag{21.1.7}$$

This form of the EPDiff equation also emphasises its nonlocality since the velocity is obtained from the momentum density by convolution against the Green's function G of the operator Q_{op}. Thus, $\mathbf{u} = G*\mathbf{m}$ with $Q_{\mathrm{op}}G = \delta(\mathbf{x})$, the Dirac measure. We may check that this "characteristic form" of EPDiff recovers its Eulerian form by computing directly

$$\frac{d}{dt}\left(\mathbf{m}\cdot d\mathbf{x}\otimes dV\right)$$

$$= \frac{d\mathbf{m}}{dt}\cdot d\mathbf{x}\otimes dV + \mathbf{m}\cdot d\frac{d\mathbf{x}}{dt}\otimes dV + \mathbf{m}\cdot d\mathbf{x}\otimes\left(\frac{d}{dt}dV\right)$$

$$\text{along } \frac{d\mathbf{x}}{dt} = \mathbf{u} = G*\mathbf{m}$$

$$= \left(\frac{\partial}{\partial t}\mathbf{m} + \mathbf{u}\cdot\nabla\mathbf{m} + \nabla\mathbf{u}^T\cdot\mathbf{m} + \mathbf{m}(\operatorname{div}\mathbf{u})\right)\cdot d\mathbf{x}\otimes dV = 0.$$

Exercise. Show that EPDiff may be written as

$$\left(\frac{\partial}{\partial t} + \mathcal{L}_{\mathbf{u}}\right)\left(\mathbf{m}\cdot d\mathbf{x}\otimes dV\right) = 0, \tag{21.1.8}$$

where $\mathcal{L}_{\mathbf{u}}$ is the Lie derivative with respect to the vector field with components $\mathbf{u} = G*\mathbf{m}$.

Hint: How does this Lie-derivative form of EPDiff in (21.1.8) differ from its characteristic form (21.1.7)? ★

EPDiff may also be written equivalently in terms of the operators div, grad and curl in 2D and 3D as

$$\frac{\partial}{\partial t}\mathbf{m} - \mathbf{u} \times \operatorname{curl} \mathbf{m} + \nabla(\mathbf{u} \cdot \mathbf{m}) + \mathbf{m}(\operatorname{div} \mathbf{u}) = 0. \qquad (21.1.9)$$

Thus, for example, its numerical solution would require an algorithm which has the capability to deal with the distinctions and relationships among the operators div, grad and curl.

21.2   Deriving the n-dimensional EPDiff equation as geodesic flow

Let's derive the EPDiff equation (21.1.6) by following the proof of the EP reduction theorem leading to the Euler–Poincaré equations for right invariance in the form (21.1.2). Following this calculation for the current case yields

$$\delta \int_a^b l(\mathbf{u})dt = \int_a^b \left\langle \frac{\delta l}{\delta \mathbf{u}}, \delta \mathbf{u} \right\rangle dt = \int_a^b \left\langle \frac{\delta l}{\delta \mathbf{u}}, \dot{\mathbf{v}} - \operatorname{ad}_{\mathbf{u}} \mathbf{v} \right\rangle dt$$

$$= \int_a^b \left\langle \frac{\delta l}{\delta \mathbf{u}}, \dot{\mathbf{v}} \right\rangle dt - \int_a^b \left\langle \frac{\delta l}{\delta \mathbf{u}}, \operatorname{ad}_{\mathbf{u}} \mathbf{v} \right\rangle dt$$

$$= - \int_a^b \left\langle \frac{d}{dt}\frac{\delta l}{\delta \mathbf{u}} + \operatorname{ad}_{\mathbf{u}}^* \frac{\delta l}{\delta \mathbf{u}}, \mathbf{v} \right\rangle dt,$$

where $\langle \cdot, \cdot \rangle$ is the pairing between elements of the Lie algebra and its dual. In our case, this is the L^2 pairing, for example,

$$\left\langle \frac{\delta l}{\delta \mathbf{u}}, \delta \mathbf{u} \right\rangle = \int \frac{\delta l}{\delta u^i} \delta u^i \, d^n x.$$

This pairing allows us to compute the coordinate form of the EPDiff equation explicitly as

$$\int_a^b \left\langle \frac{\delta l}{\delta \mathbf{u}}, \delta \mathbf{u} \right\rangle dt$$

$$= \int_a^b dt \int \frac{\delta l}{\delta u^i} \left(\frac{\partial v^i}{\partial t} + u^j \frac{\partial v^i}{\partial x^j} - v^j \frac{\partial u^i}{\partial x^j} \right) d^n x$$

$$= -\int_a^b dt \int \left\{ \frac{\partial}{\partial t} \frac{\delta l}{\delta u^i} + \frac{\partial}{\partial x^j} \left(\frac{\delta l}{\delta u^i} u^j \right) + \frac{\delta l}{\delta u^j} \frac{\partial u^j}{\partial x^i} \right\} v^i \, d^n x.$$

Substituting $\mathbf{m} = \delta l / \delta \mathbf{u}$ now recovers the coordinate forms for the coadjoint action of vector fields in (21.1.5) and the EPDiff equation itself in (21.1.6). When $\ell[\mathbf{u}] = \frac{1}{2} \|\mathbf{u}\|^2$, EPDiff describes geodesic motion on the diffeomorphisms with respect to the norm $\|\mathbf{u}\|$.

Lemma 21.2.1 *In Step IIB of the proof of the Euler–Poincaré reduction theorem (that is, (ii) $\Longleftrightarrow$ (iv) for an arbitrary Lie group), a certain formula for the variations for time-dependent vector fields was employed. That formula was employed again in the calculation above as*

$$\delta \mathbf{u} = \dot{\mathbf{v}} - \mathrm{ad}_{\mathbf{u}} \mathbf{v}. \tag{21.2.1}$$

This formula may be rederived as follows in the present context. We write $\mathbf{u} = \dot{g} g^{-1}$ and $\mathbf{v} = g' g^{-1}$ in natural notation and express the partial derivatives $\dot{g} = \partial g / \partial t$ and $g' = \partial g / \partial \epsilon$ using the right translations as

$$\dot{g} = \mathbf{u} \circ g \quad and \quad g' = \mathbf{v} \circ g.$$

To compute the mixed partials, consider the chain rule for, say, $\mathbf{u}(g(t, \epsilon)\mathbf{x}_0)$ and set $\mathbf{x}(t, \epsilon) = g(t, \epsilon) \cdot \mathbf{x}_0$. Then,

$$\mathbf{u}' = \frac{\partial \mathbf{u}}{\partial \mathbf{x}} \cdot \frac{\partial \mathbf{x}}{\partial \epsilon} = \frac{\partial \mathbf{u}}{\partial \mathbf{x}} \cdot g'(t, \epsilon)\mathbf{x}_0 = \frac{\partial \mathbf{u}}{\partial \mathbf{x}} \cdot g' g^{-1} \mathbf{x} = \frac{\partial \mathbf{u}}{\partial \mathbf{x}} \cdot \mathbf{v}(\mathbf{x}).$$

The chain rule for $\dot{\mathbf{v}}$ gives a similar formula with $\mathbf{u}$ and $\mathbf{v}$ exchanged. Thus, the chain rule gives two expressions for the mixed partial derivative $\dot{g}'$ as

$$\dot{g}' = \mathbf{u}' = \nabla \mathbf{u} \cdot \mathbf{v} \quad and \quad g'^{\cdot} = \dot{\mathbf{v}} = \nabla \mathbf{v} \cdot \mathbf{u}.$$

The difference between the mixed partial derivatives then implies the desired formula (21.2.1) since

$$\mathbf{u}' - \dot{\mathbf{v}} = \nabla \mathbf{u} \cdot \mathbf{v} - \nabla \mathbf{v} \cdot \mathbf{u} = -[\mathbf{u}, \mathbf{v}] = -\mathrm{ad}_{\mathbf{u}} \mathbf{v}.$$

22

EPDIFF SOLUTION BEHAVIOUR IN 1D

Contents

What is this lecture about? This lecture explores the construction of solutions for geodesic motion on the manifold of smooth invertible maps (diffeomorphisms).

In this lecture, we discuss the solutions of EPDiff for pressureless compressible geodesic motion in one spatial dimension. This is the

EPDiff equation in 1D, [1]

$$\partial_t m + \text{ad}_u^* \, m = 0, \quad \text{or, equivalently,} \qquad (22.0.1)$$

$$\partial_t m + u m_x + 2 u_x m = 0, \quad \text{with } m = Q_{\text{op}} u. \qquad (22.0.2)$$

The EPDiff equation describes geodesic motion on the diffeo-morphism group with respect to a family of metrics for the fluid velocity $u(t, x)$, with notation

$$m \;=\; \frac{\delta \ell}{\delta u} = Q_{\text{op}} u \quad \text{for a kinetic-energy Lagrangian} \quad (22.0.3)$$

$$\ell(u) = \tfrac{1}{2} \int u \, Q_{\text{op}} u \, dx = \tfrac{1}{2} \|u\|^2. \qquad (22.0.4)$$

In 1D, Q_{op} in equation (22.0.3) is a positive, symmetric operator that defines the kinetic energy metric for the velocity.

The EPDiff equation (22.0.2) is written in terms of the variable $m = \delta \ell / \delta u$. It is appropriate to call this variational derivative m because it is the momentum density associated with the fluid veloc-ity u.

Physically, the first nonlinear term in the EPDiff equation (22.0.2) is fluid transport.

The coefficient 2 arises in the second nonlinear term because, in 1D, two of the summands in $\text{ad}_u^* \, m = u m_x + 2 u_x m$ are the *same*, cf. equation (21.1.5).

The momentum is expressed in terms of the velocity by $m = \delta \ell / \delta u = Q_{\text{op}} u$. Equivalently, for solutions that vanish at spatial

---

[1] A 1-form density in one dimension (1D) takes the form $m (dx)^2$ and the Euler–Poincaré (EP) equation is given by

$$\frac{d}{dt} \left(m \, (dx)^2 \right) = \frac{dm}{dt} (dx)^2 + 2m \, (du)(dx) = 0$$

$$\text{with} \quad \frac{d}{dt} dx = du = u_x dx \quad \text{and} \quad u = G * m,$$

where $G * m$ denotes convolution with a function G on the real line:

infinity, one may think of the velocity as being obtained from the convolution

$$u(x) = G * m(x) = \int G(x-y)m(y)\, dy, \qquad (22.0.5)$$

where G is the Green's function for the operator Q_{op} on the real line.

The operator Q_{op} and its Green's function G are chosen to be even under reflection, $G(-x) = G(x)$, so that u and m have the same parity. Moreover, the EPDiff equation (22.0.2) conserves the total momentum $M = \int m(y)\, dy$, for any even Green's function.

Exercise. Show that equation (22.0.2) conserves $M = \int m(y)\, dy$ for any even Green's function $G(-x) = G(x)$, for either periodic or homogeneous boundary conditions. ★

The travelling wave solutions of 1D EPDiff when Green's function G is chosen to be even under reflection are the "pulsons,"

$$u(x,t) = c\, G(x-ct).$$

Exercise. Prove the statement that the travelling wave solutions of 1D EPDiff are pulsons when Green's function is even. What role is played in the solution by the Green's function being even?

Hint: Evaluate the derivative of an even function at $x = 0$. ★

See Fringer and Holm [FrHo2001] and references therein for further discussions and numerical simulations of the pulson solutions of the 1D EPDiff equation.

22.1 Pulsons

The EPDiff equation (22.0.2) on the real line has the remarkable property that its solutions *collectivise*[2] into the finite-dimensional solutions of the "N-pulson" form, which was discovered for a special form of G by Camassa and Holm [CaHo1993] and was then extended for *any* even G by Fringer and Holm [FrHo2001]:

$$u(x,t) = \sum_{i=1}^{N} p_i(t)\, G(x - q_i(t)). \tag{22.1.1}$$

Since $G(x)$ is the Green's function for the operator Q_{op}, the corresponding solution for the momentum $m = Q_{\mathrm{op}} u$ is given by a sum of delta functions,

$$m(x,t) = \sum_{i=1}^{N} p_i(t)\, \delta(x - q_i(t)). \tag{22.1.2}$$

Thus, the time-dependent "collective coordinates" $q_i(t)$ and $p_i(t)$ are the positions and velocities of the N pulses in this solution. These parameters satisfy the finite-dimensional geodesic motion equations obtained as canonical Hamiltonian equations,

$$\dot{q}_i = \frac{\partial H_N}{\partial p_i} = \sum_{j=1}^{N} p_j\, G(q_i - q_j), \tag{22.1.3}$$

$$\dot{p}_i = -\frac{\partial H_N}{\partial q_i} = -p_i \sum_{j=1}^{N} p_j\, G'(q_i - q_j), \tag{22.1.4}$$

in which the Hamiltonian is given by the quadratic form

$$H_N = \tfrac{1}{2} \sum_{i,j=1}^{N} p_i\, p_j\, G(q_i - q_j). \tag{22.1.5}$$

---

[2]See Guillemin and Sternberg [GuSt1984] for discussions of the concept of collective variables for Hamiltonian theories. We will discuss the collectivisation for the EPDiff equation later from the viewpoint of momentum maps.

Remark 22.1.1 In a certain sense, equations (22.1.3)–(22.1.4) comprise the analogue for the peakon momentum relation (22.1.2) of the "symmetric generalised rigid-body equations" in (9.5.1).   □

Thus, the canonical equations for the Hamiltonian H_N describe the nonlinear collective interactions of the N-pulson solutions of the EPDiff equation (22.0.2) as finite-dimensional geodesic motion of a particle on an N-dimensional surface whose co-metric is

$$G^{ij}(q) = G(q_i - q_j). \qquad (22.1.6)$$

Fringer and Holm [FrHo2001] showed numerically that the N-pulson solutions describe the emergent patterns in the solution of the initial-value problem for the EPDiff equation (22.0.2) with spatially confined initial conditions.

Exercise. Equations (22.1.3)–(22.1.4) describe geodesic motion.

1. Write the Lagrangian and Euler–Lagrange equations for this motion.
2. Solve equations (22.1.3)–(22.1.4) for $N = 2$ when $\lim_{|x| \to \infty} G(x) = 0$.
   a. Why should the solution be described as exchange of momentum in elastic collisions?
   b. Consider both head-on and overtaking collisions.
   c. Consider the antisymmetric case when the total momentum vanishes.   ★

22.1.1   Integrability

Calogero and Francoise [Ca1995, CaFr1996] found that for any finite number N, the Hamiltonian equations for H_N in (22.1.5) are completely integrable in the Liouville sense[3] for

$$G \equiv G_1(x) = \lambda + \mu \cos(\nu x) + \mu_1 \sin(\nu|x|)$$

$$\text{and} \quad G \equiv G_2(x) = \alpha + \beta|x| + \gamma x^2,$$

with λ, μ, μ_1, ν, and α, β, γ being arbitrary constants, such that λ and μ are real and μ_1 and ν both real or both imaginary.[4] Particular cases of G_1 and G_2 are the peakons $G_1(x) = e^{-|x|/\alpha}$ of Camassa and Holm [CaHo1993] and the compactons $G_2(x) = \max(1 - |x|, 0)$ of the Hunter–Saxton equation (see Hunter and Zheng [HuZh1994]). The latter is the EPDiff equation (22.0.2), with $\ell(u) = \frac{1}{2} \int u_x^2 \, dx$ and thus $m = -u_{xx}$.

22.1.2   Lie–Poisson Hamiltonian form of EPDiff

In terms of m, the conserved-energy Hamiltonian for the EPDiff equation (22.0.2) is obtained by Legendre-transforming the kinetic-energy Lagrangian as

$$h = \left\langle \frac{\delta\ell}{\delta u}, u \right\rangle - \ell(u).$$

Thus, the Hamiltonian depends on m as

$$h(m) = \frac{1}{2} \int m(x) G(x - y) m(y) \, dx dy,$$

which also reveals the geodesic nature of the EPDiff equation (22.0.2) and the role of $G(x)$ in the kinetic energy metric on the Hamiltonian side.

The corresponding *Lie–Poisson bracket* for EPDiff as a Hamiltonian evolution equation is given by

$$\partial_t m = \{m, h\} = -\operatorname{ad}^*_{\delta h/\delta m} m = -(\partial m + m\partial)\frac{\delta h}{\delta m} \quad \text{and} \quad \frac{\delta h}{\delta m} = u,$$

---

[3]A Hamiltonian system is integrable in the Liouville sense if the number of compatible independent constants of motion in involution is the same as the number of its degrees of freedom.

[4]This choice of the constants keeps H_N real in (22.1.5).

which recovers the starting equation and indicates some of its connections with fluid equations on the Hamiltonian side. For any two smooth functionals f and h of m in the space for which the solutions of EPDiff exist, this Lie–Poisson bracket may be expressed as

$$\{f, h\} = -\int \frac{\delta f}{\delta m}(\partial m + m\partial)\frac{\delta h}{\delta m}\,dx = -\int m\left[\frac{\delta f}{\delta m}, \frac{\delta h}{\delta m}\right]dx,$$

where $[\cdot, \cdot]$ denotes the Lie algebra bracket of vector fields. That is,

$$\left[\frac{\delta f}{\delta m}, \frac{\delta h}{\delta m}\right] = \frac{\delta f}{\delta m}\partial\frac{\delta h}{\delta m} - \frac{\delta h}{\delta m}\partial\frac{\delta f}{\delta m}.$$

Exercise. What is the Casimir for this Lie–Poisson bracket? What does it mean from the viewpoint of coadjoint orbits?   ★

22.2 Peakons

The case $G(x) = e^{-|x|/\alpha}$ with a constant length scale α is the Green's function for which the operator in the kinetic-energy Lagrangian (22.0.3) is $Q_{op} = 1 - \alpha^2\partial_x^2$. For this (Helmholtz) operator Q_{op}, the Lagrangian and corresponding kinetic-energy norm are given by

$$\ell[u] = \tfrac{1}{2}\|u\|^2 = \tfrac{1}{2}\int uQ_{op}u\,dx = \tfrac{1}{2}\int u^2 + \alpha^2 u_x^2\,dx, \quad \text{for } \lim_{|x|\to\infty} u = 0.$$

This Lagrangian is the H^1 norm of the velocity in 1D. In this case, the EPDiff equation (22.0.2) is also the zero-dispersion limit of the completely integrable CH equation for unidirectional shallow water waves, first derived by Camassa and Holm [CaHo1993],

$$m_t + um_x + 2mu_x = -c_0 u_x + \gamma u_{xxx}, \quad m = u - \alpha^2 u_{xx}. \quad (22.2.1)$$

This equation describes shallow water dynamics as completely integrable soliton motion at quadratic order in the asymptotic expansion for unidirectional shallow water waves on a free surface under gravity. See Dullin, Gottwald and Holm [DuGoHo2001, DuGoHo2003, DuGoHo2004] for more details and explanations of this asymptotic expansion for unidirectional shallow water waves to quadratic order.

Because of the relation $m = u - \alpha^2 u_{xx}$, equation (22.2.1) is non-local. In other words, it is an integral-partial differential equation. In fact, after writing equation (22.2.1) in the equivalent form,

$$(1 - \alpha^2 \partial^2)(u_t + u u_x) = -\partial\left(u^2 + \tfrac{1}{2}\alpha^2 u_x^2\right) - c_0 u_x + \gamma u_{xxx}, \quad (22.2.2)$$

one sees the interplay between local and nonlocal linear dispersion in its phase velocity relation,

$$\frac{\omega}{k} = \frac{c_0 - \gamma k^2}{1 + \alpha^2 k^2}, \quad (22.2.3)$$

for waves with frequency ω and wave number k linearised around $u = 0$. For $\gamma/c_0 < 0$, short waves and long waves travel in the same direction. Long waves travel faster than short ones (as required in shallow water) provided $\gamma/c_0 > -\alpha^2$. Then, the phase velocity lies in the interval $\omega/k \in (-\gamma/\alpha^2, c_0]$.

The famous Korteweg–de Vries (KdV) soliton equation,

$$u_t + 3 u u_x = -c_0 u_x + \gamma u_{xxx}, \quad (22.2.4)$$

emerges at *linear* order in the asymptotic expansion for shallow water waves, in which one takes $\alpha^2 \to 0$ in (22.2.2) and (22.2.3). In the KdV equation, the parameters c_0 and γ are seen as deformations of the *Riemann equation*

$$u_t + 3 u u_x = 0.$$

The parameters c_0 and γ represent linear wave dispersion, which modifies and eventually balances the tendency for nonlinear waves to steepen and break. The parameter α, which introduces nonlocality, also regularises this nonlinear tendency, even in the absence of c_0 and γ.

22.3   Integrability of the CH equation

22.3.1   The bi-Hamiltonian property of CH

The CH equation (22.2.1) may be written in bi-Hamiltonian form as

$$m_t = -(\partial - \partial^3)\frac{\delta H_2[m]}{\delta m} = -(2\kappa\partial + \partial m + m\partial)\frac{\delta H_1[m]}{\delta m}, \quad (22.3.1)$$

where the two Hamiltonians are given by

$$H_1[m] = \tfrac{1}{2}\int mu\,dx \quad \text{and} \quad H_2[m] = \tfrac{1}{2}\int (u^3 + uu_x^2 + 2\kappa u^2)dx. \quad (22.3.2)$$

The integration is over the real line for functions that decay sufficiently rapidly as $|x| \to \infty$. (The integration is over one period for periodic functions.)

22.3.2   The multi-Hamiltonian structure of CH

By Magri's theorem [Ma1978], the bi-Hamiltonian property of CH in (22.3.1) implies an infinite sequence of conservation laws for it, known as a *multi-Hamiltonian structure* $H_n[m]$, $n = 0, \pm 1, \pm 2, \dots$, such that

$$(\partial - \partial^3)\frac{\delta H_n[m]}{\delta m} = (2\kappa\partial + \partial m + m\partial)\frac{\delta H_{n-1}[m]}{\delta m}. \quad (22.3.3)$$

22.3.3   The Lax pair for CH

Its bi-Hamiltonian property also implies that the CH equation (22.2.1) admits a *Lax pair representation*, given by

$$\Psi_{xx} = \left(\frac{1}{4} + \lambda(m + \kappa)\right)\Psi, \quad (22.3.4)$$

$$\Psi_t = \left(\frac{1}{2\lambda} - u\right)\Psi_x + \frac{u_x}{2}\Psi + \gamma\Psi, \quad (22.3.5)$$

where κ and γ are arbitrary real constants and the eigenvalue λ is independent of time. The compatibility of the Lax pair ($\Psi_{xxt} = \Psi_{txx}$) for constant λ means that the eigenvalue equation in (22.3.3) is *isospectral*, i.e., its spectrum is invariant under the flow of the CH equation. The spectrum of (22.3.3) turns out to represent the speeds of the soliton solutions at late times when they are sufficiently separated. Remarkably, when $\kappa = 0$, the spectrum of the eigenvalue equation in (22.3.3) is purely *discrete*, which means the solution in this case comprises only peakons.

Many papers have been written to explore various features of the original CH equation. See Ref. [LuSz2022] for a recent brief chronological discussion of the exploration of singular peakon solutions.

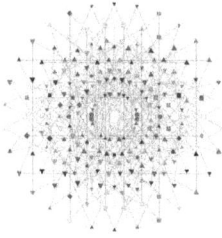

23

DIFFEONS: SINGULAR MOMENTUM SOLUTIONS OF THE EPDIFF EQUATION FOR GEODESIC MOTION IN HIGHER DIMENSIONS

Contents

What is this lecture about? This lecture extends the one-dimensional (1D) singular geodesic solutions of EPDiff to higher dimensions.

As an example of the Euler–Poincaré (EP) theory in higher dimensions, we shall generalise the 1D pulson solutions of the previous section to n dimensions. The corresponding singular momentum solutions of the EPDiff equation in higher dimensions are called "diffeons."

23.1 n-Dimensional EPDiff equation

Eulerian geodesic motion of a fluid in n dimensions is generated as an EP equation via Hamilton's principle, when the Lagrangian is given by the kinetic energy. The kinetic energy defines a norm $\|\mathbf{u}\|^2$ for the Eulerian fluid velocity, $\mathbf{u}(\mathbf{x}, t)\colon R^n \times R^1 \to R^n$. As mentioned earlier, the choice of the kinetic energy as a positive functional of fluid velocity $\mathbf{u}$ is a modelling step that depends on the physics of the problem being studied. Following our earlier procedure, as in equations (21.1.1) and (21.1.2), we shall choose the Lagrangian,

$$\|\mathbf{u}\|^2 = \int \mathbf{u} \cdot Q_{\mathrm{op}} \mathbf{u} \, d^n x = \int \mathbf{u} \cdot \mathbf{m} \, d^n x, \qquad (23.1.1)$$

so that the positive-definite, symmetric operator Q_{op} defines the norm $\|\mathbf{u}\|$ for appropriate boundary conditions, and the EPDiff equation for Eulerian geodesic motion of a fluid emerges:

$$\frac{d}{dt} \frac{\delta \ell}{\delta \mathbf{u}} + \mathrm{ad}^*_{\mathbf{u}} \frac{\delta \ell}{\delta \mathbf{u}} = 0, \quad \text{with } \ell[\mathbf{u}] = \tfrac{1}{2} \|\mathbf{u}\|^2. \qquad (23.1.2)$$

23.1.1   Legendre transforming to the Hamiltonian side

The corresponding Legendre transform yields the following invertible relations between momentum and velocity:

$$\mathbf{m} = Q_{\mathrm{op}} \mathbf{u} \quad \text{and} \quad \mathbf{u} = G * \mathbf{m}, \qquad (23.1.3)$$

where G is the *Green's function* for the operator Q_{op}, assuming appropriate boundary conditions (on $\mathbf{u}$) that allow the inversion of the operator Q_{op} to determine $\mathbf{u}$ from $\mathbf{m}$.

The corresponding *Hamiltonian* is,

$$h[\mathbf{m}] = \langle \mathbf{m}, \mathbf{u} \rangle - \tfrac{1}{2}\|\mathbf{u}\|^2 = \tfrac{1}{2}\int \mathbf{m} \cdot G * \mathbf{m} \, d^n x \equiv \tfrac{1}{2}\|\mathbf{m}\|^2, \quad (23.1.4)$$

which also defines a norm $\|\mathbf{m}\|$ via a convolution kernel G that is symmetric and positive when the Lagrangian $\ell[\mathbf{u}]$ is a norm. As expected, the norm $\|\mathbf{m}\|$ given by the Hamiltonian $h[\mathbf{m}]$ specifies the velocity $\mathbf{u}$ in terms of its Legendre-dual momentum $\mathbf{m}$ by the variational operation

$$\mathbf{u} = \frac{\delta h}{\delta \mathbf{m}} = G * \mathbf{m} \equiv \int G(\mathbf{x} - \mathbf{y}) \, \mathbf{m}(\mathbf{y}) \, d^n y. \quad (23.1.5)$$

We shall choose the kernel $G(\mathbf{x} - \mathbf{y})$ to be translation-invariant (so that Noether's theorem implies the total momentum $\mathbf{M} = \int \mathbf{m} \, d^n x$ is conserved) and symmetric under spatial reflections (so that $\mathbf{u}$ and $\mathbf{m}$ have the same parity).

After the Legendre transformation (23.1.4), the EPDiff equation (23.1.2) appears in its equivalent *Lie–Poisson Hamiltonian form*, which is reminiscent of the EPSO(3) equation for rigid-body motion since $\delta h / \delta \mathbf{m}$ is a vector field,

$$\frac{\partial}{\partial t}\mathbf{m} = \{\mathbf{m}, h\} = -\operatorname{ad}^*_{\delta h/\delta \mathbf{m}} \mathbf{m} = -\operatorname{ad}^*_{G*\mathbf{m}} \mathbf{m}. \quad (23.1.6)$$

Here, the operation $\{\cdot, \cdot\}$ denotes the Lie–Poisson bracket dual to the (right) action of vector fields among themselves by vector-field commutation

$$\{f, h\} = -\left\langle \mathbf{m}, \left[\frac{\delta f}{\delta \mathbf{m}}, \frac{\delta h}{\delta \mathbf{m}}\right] \right\rangle.$$

For more details and additional background concerning the relation of classical EP theory to Lie–Poisson Hamiltonian equations, see Holm, Marsden and Ratiu [MaRa1994, HoMaRa1998a].

In a moment, we will also consider the momentum maps for EPDiff.

23.2   Diffeons: n-Dimensional analogues of pulsons for the EPDiff equation

The momentum for the 1D pulson solutions (22.1.2) on the real line is supported at points via the Dirac delta measures in its solution *ansatz*:

$$m(x,t) = \sum_{i=1}^{N} p_i(t)\, \delta\big(x - q_i(t)\big), \quad m \in R^1. \tag{23.2.1}$$

We shall develop n-dimensional analogues of these 1D pulson solutions for the Euler–Poincaré equation (21.1.9) by generalising this solution *ansatz* to allow measure-valued n-dimensional vector solutions $\mathbf{m} \in R^n$ for which the Euler–Poincaré momentum is supported on co-dimension-k subspaces R^{n-k} with integer $k \in [1, n]$. For example, one may consider the two-dimensional (2D) vector momentum $\mathbf{m} \in R^2$ in the plane that is supported on 1D curves (momentum fronts). Likewise, in three dimensions (3D), one could consider 2D momentum surfaces (sheets), 1D momentum filaments, etc. The corresponding vector momentum *ansatz* that we shall use is the following, cf. the pulson solutions (23.2.1):

$$\mathbf{m}(\mathbf{x},t) = \sum_{i=1}^{N} \int \mathbf{P}_i(s,t)\, \delta\big(\mathbf{x} - \mathbf{Q}_i(s,t)\big)ds, \quad \mathbf{m} \in R^n. \tag{23.2.2}$$

Here, $\mathbf{P}_i, \mathbf{Q}_i \in R^n$ for $i = 1, 2, \ldots, N$. For example, when $n - k = 1$, so that $s \in R^1$ is 1D, the delta function in solution (23.2.2) supports an evolving family of vector-valued curves, called *momentum filaments*. (For simplicity of notation, we suppress the implied subscript i in the arclength s for each $\mathbf{P}_i$ and $\mathbf{Q}_i$.) The Legendre-dual relations (23.1.3) imply that the velocity corresponding to the momentum filament *ansatz* (23.2.2) is

$$\mathbf{u}(\mathbf{x},t) = G * \mathbf{m} = \sum_{j=1}^{N} \int \mathbf{P}_j(s',t)\, G\big(\mathbf{x} - \mathbf{Q}_j(s',t)\big)ds'. \tag{23.2.3}$$

Just as for the 1D case of the pulsons, we shall show that substitution of the n-dimensional solution *ansatz* (23.2.2) and (23.2.3) into the EPDiff equation (21.1.6) produces canonical geodesic Hamiltonian equations for the n-dimensional vector parameters $\mathbf{Q}_i(s,t)$ and $\mathbf{P}_i(s,t)$, $i = 1, 2, \ldots, N$.

23.2.1 Canonical Hamiltonian dynamics of diffeon filaments in R^n

For definiteness in what follows, we consider the example of momentum filaments $\mathbf{m} \in R^n$ supported on 1D space curves in R^n, so that $s \in R^1$ is the arclength parameter of one of these curves. This solution *ansatz* is reminiscent of the Biot–Savart law for vortex filaments, although the flow is not incompressible. The dynamics of momentum surfaces, for $s \in R^k$ with $k < n$, follow a similar analysis.

Substituting the momentum filament *ansatz* (23.2.2) for $s \in R^1$ and its corresponding velocity (23.2.3) into the Euler–Poincaré equation (21.1.6) and then integrating against a smooth test function $\phi(\mathbf{x})$ implies the following canonical equations (denoting explicit summation on $i, j \in 1, 2, \ldots N$):

$$\frac{\partial}{\partial t}\mathbf{Q}_i(s,t) = \sum_{j=1}^{N} \int \mathbf{P}_j(s',t)\, G(\mathbf{Q}_i(s,t) - \mathbf{Q}_j(s',t)))\, ds' = \frac{\delta H_N}{\delta \mathbf{P}_i},$$

$$(23.2.4)$$

$$\frac{\partial}{\partial t}\mathbf{P}_i(s,t) = -\sum_{j=1}^{N} \int \left(\mathbf{P}_i(s,t)\cdot\mathbf{P}_j(s',t)\right)$$

$$\times \frac{\partial}{\partial \mathbf{Q}_i(s,t)} G\big(\mathbf{Q}_i(s,t) - \mathbf{Q}_j(s',t)\big)\, ds'$$

$$= -\frac{\delta H_N}{\delta \mathbf{Q}_i}, \quad \text{(sum on } j, \text{ no sum on } i). \qquad (23.2.5)$$

The dot product $\mathbf{P}_i \cdot \mathbf{P}_j$ denotes the inner, or scalar, product of the two vectors $\mathbf{P}_i$ and $\mathbf{P}_j$ in R^n. Thus, the solution *ansatz* (23.2.2) yields a closed set of *integro-partial-differential equations (IPDEs)* given by (23.2.4) and (23.2.5) for the vector parameters $\mathbf{Q}_i(s,t)$ and $\mathbf{P}_i(s,t)$, with $i = 1,2\ldots N$. These equations are generated canonically by the following Hamiltonian function $H_N\colon (R^n \times R^n)^{\otimes N} \to R$:

$$H_N = \tfrac{1}{2}\iint \sum_{i,j=1}^{N} \big(\mathbf{P}_i(s,t)\cdot\mathbf{P}_j(s',t)\big)\,G\big(\mathbf{Q}_i(s,t)-\mathbf{Q}_j(s',t)\big)\,ds\,ds'.$$

(23.2.6)

This Hamiltonian arises by substituting the momentum *ansatz* (23.2.2) into the Hamiltonian (23.1.4) obtained from the Legendre transformation of the Lagrangian corresponding to the kinetic-energy norm of the fluid velocity. Thus, the evolutionary IPDE system (23.2.4) and (23.2.5) represents canonically Hamiltonian geodesic motion on the space of curves in R^n with respect to the co-metric given on these curves in (23.2.6). The Hamiltonian $H_N = \tfrac{1}{2}\|\mathbf{P}\|^2$ in (23.2.6) defines the norm $\|\mathbf{P}\|$ in terms of this co-metric that combines convolution using Green's function G and sum over filaments with the scalar product of momentum vectors in R^n.

Remark 23.2.1 Note the Lagrangian property of the s coordinate since

$$\frac{\partial}{\partial t}\mathbf{Q}_i(s,t) = \mathbf{u}(\mathbf{Q}_i(s,t),t). \qquad \square$$

23.3 Singular solution momentum map J_{Sing} for diffeons

The diffeon momentum filament *ansatz* (23.2.2) reduces and *collectivises* the solution of the geodesic EP PDE (21.1.6) in $n+1$ dimensions into the system (23.2.4) and (23.2.5) of $2N$ canonical evolutionary IPDEs. One can summarise the mechanism through which this process occurs by saying that the map that implements the canonical $(\mathbf{Q},\mathbf{P})$ variables in terms of singular solutions is a (cotangent bundle) momentum map. Such momentum maps are Poisson maps; therefore, the canonical Hamiltonian nature of the dynamical equations for $(\mathbf{Q}, \mathbf{P})$ fits into a general theory which also provides a framework for suggesting other avenues of investigation.

Theorem 23.3.1 *The momentum ansatz (23.2.2) for measure-valued solutions of the* EPDiff *equation (21.1.6) defines an equivariant momentum map*

$$\mathbf{J}_{\text{Sing}} \colon T^* \operatorname{Emb}(S, \mathbb{R}^n) \to \mathfrak{X}^*(\mathbb{R}^n)$$

that is called the singular solution momentum map by Holm and Marsden [HoMa2005].

We shall explain the notation used in the theorem's statement in the course of its proof. Now, however, we only note that the sense of "defines" is that the momentum solution *ansatz* (23.2.2) expressing **m** (a vector function of spatial position **x**) in terms of **Q** and **P** (which are functions of s) can be regarded as a map from the space of $(\mathbf{Q}(s), \mathbf{P}(s))$ to the space of **m**'s. This will turn out to be the Lagrange-to-Euler map for the fluid description of the singular solutions. The proof follows Holm and Marsden [HoMa2005].

Proof. For simplicity and without loss of generality, let us take $N = 1$ and thus suppress the index a. That is, we shall take the case of an isolated singular solution. As the proof will show, this is not a real restriction.

To set the notation, fix a k-dimensional manifold S with a given volume element and whose points are denoted $s \in S$. Let $\operatorname{Emb}(S, \mathbb{R}^n)$ denote the set of smooth embeddings $\mathbf{Q} \colon S \to \mathbb{R}^n$. (If the EPDiff equations are taken on a manifold M, replace $\mathbb{R}^n$ with M.) Under appropriate technical conditions, which we shall only treat formally here, $\operatorname{Emb}(S, \mathbb{R}^n)$ is a smooth manifold. (See, for example, Ebin and Marsden [EbMa1970] and Marsden and Hughes [MaHu1994] for a discussion and references.)

The tangent space $T_{\mathbf{Q}} \operatorname{Emb}(S, \mathbb{R}^n)$ to $\operatorname{Emb}(S, \mathbb{R}^n)$ at the point $\mathbf{Q} \in \operatorname{Emb}(S, \mathbb{R}^n)$ is given by the space of *material velocity fields*, namely the linear space of maps $\mathbf{V} \colon S \to \mathbb{R}^n$ that are vector fields over the map $\mathbf{Q}$. The dual space to this space will be identified with the space of 1-form densities over $\mathbf{Q}$, which we shall regard as maps

P: $S \rightarrow (\mathbb{R}^n)^*$. In summary, the cotangent bundle $T^* \operatorname{Emb}(S, \mathbb{R}^n)$ is identified with the space of pairs of maps $(\mathbf{Q}, \mathbf{P})$.

This identification gives us the domain space for the singular solution momentum map. Now, we consider the action of the symmetry group. Consider the group $\mathfrak{G} = \operatorname{Diff}$ of diffeomorphisms of the space $\mathfrak{S}$ in which the EPDiff equations are operating, concretely in our case $\mathbb{R}^n$. Let it act on $\mathfrak{S}$ by composition on the *left*. Namely, for $\eta \in \operatorname{Diff}(\mathbb{R}^n)$, we let

$$\eta \cdot \mathbf{Q} = \eta \circ \mathbf{Q}. \tag{23.3.1}$$

Now, lift this action to the cotangent bundle $T^* \operatorname{Emb}(S, \mathbb{R}^n)$ in the standard way (see, for instance, Marsden and Ratiu [MaRa1994] for this construction). This lifted action is a symplectic (and hence Poisson) action and has an equivariant momentum map. *We claim that this momentum map is precisely given by the ansatz (23.2.2).*

To see this, one only needs to recall and then apply the general formula for the momentum map associated with an action of a general Lie group $\mathfrak{G}$ on a configuration manifold Q and cotangent lifted to T^*Q.

First, let us recall the general formula. Namely, the momentum map is the map $\mathbf{J} \colon T^*Q \rightarrow \mathfrak{g}^*$ ($\mathfrak{g}^*$ denotes the dual of the Lie algebra $\mathfrak{g}$ of $\mathfrak{G}$) defined by

$$\mathbf{J}(\alpha_q) \cdot \xi = \langle \alpha_q, \xi_Q(q) \rangle, \tag{23.3.2}$$

where $\alpha_q \in T_q^*Q$ and $\xi \in \mathfrak{g}$, with ξ_Q being the infinitesimal generator of the action of $\mathfrak{G}$ on Q associated with the Lie algebra element ξ, and where $\langle \alpha_q, \xi_Q(q) \rangle$ is the natural pairing of an element of T_q^*Q with an element of T_qQ.

Now, we apply this formula to the special case in which the group $\mathfrak{G}$ is the diffeomorphism group $\operatorname{Diff}(\mathbb{R}^n)$, the manifold Q is $\operatorname{Emb}(S, \mathbb{R}^n)$ and where the action of the group on $\operatorname{Emb}(S, \mathbb{R}^n)$ is given by relation (23.3.1). The sense in which the Lie algebra of $\mathfrak{G} = \operatorname{Diff}$ is the space $\mathfrak{g} = \mathfrak{X}$ of vector fields is well understood. Hence, its dual $\mathfrak{g}^* = \mathfrak{X}^*$ is naturally regarded as the space of 1-form densities. The momentum map is thus a map, $\mathbf{J} \colon T^* \operatorname{Emb}(S, \mathbb{R}^n) \rightarrow \mathfrak{X}^*$.

With $\mathbf{J}$ given by (23.3.2), we only need to work out this formula. First, we shall work out the infinitesimal generators. Let $X \in \mathfrak{X}$ be a Lie algebra element. By differentiating the action (23.3.1) with respect to η in the direction of X at the identity element, we find that the infinitesimal generator is given by

$$X_{\mathrm{Emb}(S,\mathbb{R}^n)}(\mathbf{Q}) = X \circ \mathbf{Q}.$$

Thus, taking α_q to be the cotangent vector $(\mathbf{Q}, \mathbf{P})$, equation (23.3.2) gives

$$\langle \mathbf{J}(\mathbf{Q}, \mathbf{P}), X \rangle = \langle (\mathbf{Q}, \mathbf{P}), X \circ \mathbf{Q} \rangle$$

$$= \int_S P_i(s) X^i(\mathbf{Q}(s)) d^k s.$$

On the other hand, note that the right-hand side of (23.2.2) (again with the index a suppressed, and with t suppressed as well) when paired with the Lie algebra element X is

$$\left\langle \int_S \mathbf{P}(s)\, \delta\, (\mathbf{x} - \mathbf{Q}(s))\, d^k s, X \right\rangle$$

$$= \int_{\mathbb{R}^n} \int_S \left(P_i(s)\, \delta\, (\mathbf{x} - \mathbf{Q}(s))\, d^k s \right) X^i(\mathbf{x}) d^n x$$

$$= \int_S P_i(s) X^i(\mathbf{Q}(s)) d^k s.$$

This shows that the expression given by (23.2.2) is equal to $\mathbf{J}$, and so the result is proved. ∎

The proof has shown the following basic fact.

Corollary 23.3.1 *The singular solution momentum map defined by the singular solution ansatz (23.2.2), namely,*

$$\mathbf{J}_{\mathrm{Sing}} \colon T^* \operatorname{Emb}(S, \mathbb{R}^n) \to \mathfrak{X}(\mathbb{R}^n)^*,$$

is a Poisson map from the canonical Poisson structure on $T^* \operatorname{Emb}(S, \mathbb{R}^n)$ to the Lie–Poisson structure on $\mathfrak{X}(\mathbb{R}^n)^*$.

This is perhaps the most basic property of the singular solution momentum map. Some of its more sophisticated properties are outlined by Holm and Marsden [HoMa2005].

23.3.1 Pulling back the equations

Since the solution *ansatz* (23.2.2) has been shown in the preceding Corollary to be a Poisson map, the pull-back of the Hamiltonian from $\mathfrak{X}^*$ to $T^* \operatorname{Emb}(S, \mathbb{R}^n)$ gives the equations of motion on the latter space that project to the equations on $\mathfrak{X}^*$.

Thus, the basic fact that the momentum map $\mathbf{J}_{\text{Sing}}$ is Poisson explains why the functions $\mathbf{Q}^a(s,t)$ and $\mathbf{P}^a(s,t)$ satisfy canonical Hamiltonian equations.

Note that the coordinate $s \in \mathbb{R}^k$ that labels these functions is a "Lagrangian coordinate" in the sense that it does not evolve in time but rather labels the solution.

In terms of the pairing

$$\langle \cdot, \cdot \rangle : \mathfrak{g}^* \times \mathfrak{g} \to \mathbb{R} \tag{23.3.3}$$

between the Lie algebra $\mathfrak{g}$ (vector fields in $\mathbb{R}^n$) and its dual $\mathfrak{g}^*$ (1-form densities in $\mathbb{R}^n$), the following relation holds for measure-valued solutions under the momentum map (23.2.2):

$$\langle \mathbf{m}, \mathbf{u} \rangle = \int \mathbf{m} \cdot \mathbf{u}\, d^n\mathbf{x}, \quad L^2 \text{ pairing for } \mathbf{m}, \mathbf{u} \in \mathbb{R}^n,$$

$$= \iint \sum_{a,b=1}^{N} \left(\mathbf{P}^a(s,t) \cdot \mathbf{P}^b(s',t) \right) G\left(\mathbf{Q}^a(s,t) - \mathbf{Q}^b(s',t) \right) ds\, ds'$$

$$= \int \sum_{a=1}^{N} \mathbf{P}^a(s,t) \cdot \frac{\partial \mathbf{Q}^a(s,t)}{\partial t}\, ds$$

$$\equiv \langle\!\langle \mathbf{P}, \dot{\mathbf{Q}} \rangle\!\rangle, \tag{23.3.4}$$

which is the natural pairing between the points $(\mathbf{Q}, \mathbf{P}) \in T^* \operatorname{Emb}(S, \mathbb{R}^n)$ and $(\mathbf{Q}, \dot{\mathbf{Q}}) \in T \operatorname{Emb}(S, \mathbb{R}^n)$. The momentum map relation (23.3.4) corresponds to preservation of the action of the Lagrangian $\ell[\mathbf{u}]$ under the cotangent lift of $\operatorname{Diff}(\mathbb{R}^n)$.

When the Hamiltonian $H[\mathbf{m}]$ defined on the dual of the Lie algebra $\mathfrak{g}^*$ is pulled back to $T^*, \operatorname{Emb}(S, \mathbb{R}^n)$, one recovers

$$H[\mathbf{m}] \equiv \tfrac{1}{2} \langle \mathbf{m}, G * \mathbf{m} \rangle = \tfrac{1}{2} \langle\!\langle \mathbf{P}, G * \mathbf{P} \rangle\!\rangle \equiv H_N[\mathbf{P}, \mathbf{Q}]. \tag{23.3.5}$$

In summary, in concert with the Poisson nature of the singular solution momentum map, we see that the singular solutions in terms of $\mathbf{Q}$ and $\mathbf{P}$ satisfy Hamiltonian equations and also define an invariant solution set for the EPDiff equations. In fact:

This invariant solution set is a special coadjoint orbit for the diffeomorphism group, as we shall discuss in the following lecture.

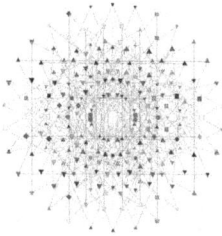

24

THE GEOMETRY OF THE MOMENTUM MAP

Contents

> **What is this lecture about?** This lecture explores the geometry of the singular solution momentum map discussed earlier in a little more detail. The treatment is formal, in the sense that there are a number of technical issues in the infinite-dimensional case that will be left open. We mention a few of these as we proceed.

24.1 Coadjoint orbits

We claim that *the image of the singular solution momentum map is a coadjoint orbit in $\mathfrak{X}^*$.* This means that (modulo some issues of connectedness and smoothness, which we do not consider here) the solution *ansatz* given by (23.2.2) defines a coadjoint orbit in the space of all 1-form densities, regarded as the dual of the Lie algebra of the diffeomorphism group. These coadjoint orbits should be thought of as singular orbits — that is, due to their special nature, they are not generic.

Recognising them as coadjoint orbits is one way of gaining further insight into why the singular solutions form dynamically invariant sets — it is a general fact that coadjoint orbits in $\mathfrak{g}^*$ are *symplectic submanifolds* of the Lie–Poisson manifold $\mathfrak{g}^*$ (in our case $\mathfrak{X}(\mathbb{R}^n)^*$) and, correspondingly, are dynamically invariant for any Hamiltonian system on $\mathfrak{g}^*$.

The idea of the proof of our claim is simply the following: whenever one has an equivariant momentum map $\mathbf{J} \colon P \to \mathfrak{g}^*$ for the action of a group G on a symplectic or Poisson manifold P, and that action is transitive, then the image of $\mathbf{J}$ is an orbit (or at least a piece of an orbit). This general result, due to Kostant, is stated more precisely by Marsden and Ratiu [MaRa1994, Theorem 14.4.5]. In rough terms, the reason that transitivity holds in our case is because one can "move the images of the manifolds S around at will with arbitrary velocity fields" using the diffeomorphisms of $\mathbb{R}^n$.

24.2   The momentum map $\mathbf{J}_S$ and the Kelvin circulation theorem

The momentum map $\mathbf{J}_{\mathrm{Sing}}$ involves $\mathrm{Diff}(\mathbb{R}^n)$, the left action of the diffeomorphism group on the space of embeddings $\mathrm{Emb}(S, \mathbb{R}^n)$ by smooth maps of the target space $\mathbb{R}^n$, namely,

$$\mathrm{Diff}(\mathbb{R}^n) \colon \mathbf{Q} \cdot \eta = \eta \circ \mathbf{Q}, \qquad (24.2.1)$$

where we recall that $\mathbf{Q} \colon S \to \mathbb{R}^n$. As above, the cotangent bundle $T^* \mathrm{Emb}(S, \mathbb{R}^n)$ is identified with the space of pairs of maps $(\mathbf{Q}, \mathbf{P})$, with $\mathbf{Q} \colon S \to \mathbb{R}^n$ and $\mathbf{P} \colon S \to T^*\mathbb{R}^n$.

However, there is another momentum map $\mathbf{J}_S$ associated with the *right action* of the diffeomorphism group of S on the embeddings $\mathrm{Emb}(S, \mathbb{R}^n)$ by smooth maps of the "Lagrangian labels" S (fluid particle relabelling by $\eta \colon S \to S$). This action is given by

$$\mathrm{Diff}(S) \colon \mathbf{Q} \cdot \eta = \mathbf{Q} \circ \eta. \qquad (24.2.2)$$

The infinitesimal generator of this right action is

$$X_{\mathrm{Emb}(S,\mathbb{R}^n)}(\mathbf{Q}) = \frac{d}{dt}\bigg|_{t=0} \mathbf{Q} \circ \eta_t = T\mathbf{Q} \circ X, \tag{24.2.3}$$

where $X \in \mathfrak{X}$ is tangent to the curve η_t at $t = 0$. Thus, again taking $N = 1$ (so we suppress the index a) and also letting α_q in the momentum map formula (23.3.2) be the cotangent vector $(\mathbf{Q}, \mathbf{P})$, one computes $\mathbf{J}_S$:

$$\begin{aligned}
\langle \mathbf{J}_S(\mathbf{Q}, \mathbf{P}), X \rangle &= \langle (\mathbf{Q}, \mathbf{P}), T\mathbf{Q} \cdot X \rangle \\
&= \int_S P_i(s) \frac{\partial Q^i(s)}{\partial s^m} X^m(s) \, d^k s \\
&= \int_S X \Big(\mathbf{P}(s) \cdot d\mathbf{Q}(s) \Big) d^k s \\
&= \left(\int_S \mathbf{P}(s) \cdot d\mathbf{Q}(s) \otimes d^k s, X(s) \right) \\
&= \langle \mathbf{P} \cdot d\mathbf{Q}, X \rangle.
\end{aligned}$$

Consequently, the momentum map formula (23.3.2) yields, cf. Exercise 20.2,

$$\mathbf{J}_S(\mathbf{Q}, \mathbf{P}) = \mathbf{P} \cdot d\mathbf{Q}, \tag{24.2.4}$$

with the indicated pairing of the 1-form density $\mathbf{P} \cdot d\mathbf{Q}$ with the vector field X.

We have set things up so that the following is true.

Proposition 24.2.1 *The momentum map $\mathbf{J}_S$ is preserved by evolution equations (23.2.4)–(23.2.5) for $\mathbf{Q}$ and $\mathbf{P}$.*

Proof. It is enough to note that the Hamiltonian H_N in equation (23.2.6) is invariant under the cotangent lift of the action of $\mathrm{Diff}(S)$; it merely amounts to the invariance of the integral over S under reparametrisation, that is, the change of variables formula. Keep in mind that $\mathbf{P}$ includes a density factor. ∎

Remark 24.2.1

1. This result is similar to the Kelvin–Noether theorem for circulation Γ of an ideal fluid, which may be written as

$$\Gamma = \oint_{c(s)} D(s)^{-1}\mathbf{P}(s) \cdot d\mathbf{Q}(s)$$

   for each Lagrangian circuit $c(s)$, where D is the mass density and $\mathbf{P}$ is again the canonical momentum density. This similarity should come as no surprise because the Kelvin–Noether theorem for ideal fluids arises from the invariance of Hamilton's principle under fluid parcel relabelling by the *same* right action of the diffeomorphism group, as in (24.2.2).

2. Note that, being an equivariant momentum map, the map $\mathbf{J}_S$, as with $\mathbf{J}_{\text{Sing}}$, is also a Poisson map. That is, substituting the canonical Poisson bracket into relation (24.2.4), i.e., the relation $\mathbf{M}(\mathbf{x}) = \sum_i P_i(\mathbf{x})\nabla Q^i(\mathbf{x})$, yields the Lie–Poisson bracket on the space of $\mathbf{M}$'s. We use the different notations $\mathbf{m}$ and $\mathbf{M}$ because these quantities are analogous to the body and spatial angular momentum for rigid body mechanics. In fact, the quantity $\mathbf{m}$ given by the solution *Ansatz*, specifically, $\mathbf{m} = \mathbf{J}_{\text{Sing}}(\mathbf{Q},\mathbf{P})$ gives the singular solutions of the EPDiff equations, while $\mathbf{M}(\mathbf{x}) = \mathbf{J}_S(\mathbf{Q},\mathbf{P}) = \sum_i P_i(\mathbf{x})\nabla Q^i(\mathbf{x})$ is a conserved quantity.

3. In the language of fluid mechanics, the expression of $\mathbf{M}$ in terms of $(\mathbf{Q},\mathbf{P})$ is an example of a *Clebsch representation*, which expresses the solution of the EPDiff equations in terms of canonical variables that evolve by standard canonical Hamilton equations. This has been known in the case of fluid mechanics for more than 100 years. For modern discussions of the Clebsch representation for ideal fluids, see, for example, Holm and Kupershmidt [HoKu1983] and Marsden and Weinstein [MaWe1983a].

4. One more remark is in order, namely the special case in which $S = M$ is of course allowed. In this case, $\mathbf{Q}$ corresponds

to the map η itself, and $\mathbf{P}$ simply corresponds to its conjugate momentum. The quantity $\mathbf{m}$ corresponds to the spatial (dynamic) momentum density (that is, the right translation of $\mathbf{P}$ to the identity), while $\mathbf{M}$ corresponds to the conserved "body" momentum density (that is, the left translation of $\mathbf{P}$ to the identity). □

24.3   Brief summary

$\mathrm{Emb}(S, \mathbb{R}^n)$ admits two group actions. These are the group $\mathrm{Diff}(S)$ of diffeomorphisms of S, which acts by composition on the *right*, and the group $\mathrm{Diff}(\mathbb{R}^n)$, which acts by composition on the *left*. The group $\mathrm{Diff}(\mathbb{R}^n)$ acting from the left produces the singular solution momentum map $\mathbf{J}_{\mathrm{Sing}}$. The action of $\mathrm{Diff}(S)$ from the right produces the conserved momentum map $\mathbf{J}_S \colon T^* \mathrm{Emb}(S, \mathbb{R}^n) \to \mathfrak{X}(S)^*$. We now assemble both momentum maps into one figure as follows [HoMa2005]:

$$
\begin{array}{ccc}
 & T^* \mathrm{Emb}(S, M) & \\
\mathbf{J}_{\mathit{Sing}} \swarrow & & \searrow \mathbf{J}_S \\
\mathfrak{X}(M)^* & & \mathfrak{X}(S)^*
\end{array}
$$

For a full discussion of this dual pair for continuum dynamics, see Ref. [GaVi2012]. We turn next to the Euler–Poincaré framework for continuum dynamics.

Part III

Euler–Poincaré Framework of Continuum Partial Differential Equations

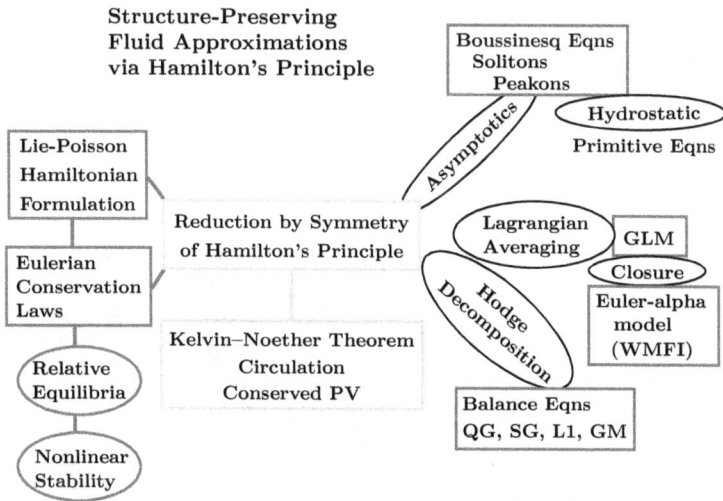

Structure-Preserving
Fluid Approximations
via Hamilton's Principle

Boussinesq Eqns
Solitons
Peakons

Hydrostatic
Primitive Eqns

Lie-Poisson
Hamiltonian
Formulation

Reduction by Symmetry
of Hamilton's Principle

Lagrangian
Averaging

GLM

Closure

Asymptotics

Eulerian
Conservation
Laws

Relative
Equilibria

Kelvin–Noether Theorem
Circulation
Conserved PV

*Hodge
Decomposition*

Euler-alpha
model
(WMFI)

Balance Eqns
QG, SG, L1, GM

Nonlinear
Stability

Structure-preserving fluid approximations via Hamilton's principle.

25

EULER–POINCARÉ FRAMEWORK OF FLUID DYNAMICS

Contents

What is this lecture about? This lecture assembles the Euler–Poincaré (EP) framework of the semidirect-product Lie–Poisson structure of ideal fluid dynamics step by step from the set up to the Kelvin–Noether theorem.

25.1   Set up and basic assumptions

Almost all fluid models of interest admit the following general assumptions. These assumptions form the basis of the EP theorem for continua, which we shall state later in this section, after introducing the notation necessary for dealing geometrically with the reduction of Hamilton's principle from the material (or Lagrangian) picture of fluid dynamics to the spatial (or Eulerian) picture. This theorem was first stated and proved by Holm, Marsden and Ratiu [HoMaRa1998a], to which we refer for additional details, as well as for abstract definitions and proofs.

25.1.1   Basic assumptions underlying the EP theorem for continua

- There is a *right* representation of a Lie group G on the vector space V and G acts in the natural way on the *right* on $TG \times V^*$: $(U_g, a)h = (U_g h, ah)$.

- The Lagrangian function $L \colon TG \times V^* \to \mathbb{R}$ is right G-invariant under the isotropy group of $a_0 \in V^*$.[1]

- In particular, if $a_0 \in V^*$, define the Lagrangian $L_{a_0} \colon TG \to \mathbb{R}$ by $L_{a_0}(U_g) = L(U_g, a_0)$. Then, L_{a_0} is right invariant under the lift to TG of the right action of G_{a_0} on G, where G_{a_0} is the isotropy group of a_0.

- Right G-invariance of L permits one to define the Lagrangian on the Lie algebra $\mathfrak{g}$ of the group G. Namely, $\ell \colon \mathfrak{g} \times V^* \to \mathbb{R}$ is defined by

$$\ell(u, a) = L\big(U_g g^{-1}(t), a_0 g^{-1}(t)\big) = L(U_g, a_0), \qquad (25.1.1)$$

where $u = U_g g^{-1}(t)$ and $a = a_0 g^{-1}(t)$. Conversely, this relation defines for any $\ell \colon \mathfrak{g} \times V^* \to \mathbb{R}$ a right G-invariant function $L \colon TG \times V^* \to \mathbb{R}$.

---

[1]For fluid dynamics, right G-invariance of the Lagrangian function L is traditionally called "particle relabeling symmetry."

- For a curve $g(t) \in G$, let $u(t) := \dot{g}(t)g(t)^{-1}$, and define the curve $a(t)$ as the unique solution of the linear differential equation with time-dependent coefficients $\dot{a}(t) = -a(t)u(t)$, where the action of an element of the Lie algebra $u \in \mathfrak{g}$ on an advected quantity $a \in V^*$ is denoted by concatenation from the right. The solution with initial condition $a(0) = a_0 \in V^*$ can be written as $a(t) = a_0 g(t)^{-1}$.

25.2 Notation for reduction of Hamilton's principle by symmetries

- Let $\mathfrak{g}(\mathcal{D})$ denote the space of vector fields on $\mathcal{D}$ of some fixed differentiability class. These vector fields are endowed with the *Lie bracket* given in components by (summing on repeated indices)

$$[\mathbf{u}, \mathbf{v}]^i = u^j \frac{\partial v^i}{\partial x^j} - v^j \frac{\partial u^i}{\partial x^j}. \tag{25.2.1}$$

The notation $\mathrm{ad}_{\mathbf{u}} \mathbf{v} := [\mathbf{u}, \mathbf{v}]$ formally denotes the adjoint action of the *right* Lie algebra of $\mathrm{Diff}(\mathcal{D})$ on itself.

- Identify the Lie algebra of vector fields $\mathfrak{g}$ with its dual $\mathfrak{g}^*$ by using the L^2 pairing

$$\langle \mathbf{u}, \mathbf{v} \rangle = \int_{\mathcal{D}} \mathbf{u} \cdot \mathbf{v} \, dV. \tag{25.2.2}$$

- Let $\mathfrak{g}(\mathcal{D})^*$ denote the geometric dual space of $\mathfrak{g}(\mathcal{D})$, that is, $\mathfrak{g}(\mathcal{D})^* := \Lambda^1(\mathcal{D}) \otimes \mathrm{Den}(\mathcal{D})$. This is the space of 1-form densities on $\mathcal{D}$. If $\mathbf{m} \otimes dV \in \Lambda^1(\mathcal{D}) \otimes \mathrm{Den}(\mathcal{D})$, then the pairing of $\mathbf{m} \otimes dV$ with $\mathbf{u} \in \mathfrak{g}(\mathcal{D})$ is given by the L^2 pairing,

$$\langle \mathbf{m} \otimes dV, \mathbf{u} \rangle = \int_{\mathcal{D}} \mathbf{m} \cdot \mathbf{u} \, dV \tag{25.2.3}$$

where $\mathbf{m} \cdot \mathbf{u}$ is the standard contraction of a 1-form $\mathbf{m}$ with a vector field $\mathbf{u}$.

- For $\mathbf{u} \in \mathfrak{g}(\mathcal{D})$ and $\mathbf{m} \otimes dV \in \mathfrak{g}(\mathcal{D})^*$, the dual of the adjoint representation is defined by

$$
\langle \mathrm{ad}_{\mathbf{u}}^*(\mathbf{m} \otimes dV), \mathbf{v} \rangle = - \int_{\mathcal{D}} \mathbf{m} \cdot \mathrm{ad}_{\mathbf{u}} \mathbf{v} \, dV = - \int_{\mathcal{D}} \mathbf{m} \cdot [\mathbf{u}, \mathbf{v}] \, dV
$$

(25.2.4)

and its expression is

$$
\mathrm{ad}_{\mathbf{u}}^*(\mathbf{m} \otimes dV) = (\pounds_{\mathbf{u}} \mathbf{m} + (\mathrm{div}_{dV} \mathbf{u})\mathbf{m}) \otimes dV = \pounds_{\mathbf{u}}(\mathbf{m} \otimes dV),
$$

(25.2.5)

where $\mathrm{div}_{dV} \mathbf{u}$ is the divergence of $\mathbf{u}$ relative to the measure dV, that is, $\pounds_{\mathbf{u}} dV = (\mathrm{div}_{dV} \mathbf{u}) dV$. Hence, $\mathrm{ad}_{\mathbf{u}}^*$ coincides with the Lie derivative $\pounds_{\mathbf{u}}$ for 1-form densities.

- If $\mathbf{u} = u^j \partial/\partial x^j$, $\mathbf{m} = m_i dx^i$, then the 1-form factor in the preceding formula for $\mathrm{ad}_{\mathbf{u}}^*(\mathbf{m} \otimes dV)$ has the *coordinate expression*

$$
\left(\mathrm{ad}_{\mathbf{u}}^* \mathbf{m} \right)_i dx^i = \left(u^j \frac{\partial m_i}{\partial x^j} + m_j \frac{\partial u^j}{\partial x^i} + (\mathrm{div}_{dV} \mathbf{u}) m_i \right) dx^i
$$

(25.2.6)

$$
= \left(\frac{\partial}{\partial x^j} (u^j m_i) + m_j \frac{\partial u^j}{\partial x^i} \right) dx^i.
$$

(25.2.7)

The last equality assumes that the divergence is taken relative to the standard measure $dV = d^n\mathbf{x}$ in $\mathbb{R}^n$. (On a Riemannian manifold, the metric divergence needs to be used.)

25.3 Conventions and terminology in continuum mechanics

Throughout the rest of the lecture notes, we shall follow Holm, Marsden and Ratiu [HoMaRa1998a] in using the conventions and terminology for the standard quantities in continuum mechanics.

Definition 25.3.1 *Elements of $\mathcal{D}$ representing the material particles of the system are denoted by X; their coordinates X^A, $A = 1, \ldots, n$ may thus be regarded as the **particle labels**.*

1. A **configuration**, which we typically denote by η, or g, is an element of $\mathrm{Diff}(\mathcal{D})$.

2. A **motion**, denoted as η_t or alternatively as $g(t)$, is a time-dependent curve in $\mathrm{Diff}(\mathcal{D})$.

Definition 25.3.2 The **Lagrangian**, or **material velocity** $\mathbf{U}(X,t)$ of the continuum along the motion η_t or $g(t)$ is defined by taking the time derivative of the motion keeping the particle labels X fixed:

$$\mathbf{U}(X,t) := \frac{d\eta_t(X)}{dt} := \left.\frac{\partial}{\partial t}\right|_X \eta_t(X) := \dot{g}(t) \cdot X.$$

These are convenient shorthand notations for the time derivative at a fixed Lagrangian coordinate X.

Consistent with this definition of material velocity, the tangent space to $\mathrm{Diff}(\mathcal{D})$ at $\eta \in \mathrm{Diff}(\mathcal{D})$ is given by

$$T_\eta \mathrm{Diff}(\mathcal{D}) = \{\mathbf{U}_\eta \colon \mathcal{D} \to T\mathcal{D} \mid \mathbf{U}_\eta(X) \in T_{\eta(X)}\mathcal{D}\}.$$

The elements of $T_\eta \mathrm{Diff}(\mathcal{D})$ are usually thought of as vector fields on $\mathcal{D}$ covering η. The tangent lift of right translations on $T\mathrm{Diff}(\mathcal{D})$ by $\varphi \in \mathrm{Diff}(\mathcal{D})$ is given by

$$\mathbf{U}_\eta \varphi := T_\eta R_\varphi(\mathbf{U}_\eta) = \mathbf{U}_\eta \circ \varphi.$$

Definition 25.3.3 During a motion η_t or $g(t)$, the particle labeled by X describes a path in $\mathcal{D}$, whose points

$$x(X,t) := \eta_t(X) := g(t) \cdot X,$$

are called the **Eulerian** or **spatial points** of this path, which is also called the **Lagrangian trajectory** because a Lagrangian fluid parcel follows this path in space. The derivative $\mathbf{u}(x,t)$ of this path, evaluated at a fixed Eulerian point x, is called the **Eulerian** or **spatial velocity** of the system:

$$\mathbf{u}(x,t) := \mathbf{u}(\eta_t(X),t) := \mathbf{U}(X,t) := \left.\frac{\partial}{\partial t}\right|_X \eta_t(X)$$

$$:= \dot{g}(t) \cdot X := \dot{g}(t)g^{-1}(t) \cdot x.$$

Thus, the Eulerian velocity **u** *is a time-dependent vector field on* $\mathcal{D}$, *denoted as* $\mathbf{u}_t \in \mathfrak{g}(\mathcal{D})$, *where* $\mathbf{u}_t(x) := \mathbf{u}(x,t)$. *We also have the fundamental relationships*

$$\mathbf{U}_t = \mathbf{u}_t \circ \eta_t \quad \text{and} \quad \mathbf{u}_t = \dot{g}(t)g^{-1}(t),$$

where we denote $\mathbf{U}_t(X) := \mathbf{U}(X,t)$.

Definition 25.3.4 *The* **representation space** V^* *of* $\mathrm{Diff}(\mathcal{D})$ *in continuum mechanics is often some subspace of the tensor field densities on* $\mathcal{D}$, *denoted as* $\mathfrak{T}(\mathcal{D}) \otimes \mathrm{Den}(\mathcal{D})$, *and the representation is given by a pull-back. It is thus a* **right** *representation of* $\mathrm{Diff}(\mathcal{D})$ *on* $\mathfrak{T}(\mathcal{D}) \otimes \mathrm{Den}(\mathcal{D})$. *The right action of the Lie algebra* $\mathfrak{g}(\mathcal{D})$ *on* V^* *is denoted as* **concatenation from the right**. *That is, we denote*

$$a\mathbf{u} := \mathcal{L}_{\mathbf{u}}a,$$

which is the Lie derivative of the tensor field density a along the vector field **u**.

Definition 25.3.5 *The* **Lagrangian of a continuum mechanical system** *is a function*

$$L \colon T\,\mathrm{Diff}(\mathcal{D}) \times V^* \to \mathbb{R},$$

which is right invariant relative to the tangent lift of the right translation of $\mathrm{Diff}(\mathcal{D})$ *on itself and pull-back on the tensor field densities. Invariance of the Lagrangian L induces a function* $\ell \colon \mathfrak{g}(\mathcal{D}) \times V^* \to \mathbb{R}$, *given by*

$$\ell(\mathbf{u}, a) = L(\mathbf{u} \circ \eta, \eta^* a) = L(\mathbf{U}, a_0),$$

where $\mathbf{u} \in \mathfrak{g}(\mathcal{D})$ *and* $a \in V^* \subset \mathfrak{T}(\mathcal{D}) \otimes \mathrm{Den}(\mathcal{D})$, *and where* $\eta^* a$ *denotes the pull-back of a by the diffeomorphism* η *and* **u** *is the Eulerian velocity. That is,*

$$\mathbf{U} = \mathbf{u} \circ \eta \quad \text{and} \quad a_0 = \eta^* a. \tag{25.3.1}$$

The evolution of a is by right action, given by the equation

$$\dot{a} = -\mathcal{L}_{\mathbf{u}}\,a = -a\mathbf{u}. \tag{25.3.2}$$

The solution of this equation, for the initial condition a_0, is

$$a(t) = \eta_{t*} a_0 = a_0 g^{-1}(t), \tag{25.3.3}$$

where the lower star denotes the push-forward operation and η_t is the flow of $\mathbf{u} = \dot{g} g^{-1}(t)$.

Definition 25.3.6 *Advected Eulerian quantities are defined in continuum mechanics to be those variables which are Lie-transported by the flow of the Eulerian velocity field. Using this standard terminology, equation (25.3.2), or its solution (25.3.3), states that the tensor field density $a(t)$ (which may include mass density and other Eulerian quantities) is advected.*

Remark 25.3.1 (Dual tensors) As we mentioned, typically, $V^* \subset \mathfrak{T}(\mathcal{D}) \otimes \mathrm{Den}(\mathcal{D})$ for continuum mechanics. On a general manifold, tensors of a given type have natural duals. For example, symmetric covariant tensors are dual to symmetric contravariant tensor densities, the pairing being given by the integration of the natural contraction of these tensors. Likewise, k-forms are naturally dual to $(n-k)$-forms, the pairing being given by taking the integral of their wedge product.   □

Definition 25.3.7 *The diamond operation $\diamond$ between the elements of V and V^* produces an element of the dual Lie algebra $\mathfrak{g}(\mathcal{D})^*$ and is defined as*

$$\langle b \diamond a, \mathbf{w} \rangle = - \int_{\mathcal{D}} b \cdot \pounds_{\mathbf{w}} a, \tag{25.3.4}$$

*where $b \cdot \pounds_{\mathbf{w}} a$ denotes the contraction, as described above, of the elements of V and the elements of V^* and $\mathbf{w} \in \mathfrak{g}(\mathcal{D})$. (These operations do **not** depend on a Riemannian structure.)*

For a path $\eta_t \in \mathrm{Diff}(\mathcal{D})$, let $\mathbf{u}(x, t)$ be its Eulerian velocity, and consider the curve $a(t)$ with initial condition a_0 given by the equation

$$\partial_t a + \pounds_{\mathbf{u}} a = 0. \tag{25.3.5}$$

Let the Lagrangian $L_{a_0}(\mathbf{U}) := L(\mathbf{U}, a_0)$ be right invariant under $\mathrm{Diff}(\mathcal{D})$. We can now state the EP theorem for continua of Holm, Marsden and Ratiu [HoMaRa1998a].

25.4   Statement of the EP theorem for continua

Theorem 25.4.1 (EP theorem for continua) *Given   a   path η_t in* Diff$(\mathcal{D})$ *with Lagrangian velocity* $\mathbf{U}$ *and Eulerian velocity* $\mathbf{u}$*, the following are equivalent:*

1. *Hamilton's variational principle*

$$\delta \int_{t_1}^{t_2} L\left(X, \mathbf{U}_t(X), a_0(X)\right) dt = 0 \qquad (25.4.1)$$

   holds for variations $\delta\eta_t$ *vanishing at the endpoints.*

2. η_t *satisfies the Euler–Lagrange equations for* L_{a_0} *on* Diff$(\mathcal{D})$.

3. *The constrained variational principle in Eulerian coordinates*

$$\delta \int_{t_1}^{t_2} \ell(\mathbf{u}, a) \, dt = 0 \qquad (25.4.2)$$

   holds on $\mathfrak{g}(\mathcal{D}) \times V^*$ *using variations of the form*

$$\delta\mathbf{u} = \frac{\partial\mathbf{w}}{\partial t} + [\mathbf{u}, \mathbf{w}] = \frac{\partial\mathbf{w}}{\partial t} + ad_{\mathbf{u}}\mathbf{w}, \qquad \delta a = -\mathcal{L}_{\mathbf{w}}\, a, \quad (25.4.3)$$

   where $\mathbf{w}_t = \delta\eta_t \circ \eta_t^{-1}$ *vanishes at the endpoints.*

4. *The EP equations for continua*

$$\frac{\partial}{\partial t}\frac{\delta\ell}{\delta\mathbf{u}} = -\, ad_{\mathbf{u}}^*\frac{\delta\ell}{\delta\mathbf{u}} + \frac{\delta\ell}{\delta a} \diamond a = -\mathcal{L}_{\mathbf{u}}\frac{\delta\ell}{\delta\mathbf{u}} + \frac{\delta\ell}{\delta a} \diamond a \qquad (25.4.4)$$

   hold, with auxiliary equations $(\partial_t + \mathcal{L}_{\mathbf{u}})a = 0$ *for each advected quantity* $a(t)$*. The* $\diamond$ *operation defined in (25.3.4) needs to be determined on a case-by-case basis, depending on the nature of the tensor* $a(t)$*. The variation* $\mathbf{m} = \delta\ell/\delta\mathbf{u}$ *is a 1-form density, and we have used relation (25.2.5) in the last step of equation (25.4.4).*

We refer to Holm, Marsden and Ratiu [HoMaRa1998a] for the proof of this theorem in the abstract setting. We shall see some of the features of this result in the concrete setting of continuum mechanics shortly.

25.5 Discussion of the EP equations

The following string of equalities shows *directly* that (3) is equivalent to (4):

$$
\begin{aligned}
0 &= \delta \int_{t_1}^{t_2} l(\mathbf{u}, a)dt = \int_{t_1}^{t_2} \left(\frac{\delta l}{\delta \mathbf{u}} \cdot \delta \mathbf{u} + \frac{\delta l}{\delta a} \cdot \delta a \right) dt \\
&= \int_{t_1}^{t_2} \left[\frac{\delta l}{\delta \mathbf{u}} \cdot \left(\frac{\partial \mathbf{w}}{\partial t} - \mathrm{ad}_{\mathbf{u}} \, \mathbf{w} \right) - \frac{\delta l}{\delta a} \cdot \pounds_{\mathbf{w}} \, a \right] dt \\
&= \int_{t_1}^{t_2} \mathbf{w} \cdot \left[-\frac{\partial}{\partial t} \frac{\delta l}{\delta \mathbf{u}} - \mathrm{ad}_{\mathbf{u}}^* \frac{\delta l}{\delta \mathbf{u}} + \frac{\delta l}{\delta a} \diamond a \right] dt.
\end{aligned} \qquad (25.5.1)
$$

The rest of the proof follows essentially the same track as the proof of the pure EP theorem, modulo slight changes to accommodate the advected quantities.

In the absence of dissipation, most Eulerian fluid equations[2] can be written in the EP form in equation (25.4.4):

$$
\frac{\partial}{\partial t} \frac{\delta \ell}{\delta \mathbf{u}} + \mathrm{ad}_{\mathbf{u}}^* \frac{\delta \ell}{\delta \mathbf{u}} = \frac{\delta \ell}{\delta a} \diamond a, \quad \text{with } (\partial_t + \pounds_{\mathbf{u}})a = 0. \qquad (25.5.2)
$$

Equation (25.5.2) is *Newton's law*: The Eulerian time derivative of the momentum density $\mathbf{m} = \delta \ell / \delta \mathbf{u}$ (a 1-form density dual to the velocity u) is equal to the force density $(\delta \ell / \delta a) \diamond a$, with the $\diamond$ operation defined in (25.3.4). Thus, Newton's law is written in the Eulerian

---

[2]Exceptions to this statement are certain multiphase fluids and complex fluids with active internal degrees of freedom such as liquid crystals. These require a further extension, which is not discussed here.

fluid representation as[3]

$$\frac{d}{dt}\Big|_{\text{Lag}} \mathbf{m} := \left(\partial_t + \pounds_{\mathbf{u}}\right)\mathbf{m} = \frac{\delta\ell}{\delta a} \diamond a, \quad \text{with} \ \frac{d}{dt}\Big|_{\text{Lag}} a := \left(\partial_t + \pounds_{\mathbf{u}}\right)a = 0.$$
$$(25.5.3)$$

The left side of the EP equation in (25.5.3) describes the fluid's dynamics due to its kinetic energy. A fluid's kinetic energy typically defines a norm for the Eulerian fluid velocity, $KE = \frac{1}{2}\|\mathbf{u}\|^2$. The left side of the EP equation is the *geodesic* part of its evolution, with respect to this norm. See Arnold and Khesin [ArKh1998] for discussions of this interpretation of ideal incompressible flow and references to the literature. However, in a gravitational field, for example, there will also be dynamics due to potential energy. And this dynamics will be governed by the right side of the EP equation.

The right side of the EP equation in (25.5.3) modifies the geodesic motion. Naturally, the right side of the EP equation is also a geometrical quantity. The diamond operation $\diamond$ represents the dual of the Lie algebra action of vectors fields on the tensor a. Here, $\delta\ell/\delta a$ is the dual tensor under the natural pairing (usually, L^2 pairing) $\langle \cdot, \cdot \rangle$ that is induced by the variational derivative of the Lagrangian $\ell(\mathbf{u}, a)$. The diamond operation $\diamond$ is defined in terms of this pairing in (25.3.4). For the L^2 pairing, this is integration by parts of (minus) the Lie derivative in (25.3.4).

The quantity a is typically a tensor (for example, a density, a scalar, or a differential form), and we shall sum over the various

---

[3]In coordinates, a one-form density takes the form $\mathbf{m} \cdot d\mathbf{x} \otimes dV$ and the EP equation (25.4.4) is given mnemonically by

$$\frac{d}{dt}\Big|_{\text{Lag}} (\mathbf{m} \cdot d\mathbf{x} \otimes dV) = \underbrace{\frac{d\mathbf{m}}{dt}\Big|_{\text{Lag}} \cdot d\mathbf{x} \otimes dV}_{\text{Advection}} + \underbrace{\mathbf{m} \cdot d\mathbf{u} \otimes dV}_{\text{Stretching}} + \underbrace{\mathbf{m} \cdot d\mathbf{x} \otimes (\nabla \cdot \mathbf{u})dV}_{\text{Expansion}}$$

$$= \frac{\delta\ell}{\delta a} \diamond a$$

with $\frac{d}{dt}\Big|_{\text{Lag}} d\mathbf{x} := \left(\partial_t + \pounds_{\mathbf{u}}\right)d\mathbf{x} = d\mathbf{u} = \mathbf{u}_{,j} dx^j$, upon using the commutation of the Lie derivative and the exterior derivative. Compare this formula with the definition of $\text{ad}^*_u(\mathbf{m} \otimes dV)$ in equation (25.2.6).

types of tensors a that are involved in the fluid description. The second equation in (25.5.3) states that each tensor a is carried along by the Eulerian fluid velocity u. Thus, a is for fluid "attribute," and its Eulerian evolution is given by minus its Lie derivative, $-\mathcal{L}_\mathbf{u}a$. That is, a stands for the set of fluid attributes that each Lagrangian fluid parcel carries around (advects), such as its buoyancy, which is determined by its individual salt, or heat, content in ocean circulation.

Many examples of how equation (25.5.3) arises in the dynamics of continuous media are given by Holm, Marsden and Ratiu [HoMaRa1998a]. The EP form of the Eulerian fluid description in (25.5.3) is analogous to the classical dynamics of rigid bodies (and tops under gravity) in body coordinates. Rigid bodies and tops are also governed by EP equations, as Poincaré showed in a two-page paper with no references over a century ago [Po1901]. For modern discussions of the EP theory, see, for example, Marsden and Ratiu [MaRa1994] or Holm, Marsden and Ratiu [HoMaRa1998a].

Exercise. For what types of tensors a_0 can one recast the EP equations for continua (25.4.4) as geodesic motion by using a version of the Kaluza–Klein construction?   ★

25.6   Corollary: The EP theorem implies the Kelvin–Noether theorem

Corollary 25.6.1 (Kelvin–Noether circulation theorem) *Assume that* $\mathbf{u}(x, t)$ *satisfies the Euler–Poincaré equations for continua:*

$$\frac{\partial}{\partial t}\left(\frac{\delta\ell}{\delta\mathbf{u}}\right) = -\mathcal{L}_\mathbf{u}\left(\frac{\delta\ell}{\delta\mathbf{u}}\right) + \frac{\delta\ell}{\delta a} \diamond a$$

and the quantity a *satisfies the advection relation*

$$\frac{\partial a}{\partial t} + \mathcal{L}_\mathbf{u}a = 0. \tag{25.6.1}$$

Let η_t be the flow of the Eulerian velocity field $\mathbf{u}$, that is, $\mathbf{u} = (d\eta_t/dt) \circ \eta_t^{-1}$. Define the advected fluid loop $\gamma_t := \eta_t \circ \gamma_0$ and the circulation map $I(t)$ by

$$I(t) = \oint_{\gamma_t} \frac{1}{D} \frac{\delta\ell}{\delta\mathbf{u}}. \tag{25.6.2}$$

In the circulation map $I(t)$, the advected mass density D_t satisfies the push-forward relation $D_t = \eta_* D_0$. This implies the advection relation (25.6.1) with $a = D$, namely, the continuity equation

$$\partial_t D + \operatorname{div} D\mathbf{u} = 0.$$

Then, the map $I(t)$ satisfies the Kelvin circulation relation:

$$\frac{d}{dt} I(t) = \oint_{\gamma_t} \frac{1}{D} \frac{\delta\ell}{\delta a} \diamond a. \tag{25.6.3}$$

Both an abstract proof of the Kelvin–Noether circulation theorem and a proof tailored for the case of continuum mechanical systems are given by Holm, Marsden and Ratiu [HoMaRa1998a]. We provide a version of the latter in the following.

Proof. First we change variables in the expression for $I(t)$:

$$I(t) = \oint_{\gamma_t} \frac{1}{D_t} \frac{\delta l}{\delta\mathbf{u}} = \oint_{\gamma_0} \eta_t^* \left[\frac{1}{D_t} \frac{\delta l}{\delta\mathbf{u}} \right] = \oint_{\gamma_0} \frac{1}{D_0} \eta_t^* \left[\frac{\delta l}{\delta\mathbf{u}} \right].$$

Next, we use the Lie derivative formula, namely

$$\frac{d}{dt} (\eta_t^* \alpha_t) = \eta_t^* \left(\frac{\partial}{\partial t} \alpha_t + \mathcal{L}_\mathbf{u} \alpha_t \right),$$

applied to a 1-form density α_t. This formula gives

$$\frac{d}{dt} I(t) = \frac{d}{dt} \oint_{\gamma_0} \frac{1}{D_0} \eta_t^* \left[\frac{\delta l}{\delta\mathbf{u}} \right] = \oint_{\gamma_0} \frac{1}{D_0} \frac{d}{dt} \left(\eta_t^* \left[\frac{\delta l}{\delta\mathbf{u}} \right] \right)$$

$$= \oint_{\gamma_0} \frac{1}{D_0} \eta_t^* \left[\frac{\partial}{\partial t} \left(\frac{\delta l}{\delta\mathbf{u}} \right) + \mathcal{L}_\mathbf{u} \left(\frac{\delta l}{\delta\mathbf{u}} \right) \right].$$

By the Euler–Poincaré equations (25.4.4), this becomes

$$\frac{d}{dt} I(t) = \oint_{\gamma_0} \frac{1}{D_0} \eta_t^* \left[\frac{\delta l}{\delta a} \diamond a \right] = \oint_{\gamma_t} \frac{1}{D_t} \left[\frac{\delta l}{\delta a} \diamond a \right],$$

again by the change of variables formula.   ∎

Corollary 25.6.2 *Since the last expression holds for every loop γ_t, we may write it as*

$$\left(\frac{\partial}{\partial t} + \mathcal{L}_{\mathbf{u}} \right) \frac{1}{D} \frac{\delta l}{\delta \mathbf{u}} = \frac{1}{D} \frac{\delta l}{\delta a} \diamond a. \qquad (25.6.4)$$

Remark 25.6.1 The Kelvin–Noether theorem is called so here because its derivation relies on the invariance of the Lagrangian L under the particle relabelling symmetry, and Noether's theorem is associated with this symmetry. However, the result (25.6.3) is the *Kelvin circulation theorem*: the circulation integral $I(t)$ around any fluid loop (γ_t, moving with the velocity of the fluid parcels $\mathbf{u}$) is invariant under the fluid motion. These two statements are equivalent. We note that *two velocities* appear in the integrand $I(t)$: the fluid velocity $\mathbf{u}$ and $D^{-1}\delta l/\delta \mathbf{u}$. The latter velocity is the momentum density $\mathbf{m} = \delta l/\delta \mathbf{u}$ divided by the mass density D. These two velocities are the basic ingredients for performing modelling and analysis in any ideal fluid problem. One simply needs to put these ingredients together in the Euler–Poincaré theorem and its corollary, the Kelvin–Noether theorem.   □

25.7   Active advection defined via composition of maps

In the Euler–Poincaré theorem, Eulerian fluid dynamics in a certain domain $\mathcal{D}$ is derived by applying for Hamilton's principle for a Lagrangian defined on $T(G \circledS V)/G = \mathfrak{g} \circledS V$, which leads to the Euler–Poincaré equations on $T^*(G \circledS V)/G = \mathfrak{g}^* \circledS V$. One may ask, though, what happens when a degree of freedom being carried along in the frame of motion of an Euler–Poincaré flow in the Euler–Poincaré theorem is not just passively advected? Suppose instead that the advected degree of freedom has its own dynamical volition, as in the case of waves propagating through a flowing fluid.

Exercise. Formulate and analyse hybrid equations of motion on $\mathfrak{g}^* \textcircled{S} V_1 \times T^* V_2$.

Use the Hamilton–Pontryagin principle to derive the equations of motion for an Euler–Poincaré system coupled to an additional *canonical* degree of freedom whose symplectic dynamics occurs in the frame of motion of the Euler–Poincaré system.

The Hamilton–Pontryagin principle for such a system is given by

$$0 = \delta S = \delta \int_0^T \ell(u, a) + \langle m, \dot{g}g^{-1} - u \rangle$$

$$+ \langle b, a_0 g^{-1} - a \rangle \qquad (25.7.1)$$

$$+ \langle p, \partial_t q + \mathcal{L}_u q \rangle - h(q, p)\, dt,$$

for $(a, b) \in T^* V_1$ and $(q, p) \in T^* V_2$.

1. Find the momentum $m \in \Lambda^1 \otimes Den$ and $p \in T^* V_2$.

2. Find the Legendre transform to obtain the Hamiltonian in the variables (m, q, p).

3. Write the equations of motion for this system in Lie–Poisson bracket form.   ★

Answer.

1. The variations of the Lagrangian in Hamilton's principle: (25.7.1)

   $$\delta u : \quad \frac{\delta \ell}{\delta u} - m - p \diamond q = 0,$$

$$\delta m: \quad \dot{g}g^{-1} - u = 0,$$

$$\delta a: \quad \frac{\delta \ell}{\delta a} - b = 0,$$

$$\delta b: \quad a_0 g^{-1} - a = 0, \quad \Longrightarrow \quad \partial_t a + \mathcal{L}_{\dot{g}g^{-1}} a = 0,$$

$$\delta p: \quad \partial_t q + \mathcal{L}_u q - \frac{\delta h}{\delta p} = 0,$$

$$\delta q: \quad -\partial_t p + \mathcal{L}_u^T p - \frac{\delta h}{\delta q} = 0,$$

$$\delta g: \quad -(\partial_t + \mathcal{L}_{\dot{g}g^{-1}})m + b \diamond a = 0. \qquad (25.7.2)$$

Kelvin–Noether theorem. Because the mass density (D) is always among the variables a advected by the flow, the last equation in (25.7.2) implies Kelvin's theorem:

$$\frac{d}{dt} \oint_{C_t} D^{-1} m = \oint_{C_t} (\partial_t + \mathcal{L}_{\dot{g}g^{-1}})(D^{-1} m)$$

$$= \oint_{C_t} D^{-1}(b \diamond a), \qquad (25.7.3)$$

for a material loop pushed forward by the flow, as $C_t = g_* C_0$.

2. The Legendre transform produces the system's Hamiltonian and its variations as

$$h(m, a, q, p) = \langle m, u \rangle + \langle p, \partial_t q \rangle - \ell(u, a)$$
$$- \langle p, \partial_t q + \mathcal{L}_u q \rangle + h(q, p)$$
$$= \langle m, u \rangle + \langle p, -\mathcal{L}_u q \rangle$$
$$- \ell(u, a) + h(q, p)$$
$$= \langle m + p \diamond q, u \rangle - \ell(u, a) + h(q, p)$$
$$h(M, a, q, p) =: \langle M, u \rangle - \ell(u, a) + h(q, p)$$

$$\delta h(M, a, q, p) = \langle \delta M \,,\, u \rangle + \left\langle M - \frac{\delta \ell}{\delta u} \,,\, \delta u \right\rangle$$

$$- \left\langle \frac{\delta \ell}{\delta a} \,,\, \delta a \right\rangle + \left\langle \delta p \,,\, -\mathcal{L}_u q + \frac{\delta h}{\delta p} \right\rangle$$

$$+ \left\langle -\mathcal{L}_u^T p + \frac{\delta h}{\delta q} \,,\, \delta q \right\rangle, \qquad (25.7.4)$$

where the notation $M := \delta \ell / \delta u = m + p \diamond q$ for the total momentum has been introduced.

3. The equations of motion for this hybrid system take the following block-diagonal bracket form:

$$\frac{d}{dt} \begin{pmatrix} M \\ a \\ p \\ q \end{pmatrix} = - \begin{bmatrix} \partial M + M \partial & \Box \diamond a & 0 & 0 \\ \mathcal{L}_\Box a & 0 & 0 & 0 \\ 0 & 0 & 0 & 1 \\ 0 & 0 & -1 & 0 \end{bmatrix}$$

$$\times \begin{pmatrix} \partial h / \partial M = u \\ \partial h / \partial a = -\partial \ell / \partial a \\ \partial h / \partial p \\ \partial h / \partial q \end{pmatrix}, \qquad (25.7.5)$$

which comprises the sum of a Lie–Poisson bracket for the fluid and a symplectic bracket for the actively advected degree of freedom, coupled through their total momentum density, $M :=$ $m + p \diamond q$. ▲

Remark 25.7.1 This exercise combines the dynamics of mass flow and dynamical advection, e.g., for wave propagation in a moving medium. This is accomplished by coupling the Hamilton–Pontryagin constrained Lagrangian (for the flow of the medium) with a phase-space Lagrangian (for the propagation of the waves in the frame of motion of the moving medium). This coupling provides a means of describing how the reaction to wave propagation in the moving frame affects the motion of the medium. This approach also applies to complex fluids such as liquid crystals (where order-parameter dynamics takes place in a moving fluid) [Ho2002b] and to superfluids (where the superfluid velocity is measured relative to the normal fluid motion) [HoKu1982, HoKu1987]. These applications may all be recognised as examples of the composition of maps approach [HoHuSt2023]. □

26

EULER–POINCARÉ THEORY OF GEOPHYSICAL FLUID DYNAMICS

What is this lecture about? This lecture provides the geometric mechanics framework for geophysical fluid dynamics (GFD).

26.1 Variational formulae in three dimensions

We compute explicit formulae for the variations δa in the cases that the set of tensors a is drawn from a set of scalar fields and densities on $\mathbb{R}^3$. We denote this symbolically by writing

$$a \in \{b, D\,d^3x\}. \tag{26.1.1}$$

We have seen that the invariance of the set a in the Lagrangian picture under the dynamics of $\mathbf{u}$ implies in the Eulerian picture that

$$\left(\frac{\partial}{\partial t} + \mathcal{L}_{\mathbf{u}}\right) a = 0,$$

where $\mathcal{L}_{\mathbf{u}}$ denotes the Lie derivative with respect to the velocity vector field $\mathbf{u}$. Hence, for a fluid dynamical Eulerian action $\mathfrak{S} = \int dt\,\ell(\mathbf{u}; b, D)$, the advected variables b and D satisfy the following Lie-derivative relations:

$$\left(\frac{\partial}{\partial t} + \mathcal{L}_{\mathbf{u}}\right) b = 0, \quad \text{or} \quad \frac{\partial b}{\partial t} = -\,\mathbf{u} \cdot \nabla\, b, \tag{26.1.2}$$

$$\left(\frac{\partial}{\partial t} + \mathcal{L}_{\mathbf{u}}\right) D\,d^3x = 0, \quad \text{or} \quad \frac{\partial D}{\partial t} = -\,\nabla \cdot (D\mathbf{u}). \tag{26.1.3}$$

In fluid dynamical applications, the advected Eulerian variables b and $D\,d^3x$ represent the buoyancy b (or specific entropy, for the compressible case) and volume element (or mass density) $D\,d^3x$, respectively. According to Section [25.4.1], equation (25.4.3), the variations of the tensor functions a at fixed $\mathbf{x}$ and t are also given by Lie derivatives, namely $\delta a = -\mathcal{L}_{\mathbf{w}}\,a$, or

$$\delta b = -\mathcal{L}_{\mathbf{w}}\,b = -\mathbf{w}\cdot\nabla\, b,$$

$$\delta D\,d^3x = -\mathcal{L}_{\mathbf{w}}\,(D\,d^3x) = -\nabla \cdot (D\mathbf{w})\,d^3x. \tag{26.1.4}$$

Hence, Hamilton's principle (25.4.2) with this dependence yields

$$0 = \delta \int dt\,\ell(\mathbf{u}; b, D)$$

$$= \int dt \left[\frac{\delta\ell}{\delta\mathbf{u}} \cdot \delta\mathbf{u} + \frac{\delta\ell}{\delta b}\,\delta b + \frac{\delta\ell}{\delta D}\,\delta D\right]$$

$$= \int dt \left[\frac{\delta\ell}{\delta\mathbf{u}} \cdot \left(\frac{\partial\mathbf{w}}{\partial t} - \mathrm{ad}_{\mathbf{u}}\,\mathbf{w}\right) - \frac{\delta\ell}{\delta b}\,\mathbf{w} \cdot \nabla\, b - \frac{\delta\ell}{\delta D}\,(\nabla \cdot (D\mathbf{w}))\right]$$

$$= \int dt\, \mathbf{w} \cdot \left[-\frac{\partial}{\partial t}\frac{\delta\ell}{\delta\mathbf{u}} - \mathrm{ad}_{\mathbf{u}}^{*}\frac{\delta\ell}{\delta\mathbf{u}} - \frac{\delta\ell}{\delta b}\nabla b + D\,\nabla\frac{\delta\ell}{\delta D} \right]$$

$$= -\int dt\, \mathbf{w} \cdot \left[\left(\frac{\partial}{\partial t} + \mathcal{L}_{\mathbf{u}}\right)\frac{\delta\ell}{\delta\mathbf{u}} + \frac{\delta\ell}{\delta b}\nabla b - D\,\nabla\frac{\delta\ell}{\delta D} \right], \qquad (26.1.5)$$

where we have consistently dropped boundary terms arising from integrations by parts by invoking natural boundary conditions. Specifically, we may impose $\hat{\mathbf{n}} \cdot \mathbf{w} = 0$ on the boundary, where $\hat{\mathbf{n}}$ is the boundary's outward unit normal vector and $\mathbf{w} = \delta\eta_t \circ \eta_t^{-1}$ vanishes at the endpoints.

26.2 Euler–Poincaré framework for GFD

The Euler–Poincaré equations for continua (25.4.4) may now be summarised in vector form for advected Eulerian variables a in the set (26.1.1). We adopt the notational convention of the circulation map I in equations (25.6.2) and (25.6.3) that a 1-form density can be made into a 1-form (no longer a density) by dividing it by the mass density D, and we use the Lie-derivative relation for the continuity equation $(\partial/\partial t + \mathcal{L}_{\mathbf{u}})Dd^3x = 0$. Then, the Euclidean components of the Euler–Poincaré equations for continua in equation (26.1.5) are expressed in the Kelvin theorem form (25.6.4) with a slight abuse of notation as

$$\left(\frac{\partial}{\partial t} + \mathcal{L}_{\mathbf{u}}\right)\left(\frac{1}{D}\frac{\delta\ell}{\delta\mathbf{u}}\cdot d\mathbf{x}\right) + \frac{1}{D}\frac{\delta\ell}{\delta b}\nabla b \cdot d\mathbf{x} - \nabla\left(\frac{\delta\ell}{\delta D}\right)\cdot d\mathbf{x} = 0,$$

$$(26.2.1)$$

in which the variational derivatives of the Lagrangian ℓ are to be computed according to the usual physical conventions, that is, as Fréchet derivatives. Formula (26.2.1) is the Kelvin–Noether form of the equation of motion for ideal continua. Hence, we have the explicit Kelvin theorem expression, cf. equations (25.6.2) and (25.6.3),

$$\frac{d}{dt}\oint_{\gamma_t(\mathbf{u})}\frac{1}{D}\frac{\delta\ell}{\delta\mathbf{u}}\cdot d\mathbf{x} = -\oint_{\gamma_t(\mathbf{u})}\frac{1}{D}\frac{\delta\ell}{\delta b}\nabla b \cdot d\mathbf{x}\,, \qquad (26.2.2)$$

where the curve $\gamma_t(\mathbf{u})$ moves with the fluid velocity $\mathbf{u}$. Then, by Stokes' theorem, the Euler equations generate a circulation of $\mathbf{v} := (D^{-1}\delta l/\delta \mathbf{u})$ whenever the gradients ∇b and $\nabla(D^{-1}\delta l/\delta b)$ are not collinear. The corresponding *conservation of potential vorticity q* on fluid parcels is given by

$$\frac{\partial q}{\partial t} + \mathbf{u} \cdot \nabla q = 0, \quad \text{where } q = \frac{1}{D}\nabla b \cdot \text{curl}\left(\frac{1}{D}\frac{\delta \ell}{\delta \mathbf{u}}\right). \qquad (26.2.3)$$

This is also called *PV convection*. Equations (26.2.1)–(26.2.3) embody most of the panoply of equations for GFD. The vector form of equation (26.2.1) is

$$\underbrace{\left(\frac{\partial}{\partial t} + \mathbf{u} \cdot \nabla\right)\left(\frac{1}{D}\frac{\delta l}{\delta \mathbf{u}}\right) + \frac{1}{D}\frac{\delta l}{\delta u^j}\nabla u^j}_{\text{Geodesic nonlinearity: kinetic energy}}$$

$$= \underbrace{\nabla\frac{\delta l}{\delta D} - \frac{1}{D}\frac{\delta l}{\delta b}\nabla b}_{\text{Potential energy}}. \qquad (26.2.4)$$

In geophysical applications, the Eulerian variable D represents the frozen-in volume element and b is the buoyancy. In this case, *Kelvin's theorem* is

$$\frac{dI}{dt} = \iint_{S(t)} \nabla\left(\frac{1}{D}\frac{\delta l}{\delta b}\right) \times \nabla b \cdot d\mathbf{S},$$

with circulation integral

$$I = \oint_{\gamma(t)} \frac{1}{D}\frac{\delta l}{\delta \mathbf{u}} \cdot d\mathbf{x}.$$

26.3   Euler's equations for a rotating stratified ideal incompressible fluid

26.3.1   The Euler–Boussinesq Lagrangian

In the Eulerian velocity representation, we consider Hamilton's principle for fluid motion in a three-dimensional domain with action functional $S = \int l\, dt$ and Lagrangian $l(\mathbf{u}, b, D)$ given by

$$l(\mathbf{u}, b, D) = \int \rho_0 D(1+b)\left(\tfrac{1}{2}|\mathbf{u}|^2 + \mathbf{u} \cdot \mathbf{R}(\mathbf{x}) - gz\right) - p(D-1)\, d^3x,$$

(26.3.1)

where $\rho_{tot} = \rho_0 D(1+b)$ is the total mass density, ρ_0 is a dimensional constant and $\mathbf{R}$ is a given function of $\mathbf{x}$. This variations at fixed $\mathbf{x}$ and t of this Lagrangian are the following:

$$\frac{1}{D}\frac{\delta l}{\delta \mathbf{u}} = \rho_0(1+b)(\mathbf{u}+\mathbf{R}), \qquad \frac{\delta l}{\delta b} = \rho_0 D\left(\tfrac{1}{2}|\mathbf{u}|^2 + \mathbf{u}\cdot\mathbf{R} - gz\right),$$

$$\frac{\delta l}{\delta D} = \rho_0(1+b)\left(\tfrac{1}{2}|\mathbf{u}|^2 + \mathbf{u}\cdot\mathbf{R} - gz\right) - p, \qquad \frac{\delta l}{\delta p} = -(D-1). \quad (26.3.2)$$

Hence, from the Euclidean component formula (26.2.4) for Hamilton principles of this type and the fundamental vector identity

$$(\mathbf{b}\cdot\nabla)\mathbf{a} + a_j\nabla b^j = -\,\mathbf{b}\times(\nabla\times\mathbf{a}) + \nabla(\mathbf{b}\cdot\mathbf{a}), \qquad (26.3.3)$$

we find the motion equation for an Euler fluid in three dimensions:

$$\left[\frac{d\mathbf{u}}{dt}\right] \implies \frac{\partial\mathbf{u}}{\partial t} - u\times\mathrm{curl}\mathbf{R} + g\hat{\mathbf{z}} + \frac{1}{\rho_0(1+b)}\nabla p = 0, \qquad (26.3.4)$$

where $\mathrm{curl}\mathbf{R} = 2\boldsymbol{\Omega}(\mathbf{x})$ is the Coriolis parameter (that is, twice the local angular rotation frequency). In writing this equation, we have used the advection of buoyancy,

$$\frac{\partial b}{\partial t} + \mathbf{u}\cdot\nabla b = 0,$$

from equation (26.1.2). The pressure p is determined by requiring the preservation of the constraint $D = 1$, for which the continuity equation (26.1.3) implies $\mathrm{div}\,u = 0$. The Euler motion equation (26.3.4) represents Newton's law for the acceleration of a fluid due to three forces: Coriolis, gravity and pressure gradient. The dynamic balances among these three forces produce the many circulatory flows of GFD. The *conservation of potential vorticity* q on fluid parcels for these Euler GFD flows is given by

$$\frac{\partial q}{\partial t} + \mathbf{u}\cdot\nabla q = 0, \quad \text{where, on using } D = 1, \quad q = \nabla b\cdot\mathrm{curl}(\mathbf{u}+\mathbf{R}).$$

(26.3.5)

Exercise. (Semidirect-product Lie–Poisson bracket for compressible ideal fluids)

1. Compute the Legendre transform for the Lagrangian

$$l(\mathbf{u}, s, D): \mathfrak{X} \times \Lambda^0 \times \Lambda^3 \mapsto \mathbb{R}$$

whose advected variables satisfy the auxiliary equations,

$$\frac{\partial s}{\partial t} = -\mathbf{u} \cdot \nabla s, \quad \frac{\partial D}{\partial t} = -\nabla \cdot (D\mathbf{u}).$$

Here, s represents entropy per unit mass and D represents mass density, which are thermodynamic variables.

2. Compute the Hamiltonian, assuming the Legendre transform is a linear invertible operator on the velocity $\mathbf{u}$. For definiteness in computing the Hamiltonian, assume that the Lagrangian is given by

$$l(\mathbf{u}, s, D) = \int D\left(\tfrac{1}{2}|\mathbf{u}|^2 + \mathbf{u} \cdot \mathbf{R}(\mathbf{x}) - e(D, s)\right) d^3x,$$

$$(26.3.6)$$

with prescribed function $\mathbf{R}(\mathbf{x})$ and specific internal energy $e(D, s)$ satisfying the first law of thermodynamics,

$$de = \frac{p}{D^2} dD + T ds,$$

where p is pressure and T temperature.

3. Find the semidirect-product Lie–Poisson bracket for the Hamiltonian formulation of these equations.

4. Does this Lie–Poisson bracket have Casimirs? If so, what are the corresponding symmetries and momentum maps?   ★

Answer. The solution here is essentially the same as for the Euler–Boussinesq equations treated above.   ▲

26.4   Hamilton–Poincaré reduction implies Lie–Poisson fluid equations

In the Euler–Poincaré framework, one starts with a Lagrangian defined on the tangent bundle of a Lie group G,

$$L: TG \to \mathbb{R},$$

and the dynamics is given by the Euler–Lagrange equations arising from the variational principle

$$\delta \int_{t_0}^{t_1} L(g, \dot{g})dt = 0.$$

The Lagrangian L is considered left/right invariant, and because of this property, one can *reduce* the problem, obtaining a new system which is defined on the Lie algebra $\mathfrak{g}$ of G and a new set of equations, the Euler–Poincaré equations, arising from a reduced variational principle

$$\delta \int_{t_0}^{t_1} l(\xi)dt = 0,$$

where $l(\xi)$ is the reduced Lagrangian and $\xi \in \mathfrak{g}$.

Exercise. Is there a similar procedure for Hamiltonian systems? More precisely, given a Hamiltonian function

$$H: T^*G \to \mathbb{R}$$

defined on the cotangent bundle T^*G, can one perform a similar procedure of reduction and derive the equations of motion on the dual of the Lie algebra $\mathfrak{g}^*$, provided the Hamiltonian is again left/right invariant?   ★

Answer. Hamilton–Poincaré reduction gives a positive answer to this question in the context of variational principles, as is done in the Euler–Poincaré framework; we are going to explain how this procedure is performed.

More generally, we will also consider advected quantities belonging to a vector space V on which G acts, so that the Hamiltonian is written in this case as

$$H \colon T^*G \times V^* \to \mathbb{R}$$

(see Holm, Marsden and Ratiu [HoMaRa1998a, HoMaRa1998b]). The space V is regarded here to be exactly the same as in the Euler–Poincaré theory.

The equations of motion, that is, Hamilton's equations, may be derived from the following variational principle:

$$\delta \int_{t_0}^{t_1} \left\{ \langle p(t), \dot{g}(t) \rangle - H_{a_0}(g(t), p(t)) \right\} dt = 0.$$

As we know from ordinary classical mechanics, $\dot{g}(t)$ should be considered as the tangent vector to the curve $g(t)$, so that $\dot{g}(t) \in T_{g(t)}G$.   ▲

Exercise. What happens if H_{a_0} is left/right invariant?

★

Answer. It turns out that, in this case, the whole function

$$F(g, \dot{g}, p) = \langle p, \dot{g} \rangle - H_{a_0}(g, p)$$

is also invariant. The proof is straightforward once the action is specified as[1]

$$h\,(g, \dot{g}, p) = (hg, T_g L_h\,\dot{g}, T^*_{hg} L_{h^{-1}}\,p),$$

---

[1] From here until the end of this lecture, we will consider only *left invariance*. Examples of right- invariant Hamiltonians will be considered in later lectures.

where $T_g L_h: T_g G \rightarrow T_{hg} G$ is the tangent of the left translation map $L_h g = hg \in G$ at the point g and $T_{hg}^* L_{h^{-1}}: T_g^* G \rightarrow T_{hg}^* G$ is the dual of the map $T_{hg} L_{h^{-1}}: T_{hg} G \rightarrow T_g G$.

We now check that

$$\langle h\,p, h\,\dot{g}\rangle = \langle T_{hg}^* L_{h^{-1}}\,p, T_g L_h\,\dot{g}\rangle$$

$$= \langle p, T_{hg} L_{h^{-1}} \circ T_g L_h\,\dot{g}\rangle$$

$$= \langle p, T_g(L_{h^{-1}} \circ L_h)\,\dot{g}\rangle = \langle p, \dot{g}\rangle,$$

where the chain rule for the tangent map has been used. The same result holds for the right action.

Due to this invariance property, one can write the variational principle as

$$\delta \int_{t_0}^{t_1} \{\langle \mu, \xi \rangle - h(\mu, a)\}\, dt = 0,$$

with

$$\mu(t) = g^{-1}(t)\, p(t) \in \mathfrak{g}^*, \quad \xi(t) = g^{-1}(t)\, \dot{g}(t) \in \mathfrak{g},$$

$$a(t) = g^{-1}(t)\, a_0 \in V^*.$$

In particular, $a(t)$ is the solution of

$$\partial_t a(t) = -\,\xi(t)\, a_0,$$

where a Lie algebra action of $\mathfrak{g}$ on V^* is implicitly defined. In order to find the equations of motion, one calculates the variations

$$\delta \int_{t_0}^{t_1} \{\langle \mu, \xi \rangle - h(\mu, a)\}\, dt$$

$$= \int_{t_0}^{t_1} \left\{ \langle \delta\mu, \xi \rangle + \langle \mu, \delta\xi \rangle - \left\langle \delta\mu, \frac{\delta h}{\delta \mu} \right\rangle - \left\langle \delta a, \frac{\delta h}{\delta a} \right\rangle \right\} dt.$$

As in the Euler–Poincaré theorem, we use the following expressions for the variations:

$$\delta\xi = \partial_t \eta + [\xi, \eta], \quad \delta a = -\eta a.$$

Then, using the definition of the diamond operator, we find

$$
\int_{t_0}^{t_1} \left\{ \langle \delta\mu, \xi \rangle + \langle \mu, \delta\xi \rangle - \left\langle \delta\mu, \frac{\delta h}{\delta\mu} \right\rangle - \left\langle \delta a, \frac{\delta h}{\delta a} \right\rangle \right\} dt
$$

$$
= \int_{t_0}^{t_1} \left\{ \left\langle \delta\mu, \xi - \frac{\delta h}{\delta\mu} \right\rangle + \langle \mu, \dot{\eta} + \mathrm{ad}_\xi \eta \rangle \right.
$$

$$
\left. + \left\langle \eta a, \frac{\delta h}{\delta a} \right\rangle \right\} dt
$$

$$
= \int_{t_0}^{t_1} \left\{ \left\langle \delta\mu, \xi - \frac{\delta h}{\delta\mu} \right\rangle + \langle -\dot{\mu} + \mathrm{ad}_\xi^* \mu, \eta \rangle \right.
$$

$$
\left. + - \left\langle \frac{\delta h}{\delta a} \diamond a, \eta \right\rangle \right\} dt.
$$

Consequently, the $\delta\mu$ variation yields

$$
\xi = \frac{\delta h}{\delta\mu},
$$

and the equations of motion are given by

$$
\dot{\mu} = \mathrm{ad}_\xi^* \mu - \frac{\delta h}{\delta a} \diamond a,
$$

together with[2]

$$
\dot{a} = -\frac{\delta h}{\delta\mu} a.
$$

The equations of motion written on the dual Lie algebra $\mathfrak{g}$ are called *Lie–Poisson* equations.   ▲

This exercise has proven the following theorem.

Theorem 26.4.1 (Hamilton–Poincaré reduction theorem) *With the preceding notation, the following statements are equivalent:*

---

[2]Recall the concatenation notation for the Lie algebra action, in which $\frac{\delta h}{\delta\mu} a \simeq \mathcal{L}_{\delta h/\delta\mu} a$.

1. *With a_0 held fixed, the variational principle*

$$\delta \int_{t_0}^{t_1} \{ \langle p(t), \dot{g}(t) \rangle - H_{a_0}(g(t), p(t)) \} \, dt = 0$$

holds, for variations $\delta g(t)$ of $g(t)$ vanishing at the endpoints.

2. *$(g(t), p(t))$ satisfies Hamilton's equations for H_{a_0} on G.*

3. *The constrained variational principle*

$$\delta \int_{t_0}^{t_1} \{ \langle \mu(t), \xi(t) \rangle - h(\mu(t), a(t)) \} \, dt = 0$$

holds for $\mathfrak{g} \times V^*$ using variations of ξ and a of the form

$$\delta \xi = \partial_t \eta + [\xi, \eta], \quad \delta a = -\eta a$$

where $\eta(t) \in \mathfrak{g}$ vanishes at the endpoints.

4. *The Lie–Poisson equations hold on $\mathfrak{g} \times V^*$:*

$$(\partial_t \mu, \partial_t a) = \left(\mathrm{ad}_\xi^* \mu - \frac{\delta h}{\delta a} \diamond a, -\frac{\delta h}{\delta \mu} a \right).$$

Remark 26.4.1 One might have preferred to start with an invariant Hamiltonian defined on

$$T^*(G \times V) = T^* G \times V \times V^*.$$

However, as mentioned by Holm, Marsden and Ratiu [HoMaRa1998a, HoMaRa1998b], such an approach turns out to be equivalent to the treatment presented here.   □

Remark 26.4.2 (Legendre transform) Lie–Poisson equations may arise from the Euler–Poincaré setting by the Legendre transform

$$\mu = \frac{\delta l}{\delta \xi}.$$

If this is a diffeomorphism, then the Hamilton–Poincaré theorem is equivalent to the Euler–Poincaré theorem.   □

Exercise. (Lie–Poisson structure) Show that the space $\mathfrak{g}^* \times V^*$ is a Poisson manifold.   ★

Answer.

$$\partial_t F(\mu, a)$$

$$= \left\langle \partial_t \mu, \frac{\delta F}{\delta \mu} \right\rangle + \left\langle \partial_t a, \frac{\delta F}{\delta a} \right\rangle$$

$$= \left\langle \mathrm{ad}^*_{\delta H/\delta \mu} \mu - \frac{\delta H}{\delta a} \diamond a, \frac{\delta F}{\delta \mu} \right\rangle - \left\langle \frac{\delta H}{\delta \mu} a, \frac{\delta F}{\delta a} \right\rangle$$

$$= \left\langle \mu, \left[\frac{\delta H}{\delta \mu}, \frac{\delta F}{\delta \mu} \right] \right\rangle - \left\langle \frac{\delta H}{\delta a} \diamond a, \frac{\delta F}{\delta \mu} \right\rangle - \left\langle \frac{\delta H}{\delta \mu} a, \frac{\delta F}{\delta a} \right\rangle$$

$$= - \left\langle \mu, \left[\frac{\delta F}{\delta \mu}, \frac{\delta H}{\delta \mu} \right] \right\rangle - \left\langle a, \frac{\delta F}{\delta \mu} \frac{\delta H}{\delta a} - \frac{\delta H}{\delta \mu} \frac{\delta F}{\delta a} \right\rangle.$$

One may directly verify that the bracket formula

$$\{F, H\}(\mu, a) = - \left\langle \mu, \left[\frac{\delta F}{\delta \mu}, \frac{\delta H}{\delta \mu} \right] \right\rangle$$
$$- \left\langle a, \frac{\delta F}{\delta \mu} \frac{\delta H}{\delta a} - \frac{\delta H}{\delta \mu} \frac{\delta F}{\delta a} \right\rangle$$

satisfies the definition of a Poisson structure. This Lie–Poisson structure has been found in lectures discussing the simpler case without advected quantities. In particular, this calculation verifies that *any dual Lie algebra* $\mathfrak{g}^*$ *is a Poisson manifold*.   ▲

Remark 26.4.3 (Right invariance) It can be shown that for a right-invariant Hamiltonian, one has

$$\{F, H\}(\mu, a) = + \left\langle \mu, \left[\frac{\delta F}{\delta \mu}, \frac{\delta H}{\delta \mu}\right] \right\rangle + \left\langle a, \frac{\delta F}{\delta \mu}\frac{\delta H}{\delta a} - \frac{\delta H}{\delta \mu}\frac{\delta F}{\delta a} \right\rangle$$

$$(\partial_t \mu, \partial_t a) = - \left(\mathrm{ad}^*_\xi \mu - \frac{\delta h}{\delta a} \diamond a, -\frac{\delta h}{\delta \mu} a \right),$$

with all signs changed for a left-invariant Hamiltonian. $\square$

27

FIVE MORE CONTINUUM APPLICATIONS

Contents

What is this lecture about? This lecture applies deterministic geometric mechanics to derive five continuum dynamics equations of current research interest ranging from kinetic theory to nonlinear waves.

27.1 The Vlasov equation in kinetic theory

In plasma physics, a major topic is collision-less particle dynamics in N dimensions, whose primary equation, the Vlasov equation, will be derived heuristically here. In this context, a central role is held by the probability distribution on the phase space $P :=$ $f(\mathbf{q}, \mathbf{p}, t)\, d^N q \wedge d^N p$. Because the total probability is conserved, one may write the continuity equation just as one does for the conservation of total mass in the context of fluid dynamics:

$$\left(\partial_t + \mathcal{L}_u\right) P = \left(\partial_t f + \nabla_{(\mathbf{q}, \mathbf{p})} \cdot (\mathbf{u}\, f)\right) d^N q \wedge d^N p = 0,$$

where $\nabla_{(\mathbf{q}, \mathbf{p})}$ is the divergence in $2N$-dimensional phase space and $\mathbf{u}$ is a Hamiltonian vector field on phase space, which is given by the single particle motion

$$\mathbf{u} = (\dot{\mathbf{q}}, \dot{\mathbf{p}}) \in \mathfrak{X}\big(T^*\mathbb{R}^N\big).$$

If we now assume that the generic single particle undergoes Hamiltonian motion, the Hamiltonian function $h(\mathbf{q}, \mathbf{p})$ can be introduced directly by means of the single-particle Hamilton's equations:

$$\mathbf{u} := (\dot{\mathbf{q}}, \dot{\mathbf{p}}) = \left(\frac{\partial h}{\partial \mathbf{p}}, -\frac{\partial h}{\partial \mathbf{q}}\right).$$

This implies that the vector field $\mathbf{u}$ is divergence-free, $\nabla_{(\mathbf{q}, \mathbf{p})} \cdot \mathbf{u} = 0$, provided the Hessian of the particle Hamiltonian h is symmetric. Therefore, the Vlasov equation written in terms of the distribution function $f(\mathbf{q}, \mathbf{p}, t)$ emerges as

$$\partial_t f + \mathbf{u} \cdot \nabla_{(\mathbf{q}, \mathbf{p})} f = 0.$$

By expanding the Hamiltonian h as the total single-particle energy

$$h(\mathbf{q}, \mathbf{p}) = \frac{1}{2m}\mathbf{p}^2 + V(\mathbf{q}, \mathbf{p}),$$

one obtains the more common form of the Vlasov equation:

$$\frac{\partial f}{\partial t} + \frac{\mathbf{p}}{m} \cdot \frac{\partial f}{\partial \mathbf{q}} - \frac{\partial V}{\partial \mathbf{q}} \cdot \frac{\partial f}{\partial \mathbf{p}} = 0.$$

For more information about what can be done with this hybrid phase-space approach to geometric mechanics, see Ref. [Tr2010].

Exercise. Cast the evolution equations for differentiable functionals F of the Vlasov probability density f into the Lie–Poisson form. ★

Answer. First, write the Vlasov equation in terms of a generic single-particle Hamiltonian $h(\mathbf{q}, \mathbf{p})$ as

$$\partial_t f + \{f, h\} = 0,$$

and recall the canonical Poisson bracket

$$\{f, h\} = \frac{\partial f}{\partial \mathbf{q}} \cdot \frac{\partial h}{\partial \mathbf{p}} - \frac{\partial f}{\partial \mathbf{p}} \cdot \frac{\partial h}{\partial \mathbf{q}}.$$

Recall also that the canonical Poisson bracket endows the set $\mathcal{F}(T^*\mathbb{R}^N)$ of phase-space functions with a Lie algebra structure. Namely, for a right action of the symplectic diffeomorphisms on $\mathcal{F}(T^*\mathbb{R}^N)$,

$$[k, h] = \{k, h\} = \mathrm{ad}_h\, k.$$

At this point, in order to look for a Lie–Poisson equation, one calculates the coadjoint operator such that

$$\langle f, \{h, k\} \rangle = \langle f, -\mathrm{ad}_h k \rangle = \langle -\mathrm{ad}_h^* f, k \rangle = \langle \{h, f\}, k \rangle,$$

where the last equality is justified by the Leibniz property of the Poisson bracket, with the pairing defined as the L^2 pairing in phase space,

$$\langle f, g \rangle = \int f g \, d^N q \wedge d^N p.$$

This argument shows that the Vlasov equation can be written in the Lie–Poisson form

$$\partial_t f + \mathrm{ad}_h^* f = 0.$$

The corresponding Lie–Poisson bracket for the functionals F and H of the Vlasov probability density f is given by

$$\frac{dF}{dt} = \{F, H\}_{LP} = -\left\langle f, \left[\frac{\delta F}{\delta f}, \frac{\delta H}{\delta f}\right]\right\rangle$$

$$= -\left\langle \mathrm{ad}^*_{\delta H/\delta f}\, f, \frac{\delta F}{\delta f}\right\rangle.$$

To check this result, note that the Vlasov equation can now be rewritten as

$$\partial_t f = \{f, H\} = -\,\mathrm{ad}^*_{\delta H/\delta f}\, f$$

by setting $\delta H/\delta f = h(\mathbf{q}, \mathbf{p})$ and writing the functional F as

$$F(f) = \int f(\mathbf{q}', \mathbf{p}')\delta((\mathbf{q}', \mathbf{p}') - (\mathbf{q}, \mathbf{p}))\, d^N q' \wedge d^N p'$$

when performing the variation with respect to f.   ▲

27.2   Ideal barotropic compressible fluids in 3D

The reduced Lagrangian for ideal barotropic compressible fluids is written as

$$l(\mathbf{u}, D) = \int \left(\tfrac{1}{2}|\mathbf{u}|^2 - e(D)\right) Dd^3x,$$

where $|\mathbf{u}|^2 := u \lrcorner u^\flat$ with $u^\flat = \mathbf{u} \cdot d\mathbf{x}$, the velocity vector field $u \in \mathfrak{X}(M \subset \mathbb{R}^3)$ written in components as $u = \mathbf{u} \cdot \nabla$ has no normal components $u \lrcorner d^3x = \mathbf{u} \cdot \hat{\mathbf{n}}\, dS = 0$ on the boundary ∂M and Dd^3x is the advected density, which satisfies the *continuity equation*

$$\partial_t Dd^3x + \mathcal{L}_u(Dd^3x) = 0.$$

Moreover, the internal energy satisfies the barotropic first law of thermodynamics,

$$de = -p(D)d(D^{-1}) = \frac{p(D)}{D^2}\, dD$$

for the pressure function $p(D)$. The reduced Legendre transform on this Lie algebra $\mathfrak{X}(\mathbb{R}^3)$ is given by

$$m := \frac{\delta \ell}{\delta u} = \mathbf{m} \cdot d\mathbf{x} \otimes d^3 x = \mathbf{u} \cdot d\mathbf{x} \otimes D d^3 x,$$

and the Hamiltonian is then written as

$$h(\mathbf{m}, D) = \langle \mathbf{m}, \mathbf{u} \rangle - l(\mathbf{u}, D),$$

that is,

$$h(\mathbf{m}, D) = \int \frac{1}{2D} |\mathbf{m}|^2 + De(D) \, d^3 x.$$

The Lie–Poisson equations in this case arise from the general theory:

$$\partial_t m = -\mathrm{ad}^*_{\delta h / \delta m} m - \frac{\delta h}{\delta D} \diamond D,$$

$$\partial_t D d^3 x = -\mathcal{L}_{\delta h / \delta m} D d^3 x. \tag{27.2.1}$$

Earlier, we found that the coadjoint action is given by the Lie derivative. On the other hand, we may calculate the expression of the diamond operation from its definition,

$$\left\langle \frac{\delta h}{\delta D}, -\pounds_\eta D \right\rangle = \left\langle \frac{\delta h}{\delta D} \diamond D, \eta \right\rangle,$$

to find

$$\left\langle \frac{\delta h}{\delta D}, -\mathrm{div} \, D\eta \right\rangle = \left\langle D\nabla \frac{\delta h}{\delta D}, \eta \right\rangle.$$

Therefore, we have

$$\frac{\delta h}{\delta D} \diamond D = D\nabla \frac{\delta h}{\delta D},$$

where

$$\delta h / \delta D = -\frac{|\mathbf{m}|^2}{2D^2} + \left(e + \frac{p}{D} \right).$$

Substituting into the momentum equation and using the first law to find $d(e + p/D) = (1/D)dp$ yields

$$(\partial_t + \mathcal{L}_u)(\mathbf{m}/D) \cdot d\mathbf{x} = -\nabla(e + p/D - |\mathbf{u}|^2/2) \cdot d\mathbf{x}. \qquad (27.2.2)$$

Upon expanding the Lie derivative for $\mathbf{m}/D = \mathbf{u}$ and using the continuity equation for the density, this quickly becomes

$$\partial_t \mathbf{u} = -\mathbf{u} \cdot \nabla \mathbf{u} - \frac{1}{D} \nabla p,$$

which is Euler's equation for a barotropic fluid.

Remark 27.2.1 (Kelvin–Noether theorem for barotropic fluid dynamics) Conservation of the circulation integral for the barotropic fluid motion equation in (27.2.2) is derived from

$$\frac{d}{dt} \oint_{c(u)} \mathbf{u} \cdot d\mathbf{x} = \oint_{c(u)} (\partial_t + \mathcal{L}_u)(\mathbf{u} \cdot d\mathbf{x})$$

$$= -\oint_{c(u)} d(e + p/D - |\mathbf{u}|^2/2) = 0, \qquad (27.2.3)$$

where $c(u)$ denotes a material loop moving with the barotropic fluid flow and the fundamental law of calculus (the vanishing of the loop integral of the differential of a function) has been used.   □

27.3   Euler's equations for 3D ideal incompressible fluid motion

The barotropic equations for compressible flows yield Euler's equations for ideal incompressible fluid motion in three dimensions when the internal energy in the reduced Lagrangian for ideal compressible fluids is replaced by the constraint $D = 1$:

$$l(\mathbf{u}, D) = \int \frac{D}{2} |\mathbf{u}|^2 - p(D - 1) \, d^3x, \qquad (27.3.1)$$

where, again, $\mathbf{u} \in \mathfrak{X}(M \subset \mathbb{R}^3)$ is tangential on the boundary ∂M and the advected density D satisfies the continuity equation

$$\partial_t D + \operatorname{div} D\mathbf{u} = 0. \tag{27.3.2}$$

The continuity equation (27.3.2) enforces the divergence-free condition $\operatorname{div} \mathbf{u} = 0$ when evaluated under the constraint $D = 1$. The pressure p in the reduced Lagrangian in (27.3.1) is now a Lagrange multiplier, which is determined by the condition that incompressibility be preserved by the dynamics.

As for the barotropic case, the Hamiltonian is obtained from

$$h(\mathbf{m}, D) = \langle \mathbf{m}, \mathbf{u} \rangle - l(\mathbf{u}, D)$$

and found with $m := \frac{\delta \ell}{\delta u} = \mathbf{m} \cdot d\mathbf{x} \otimes d^3 x$ to be

$$h(\mathbf{m}, D) = \int \frac{1}{2D} |\mathbf{m}|^2 + p(D-1) \, d^3 x.$$

The Lie–Poisson equations in this case arise from the general theory:

$$\partial_t m = -\operatorname{ad}^*_{\delta h/\delta m} m - \frac{\delta h}{\delta D} \diamond D,$$
$$\partial_t (Dd^3 x) = -\mathcal{L}_{\delta h/\delta m} (Dd^3 x). \tag{27.3.3}$$

In Lie–Poisson matrix operator form, this is

$$\partial_t \begin{pmatrix} m_i \\ D \end{pmatrix} = - \begin{bmatrix} \partial_j m_i + m_j \partial_i & \Box \diamond D \\ \mathcal{L}_\Box D & 0 \end{bmatrix}$$
$$\times \begin{pmatrix} \delta h/\delta m_j = m^j/D = u^j \\ \delta h/\delta D = p - \frac{1}{2}|\mathbf{m}|^2/D^2 = p - \frac{1}{2}|\mathbf{u}|^2 \end{pmatrix}. \tag{27.3.4}$$

The corresponding Lie–Poisson bracket is defined on the dual[1] of the semidirect-product Lie algebra $\mathfrak{X}_{vol} \circledS \Lambda^0$ of volume-preserving vector fields $\mathfrak{X}_{vol}(\mathbb{R}^3)$ acting on functions $\Lambda^0(\mathbb{R}^3)$. Dual coordinates:

---

[1] Dual spaces are understood in the sense of L^2 pairing in the plane denoted by angle brackets $\langle \cdot, \cdot \rangle$.

the 1-form density $m \in \mathfrak{X}_{\text{vol}} \otimes \text{Den}(\mathbb{R}^3)$ is dual to $\mathfrak{X}_{\text{vol}}(\mathbb{R}^3)$; and density D is dual to scalar functions in $\mathbb{R}^3$.

The Lie–Poisson motion equation for Euler's fluid equation in (27.3.3) for an incompressible fluid is given by

$$\partial_t \mathbf{u} + \mathbf{u} \cdot \nabla \mathbf{u} + u_j \nabla u^j = -\nabla\left(p - \tfrac{1}{2}|\mathbf{u}|^2\right),$$

or, in Lie-derivative form,

$$\left(\partial_t + \mathcal{L}_u\right)(\mathbf{u} \cdot dx) = \left(\partial_t \mathbf{u} - \mathbf{u} \times \operatorname{curl} \mathbf{u} + \frac{1}{2}\nabla|\mathbf{u}|^2\right) \cdot d\mathbf{x}$$

$$= -d\left(p - \tfrac{1}{2}|\mathbf{u}|^2\right). \tag{27.3.5}$$

The differential of equation (27.3.5) and the property $d^2 = 0$ combine now to yield an equation for the vorticity $\omega := \operatorname{curl} u$:

$$\left(\partial_t + \mathcal{L}_u\right) d(\mathbf{u} \cdot dx) = \left(\partial_t + \mathcal{L}_u\right)(\operatorname{curl} u \cdot dS) = 0.$$

Expanding the Lie derivative in the previous formula and using $\operatorname{div} \omega = 0$ yields the classic vortex advection and stretching formula for the vorticity $\omega := \operatorname{curl} u$ in three dimensions:

$$\left(\partial_t + \mathcal{L}_u\right)(\operatorname{curl} u \cdot dS) = \left(\partial_t \omega + \mathbf{u} \cdot \nabla \omega - \omega \cdot \nabla \mathbf{u}\right) \cdot dS = 0.$$

Since both u and ω are divergence-free, this equation may also be written in vector form as

$$\partial_t \omega - \operatorname{curl}\left(\mathbf{u} \times \omega\right) = 0.$$

Exercise. Show that the Euler fluid equation (27.3.5) implies conservation of both the kinetic energy,

$$H := \int_D \frac{1}{2}|\mathbf{u}|^2 \, d^3x,$$

and the following interesting quantity known as the Hopf invariant, or *helicity*:

$$\Lambda := \int_D \mathbf{u} \cdot d\mathbf{x} \wedge d(\mathbf{u} \cdot d\mathbf{x}) = \int_D \mathbf{u} \cdot \omega \, d^3x.$$

The Hopf invariant (or helicity) is a topological quantity which characterises the knottedness of the lines of vorticity. It is also the *only Casimir* of the Lie–Poisson bracket for Euler's 3D fluid equations in (27.3.4).   ★

Exercise. Another potential Poisson bracket candidate for the vorticity $\omega := \mathrm{curl}\, u$ dynamics of Euler's fluid equations might be written for the real functionals Λ, F and H of vorticity ω as

$$\frac{dF}{dt} = \{F, H\} := \int_{\mathcal{D}} \mathrm{curl}\frac{\delta\Lambda}{\delta\omega} \cdot \mathrm{curl}\frac{\delta F}{\delta\omega} \times \mathrm{curl}\frac{\delta H}{\delta\omega}\, d^3x$$

$$= -\int_{\mathcal{D}} \frac{\delta F}{\delta\omega} \cdot \mathrm{curl}\left(\frac{\delta\Lambda}{\delta u} \times \frac{\delta H}{\delta u}\right) d^3x,$$

with the Hamiltonian $H = -\frac{1}{2}\int_{\mathcal{D}} \omega \cdot \Delta^{-1}\omega = \frac{1}{2}\int_{\mathcal{D}} |u|^2 d^3x$.

Would this formulation make sense as a Poisson bracket for Euler's fluid equations? What would the Casimirs be for such a Poisson bracket?   ★

27.4   Euler's 3D fluid equations for variables depending only on the horizontal coordinates (x,y)

When the fluid velocity depends on only two planar dimensions with (x, y) coordinates, its Hodge decomposition may be written as

$$\mathbf{u}(x, y, t) = \nabla^\perp\psi(x, y, t) + v(x, y, t)\widehat{\mathbf{z}}$$

$$= (-\psi_y, \psi_x, v), \quad \text{with } \mathrm{div}\,\mathbf{u} = 0, \qquad (27.4.1)$$

where the *grad-perp* operator $\nabla^\perp$ is defined as

$$\nabla^\perp := \hat{z} \times \nabla.$$

The corresponding 3D vorticity is given by

$$\boldsymbol{\omega} = \Delta\psi\,\hat{z} - \nabla^\perp v. \qquad (27.4.2)$$

The divergence-free velocity $\mathbf{u}$ is written in terms of the operator $\nabla^\perp = \hat{z} \times \nabla$ acting on the stream function ψ and on the vertical velocity component v. The vorticity equation then becomes

$$
\begin{aligned}
0 &= (\partial_t + \mathcal{L}_u)(\boldsymbol{\omega} \cdot d\mathbf{S}) \\
&= d\big(((\partial_t + \mathcal{L}_u)(v\,dz)\big) + (\partial_t + \mathcal{L}_u)(\Delta\psi\,dx \wedge dy) \qquad (27.4.3) \\
&= d\big((\partial_t v + J(\psi, v))\,dz\big) + \big(\partial_t \omega + J(\psi, \omega)\big)\,dx \wedge dy,
\end{aligned}
$$

where one defines $\omega := \Delta\psi$ and $\psi = \Delta^{-1}\omega$, as well as the notation

$$J(\psi, \omega) = [\psi, \omega] = \psi_x \omega_y - \psi_y \omega_x.$$

This calculation has applied the commutation of the Lie derivative and differential d, $d^2 z = 0$, and

$$0 = (\partial_t + \mathcal{L}_u)(dx \wedge dy) = \operatorname{div} \mathbf{u}\,dx \wedge dy = 0. \qquad (27.4.4)$$

> **Exercise.** Explain why equation (27.4.4) holds and how it enters the calculation of (27.4.3).   ★

From the result (27.4.3), we have obtained two scalar advection equations, namely,

$$v_t + [\psi, v] = 0 \quad \text{and} \quad \partial_t \omega + [\psi, \omega] = 0. \qquad (27.4.5)$$

For the kinetic-energy Hamiltonian for incompressible fluid flows given by

$$H(\omega) = \frac{1}{2}\int_{\mathbb{R}^2} |\mathbf{u}|^2 + v^2\,dxdy = \frac{1}{2}\int_{\mathbb{R}^2} \omega\Delta^{-1}\omega + v^2\,dxdy, \qquad (27.4.6)$$

one may write equations (27.4.5) in Lie–Poisson matrix form as

$$
\partial_t \begin{pmatrix} \omega \\ v \end{pmatrix} = - \begin{bmatrix} [\Box, \omega] & [\Box, v] \\ [\Box, v] & 0 \end{bmatrix} \begin{pmatrix} \delta H/\delta \omega = \psi \\ \delta H/\delta v = v \end{pmatrix} = - \begin{pmatrix} [\psi, \omega] \\ [\psi, v] \end{pmatrix}.
$$
(27.4.7)

Exercise. Derive equation (27.4.7) and write its corresponding Lie–Poisson bracket. ★

Thus, the vertical component $\omega\hat{z}$ of the 3D vorticity ω is advected actively by the 2D divergence-free velocity, and the vertical component $v\hat{z}$ of the 3D velocity is advected passively. The corresponding Lie–Poisson bracket is defined on the dual of the semidirect-product Lie algebra $\mathfrak{X}_{vol}\,\textcircled{s}\,\Lambda^2$ of volume-preserving vector fields $\mathfrak{X}_{vol}(\mathbb{R}^2)$ acting on 2-forms $\Lambda^2(\mathbb{R}^2)$, modulo exact 2-forms [MaWe1983b]. Dual coordinates: the 2-form ω is dual to 1-forms, and scalar function v is dual to densities $\mathrm{Den}(\mathbb{R}^2)$. (In $\mathbb{R}^2$, densities are also 2-forms.)

27.5   A modified quasi-geostrophic (QG) in the β-plane approximation

The quasi-geostrophic (QG) approximation is a basic tool for the analysis of meso- and large-scale motion in geophysical and astrophysical fluid dynamics [Pe2013]. Physically, QG theory applies when the motion is nearly in geostrophic balance, that is, when pressure gradients nearly balance the Coriolis force. Mathematically, in the simplest case of a barotropic fluid in a domain $\mathcal{D}$ on the plane $\mathbb{R}^2$ with coordinates (x_1, x_2), QG dynamics in the β-plane approximation is expressed by the following evolution equation for the stream function ψ of the geostrophic fluid velocity $\mathbf{u} = \hat{z} \times \nabla\psi$:

$$
\partial_t(\Delta\psi - \mathcal{F}\psi) + [\psi, \Delta\psi] + \beta\frac{\partial\psi}{\partial x_1} = 0,
$$
(27.5.1)

where ∂_t is the partial time derivative and Δ is the planar Laplacian. Also, the brackets

$$[a, b] \equiv \partial(a, b)/\partial(x_1, x_2) = J(a, b)$$

represent the Jacobi bracket (Jacobian) for functions a and b on $\mathbb{R}^2$. In other notation, β is the gradient of the Coriolis parameter, f, taken as $f = f_0 + \beta x_2$ in the β-plane approximation, with constants β and f_0. (Neglecting β gives the f-plane approximation.) The symbol $\mathcal{F}$ in the QG equation (27.5.1) denotes the rotational Froude number:

$$\mathcal{F} := \frac{Fr^2}{Ro^2} = \frac{f_0^2 L^2}{U^2} \frac{U^2}{gb_0} = L^2/L_R^2 = O(1) \quad \text{and} \quad L_R^2 = gb_0/f_0^2,$$

(27.5.2)

where $R_o = U/f_0 L$ is the Rossby number, $Fr = U/\sqrt{gb_0}$ is the Froude number, g is gravitational acceleration, b_0 is mean depth and L and U are typical lengths and velocities, respectively. Thus, the QG approximation with a *rotational* Froude number of order $O(1)$ corresponds to slow flows in a rapidly rotating frame of motion, such as the flows which occur in the ocean and atmosphere. Consequently, fluid dynamics in the regime $\mathcal{F} = O(1)$ is often called *geophysical fluid dynamics* (GFD).

The QG equation (27.5.1) may be derived from the basic equations of a rotating shallow water flow through proper scaling and subsequent asymptotic expansion in the Rossby number $Ro \ll 1$ and $\mathcal{F} = O(1)$ in the GFD regime, where the rotational Froude number $\mathcal{F}$ is given by the square of the ratio of the characteristic scale of the motion L to the deformation radius L_R defined in (27.5.2). See, e.g., Pedlosky [Pe2013], Allen and Holm [AlHo1996] and Stegner and Zeitlin [StZe1996]. In the the f-plane approximation, equation (27.5.1) may be written in terms of the potential vorticity, q, as

$$\partial_t q + \mathbf{u} \cdot \nabla q = 0, \quad q := \Delta\psi - \mathcal{F}\psi + f =: \mu + f. \qquad (27.5.3)$$

Upon augmenting the QG kinetic-energy Hamiltonian to account for the vertical component of divergence-free 3D velocity, as done in (27.4.6) for the Euler fluid equation, one has

$$
H(\mu, v) = \frac{1}{2} \int_{\mathbb{R}^2} |\nabla \psi|^2 + \mathcal{F}\psi^2 + v^2 \, dxdy
$$

$$
= \frac{1}{2} \int_{\mathbb{R}^2} \mu (\Delta - \mathcal{F})^{-1} \mu + v^2 \, dxdy.
$$

(27.5.4)

One may then write equations (27.5.3) in Lie–Poisson form as, cf. equation (27.4.7),

$$
\partial_t \begin{pmatrix} \mu \\ v \end{pmatrix} = - \begin{bmatrix} [\Box, q] & [\Box, v] \\ [\Box, v] & 0 \end{bmatrix} \begin{pmatrix} \delta H/\delta \mu = \psi \\ \delta H/\delta v = v \end{pmatrix} = - \begin{pmatrix} [\psi, q] \\ [\psi, v] \end{pmatrix}.
$$

(27.5.5)

Thus, the quantity $\mu = q - f$ with potential vorticity q is advected *actively* by the 2D divergence-free horizontal velocity $\hat{z} \times \nabla \psi$, and the vertical component v of the 3D velocity is advected *passively* as a *diagnostic tracer* for the vertical component of the QG flow.

The Lie–Poisson bracket corresponding to the Poisson matrix in (27.5.5) is

$$
\{F, H\}(\mu, q) = - \int_D \begin{pmatrix} \delta F/\delta \mu \\ \delta F/\delta v \end{pmatrix} \begin{bmatrix} [\Box, q] & [\Box, v] \\ [\Box, v] & 0 \end{bmatrix} \begin{pmatrix} \delta H/\delta \mu \\ \delta H/\delta v \end{pmatrix} dx_1 dx_2.
$$

This Lie–Poisson bracket is defined on the dual of the semidirect-product Lie algebra $\mathfrak{X}_{vol} \circledS \Lambda^2$ of volume-preserving vector fields $\mathfrak{X}_{vol}(\mathbb{R}^2)$ acting on 2-forms $(\mu, v) \in \Lambda^2(\mathbb{R}^2)$, modulo exact 2-forms [We1983a, MaWe1983b]. Dual coordinates are as follows: 2-form μ is dual to scalar functions; and scalar function v (vertical velocity) is dual to densities $\mathrm{Den}(\mathbb{R}^3)$. (In $\mathbb{R}^2$, densities are also 2-forms.) For related treatments of the geometrical/variational approach to the QG model, see, e.g., Refs. [HoMaRaWe1985, HoMaRa1998a, Vi1981, ZePa1994].

Exercise. What are the Casimirs for the Poisson matrix in equation (27.5.5)? ★

Answer. The Casimirs for the Poisson matrix in (27.5.5) are easily seen to be

$$C_{\Phi,\Psi}(\mu, v) = \int_D \Phi(v) + q\Psi(v)\, d^2x.$$

If the vertical velocity v vanishes initially, though, it will remain so, for the Hamiltonian in (27.5.4),

$$v(\mathbf{x}, 0) = 0 \rightarrow v(\mathbf{x}, t) = 0.$$

In this case, the variable v can be ignored, and the Lie–Poisson bracket simply becomes

$$\{F, H\} = -\int_D q\left[\frac{\delta F}{\delta q}, \frac{\delta H}{\delta q}\right] d^2x. \qquad (27.5.6)$$

The Casimirs have a direct effect on the potential vorticity

$$q = \mu + f = \Delta\psi - \mathcal{F}\psi + f$$

if vertical velocity v vanishes initially because the Casimirs for the Lie–Poisson bracket in (27.5.6) are

$$C_{\Phi}(q) = \int_D \Phi(q)\, d^2x, \qquad (27.5.7)$$

provided the stream function ψ is constant on the boundary, assuming a simply connected domain of flow.

▲

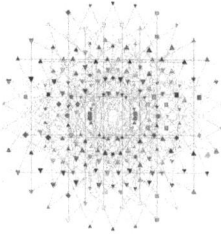

28

DISPERSIVE SHALLOW WATER (DSW) EQUATIONS IN 1D AND 2D

Contents

What is this lecture about? In this lecture, a dynamical shift in the transport velocity in equation (28.2.2) for the shallow water dynamics in two spatial dimensions is found to introduce additive second-derivative terms in one dimension, which lead to a completely integrable Hamiltonian nonlinear wave equation.

28.1 Hamilton–Pontryagin derivation of 2D DSW equations

We consider the constrained reduced Lagrangian for ideal *dispersive shallow water* (DSW) equations in two spatial dimensions, written in *Hamilton–Pontryagin* form as

$$l(\mathbf{u}, \eta) = \int_M \left(\tfrac{1}{2}\eta |\mathbf{u}|^2 - \tfrac{1}{2}g(\eta^2 + \alpha^2 |\nabla \eta|^2) \right) d^2x$$

$$+ \left\langle m,\, \dot{g}g^{-1} - u - \kappa \nabla^\sharp \log \eta \right\rangle + \left\langle \phi,\, \eta_0 g^{-1} - \eta \right\rangle,$$

(28.1.1)

where $\langle \cdot,\, \cdot \rangle$ denotes L^2 pairing. In the Lagrangian (28.1.1), the gravitational acceleration constant, g, multiplies $\eta^2/2$ in the usual potential energy. An energy penalty term $\alpha^2 |\nabla \eta|^2$ with constant positive α^2 of dimension $[L]^2$ has been added to the gravitational potential energy to control wave steepness and thereby ensure that the resulting variational equations are well posed in the sense of Hadamard. An additional transport velocity vector field $\kappa \nabla^\sharp \log \eta$ with constant κ of the dimension of diffusivity $[L]^2/[T]$ has also been included to produce dispersion. When α^2 and κ are absent, then one acquires the standard theory of shallow water dynamics.

In geometric notation, one has $|\mathbf{u}|^2 := u \lrcorner u^\flat$ with $u^\flat = \mathbf{u} \cdot d\boldsymbol{x}$, and the velocity vector field $u \in \mathfrak{X}(M \subset \mathbb{R}^2)$ (written in components as $u = \mathbf{u} \cdot \nabla$) has no normal components on the boundary ∂M. That is, $u \lrcorner d^2x = \mathbf{u} \cdot \hat{n}\, ds = 0$ on the boundary. The columnar volume of fluid ηd^2x with local depth η moves with the flow and thus satisfies the *continuity equation*

$$\left(\partial_t + \mathcal{L}_{\dot{g}g^{-1}} \right)(\eta d^2x) = 0 = \left(\partial_t \eta + \text{div}(\eta(\dot{g}g^{-1})) \right) d^2x. \quad (28.1.2)$$

In this lecture, one takes advantage of the feature discussed in Remark 5.2.1 that the Hamilton–Pontryagin approach accommodates alternative types of transport velocity $v = \dot{g}g^{-1}$ and advected quantities $a_t = a_0 g^{-1}$.

Taking variations in the Hamilton–Pontryagin action integral with constrained Lagrangian $l(\mathbf{u}, \eta)$ in (28.1.1) yields

$$0 = \delta S = \delta \int_M l(\mathbf{u}, \eta)\, dt$$

$$= \int_M \left\langle \eta u^b - m, \delta u \right\rangle + \left\langle B(\mathbf{u}, \eta) - \phi, \delta \eta \right\rangle + \left\langle \delta \phi, \eta_0 g^{-1} - \eta \right\rangle$$

$$\times \left\langle -(\partial_t + \mathcal{L}_{\dot{g}g^{-1}})m + \eta\, dB, \xi \right\rangle$$

$$+ \left\langle \delta m, \dot{g}g^{-1} - u - \kappa \nabla^\sharp \log \eta \right\rangle dt. \tag{28.1.3}$$

Here, one has the following notation:

$$\xi = \delta g g^{-1}, \quad \delta(\dot{g}g^{-1}) = (\partial_t - \mathrm{ad}_{\dot{g}g^{-1}})\xi, \quad \delta \eta = \delta(\eta_0 g^{-1}) = -\mathcal{L}_\xi \eta, \tag{28.1.4}$$

in which the ϕ-constraint relation has been used. The other constraints yield

$$m = \eta u^b = \mathbf{u} \cdot d\mathbf{x} \otimes \eta d^2 x = \mathbf{m} \cdot d\mathbf{x} \otimes d^2 x, \quad \dot{g}g^{-1} = u - \kappa \nabla^\sharp \log \eta, \tag{28.1.5}$$

and also $B(\mathbf{u}, \eta) = \phi$, where the Bernoulli function $B(\mathbf{u}, \eta)$ is given by

$$B(\mathbf{u}, \eta) := \frac{1}{2}|\mathbf{u}|^2 - g(1 - \alpha^2 \Delta)\eta - \frac{\kappa}{\eta} \mathrm{div}(\eta \mathbf{u}). \tag{28.1.6}$$

The full set of DSW equations may be written as

$$(\partial_t + \mathcal{L}_{\dot{g}g^{-1}})u^b = dB,$$
$$(\partial_t + \mathcal{L}_{\dot{g}g^{-1}})\eta d^2 x = 0. \tag{28.1.7}$$

When κ and α both vanish, equations (28.1.7) recover the familiar equations for shallow water waves.

28.2   Kelvin's theorem and potential vorticity for DSW equations

The motion equation in (28.1.7) yields the Kelvin theorem

$$\frac{d}{dt} \oint_{c(\dot{g}g^{-1})} u^\flat = \oint_{c(\dot{g}g^{-1})} \left(\partial_t + \mathcal{L}_{\dot{g}g^{-1}}\right) u^\flat = \oint_{c(\dot{g}g^{-1})} d\mathcal{B} = 0, \quad (28.2.1)$$

where $c(\dot{g}g^{-1})$ is a material loop moving with the transport velocity vector field

$$v := \dot{g}g^{-1} = u - \kappa \nabla^\sharp \log \eta, \quad (28.2.2)$$

as evaluated by the constraint relation in (28.1.5).

The Kelvin theorem for the DSW equations may also be written in terms of fluid dynamics as

$$\frac{d}{dt} \oint_{c(v)} \mathbf{u} \cdot d\mathbf{x} = \oint_{c(v)} \nabla \mathcal{B} \cdot d\mathbf{x} = 0. \quad (28.2.3)$$

The motion equation in (28.1.6) also implies

$$\left(\partial_t + \mathcal{L}_{\dot{g}g^{-1}}\right) u^\flat = d\mathcal{B}, \quad (28.2.4)$$

whose exterior derivative yields

$$\left(\partial_t + \mathcal{L}_{\dot{g}g^{-1}}\right) du^\flat = d^2 \mathcal{B} = 0. \quad (28.2.5)$$

In fluid dynamics terms with the vorticity 2-form defined by $du^\flat =: \omega d^2 x$ along with the continuity equation for the depth in (28.1.2), the previous equation implies the following advection law for the scalar potential vorticity (PV) defined as the function

$$q := \omega/\eta = (u_{2,1} - u_{1,2})/\eta,$$

$$\left(\partial_t + \mathbf{v} \cdot \nabla\right) q = \left(\partial_t + (\mathbf{u} - \kappa \nabla^\sharp \log \eta) \cdot \nabla\right) q = 0,$$

$$\text{with PV,} \quad q := \omega/\eta. \quad (28.2.6)$$

In turn, the PV advection equation (28.3.2) implies conservation by the DSW equations in (28.1.6) of the quantity

$$C_\Phi := \int_M \eta \Phi(q) \, d^2 x, \qquad (28.2.7)$$

for any differentiable function Φ.

28.3   Hamiltonian formulation of the DSW equations

By the Legendre transformation of the DSW Lagrangian in equation (28.1.1), one finds the DSW Hamiltonian

$$h(m, \eta) := \langle m, \dot{g}g^{-1} \rangle - l(u, \eta)$$

$$:= \int_M \frac{|\mathbf{m}|^2}{2\eta} - \kappa \mathbf{m} \cdot \nabla^\sharp \log(\eta) + \frac{g}{2}(\eta^2 + \alpha^2 |\nabla \eta|^2) \, d^2 x,$$

$$(28.3.1)$$

whose variational derivatives are given by

$$\frac{\delta h}{\delta \mathbf{m}} = \mathbf{m}/\eta - \kappa \nabla^\sharp \log(\eta) = \mathbf{u} - \kappa \nabla^\sharp \log(\eta),$$

$$\frac{\delta h}{\delta \eta} = -\frac{|\mathbf{m}|^2}{2\eta^2} + \frac{\kappa}{\eta} \mathrm{div}\, \mathbf{m} + g(1 - \alpha^2 \Delta)\eta = -\mathcal{B}. \qquad (28.3.2)$$

The Lie–Poisson equations in this case arise from the standard theory:

$$\partial_t m + \mathrm{ad}^*_{\delta h/\delta m}\, m = -\frac{\delta h}{\delta \eta} \diamond \eta = -\eta \nabla \frac{\delta h}{\delta \eta},$$

$$\partial_t(\eta d^2 x) + \mathcal{L}_{\delta h/\delta m}(\eta d^2 x) = 0. \qquad (28.3.3)$$

In Lie–Poisson matrix operator form, this is

$$\partial_t \begin{pmatrix} m_i \\ \eta \end{pmatrix} = - \begin{bmatrix} \partial_j m_i + m_j \partial_i & \Box \diamond \eta \\ \mathcal{L}_\Box \eta & 0 \end{bmatrix}$$

$$\times \begin{pmatrix} \delta h/\delta m_j = v^j := u^j - \kappa \partial^j \log(\eta) \\ \delta h/\delta \eta = -\mathcal{B} \end{pmatrix}. \qquad (28.3.4)$$

The variables in this Lie–Poisson operator are defined on the dual[1] of the semidirect-product Lie algebra $\mathfrak{X}⑤\Lambda^0$ of vector fields $\mathfrak{X}(M)$ acting on functions $\Lambda^0(M)$. Dual coordinates: the 1-form density $m \in \mathfrak{X}^* \otimes \mathrm{Den}(M)$ is dual to the vector fields $\mathfrak{X}(M)$; and depth $\eta \in \mathrm{Den}(M)$ is dual to scalar functions defined in domain M.

In the usual fluid dynamics notation, the system (28.3.3) becomes

$$\partial_t \mathbf{u} + (\mathbf{v} \cdot \nabla)\mathbf{u} + u_j \nabla v^j = \nabla \mathcal{B},$$

$$\partial_t \eta + \mathrm{div}(\eta \mathbf{v}) = 0,$$

(28.3.5)

where the transport velocity $\mathbf{v} = \mathbf{u} - \kappa \nabla \log \eta$ is given in (28.2.2) and the Bernoulli function $\mathcal{B}(\mathbf{u}, \eta)$ is defined in (28.1.6).

Exercise. Calculate the transformation of variables $(\mathbf{m}, \eta) \to (\mathbf{u} = \mathbf{m}/\eta, \eta)$ for the 2D version of the Lie–Poisson structure in (28.3.4), and determine the equations of motion in the variables $(\mathbf{u}, \eta)$.   ★

Answer. In two dimensions, the transformation of the Lie-Poisson operator becomes

$$\begin{bmatrix} \delta_{ki}/\eta & -m_k/\eta^2 \\ 0 & 1 \end{bmatrix} \begin{bmatrix} \partial_j m_i + m_j \partial_i & \eta \partial_i \\ \partial_j \eta & 0 \end{bmatrix} \begin{bmatrix} \delta_{jl}/\eta & 0 \\ -m_l/\eta^2 & 1 \end{bmatrix}$$

$$= \begin{bmatrix} (u_{k,l} - u_{l,k})/\eta & \partial_k \\ \partial_l & 0 \end{bmatrix}.$$

The Hamiltonian (28.3.1) in the variables $(\mathbf{u}, \eta)$ is written as

$$\tilde{h}(\mathbf{u}, \eta) := \int_M \eta \frac{|\mathbf{u}|^2}{2} - \kappa \mathbf{u} \cdot \nabla \eta + \frac{g}{2}(\eta^2 + \alpha^2 |\nabla \eta|^2) \, d^2 x,$$

(28.3.6)

---

[1]Dual spaces are defined for L^2 pairing in the planar domain M, denoted by angle brackets $\langle \cdot, \cdot \rangle$.

with variations

$$\tilde{\delta h}(\mathbf{u}, \eta) := \int_M \eta \mathbf{v} \cdot \delta\mathbf{u} + \tilde{B}\,\delta\eta\,d^2x, \qquad (28.3.7)$$

with transport velocity $\mathbf{v} := \mathbf{u} - \kappa\nabla\log\eta$ and Bernoulli function, cf. (28.1.6),

$$\tilde{B} = \frac{1}{2}|\mathbf{u}|^2 + \kappa\,\mathrm{div}\,\mathbf{u} + g(1 - \alpha^2\Delta)\eta. \qquad (28.3.8)$$

The equations of motion in the variables $(\mathbf{u}, \eta)$ in (28.3.5) can be rewritten as

$$\partial_t \begin{pmatrix} u_k \\ \eta \end{pmatrix} = -\begin{bmatrix} (u_{k,l} - u_{l,k})/\eta & \partial_k \\ \partial_l & 0 \end{bmatrix}\begin{pmatrix} \delta h/\delta u_l = \eta v^l \\ \delta h/\delta\eta = \tilde{B} \end{pmatrix},$$

$$\partial_t \begin{pmatrix} \mathbf{u} \\ \eta \end{pmatrix} = -\begin{pmatrix} -\mathbf{v}\times\mathrm{curl}\,\mathbf{u} + \nabla\tilde{B} \\ \mathrm{div}(\eta\mathbf{v}) \end{pmatrix}.$$

$$(28.3.9)$$

When κ and α both vanish, equations (28.3.9) recover the familiar equations for shallow water waves.   ▲

28.4   The BKBK equation: DSW in one dimension

In one dimension, the DSW system (28.3.5) becomes

$$\partial_t u + \partial_x(uv - B) = 0$$
$$= \partial_t u + \partial_x\left(u^2/2 + g(1 - \alpha^2\partial_x^2)\eta + \kappa u_x\right)$$
$$= \partial_t u + uu_x + g(1 - \alpha^2\partial_x^2)\eta_x + \kappa u_{xx}, \qquad (28.4.1)$$
$$\partial_t \eta + \partial_x(\eta v) = 0$$
$$= \partial_t \eta + \partial_x\left(\eta(u - \kappa\partial_x\log\eta)\right),$$

where we have used the one-dimensional version of B defined in (28.1.6) and the relation between u and v, namely, $\kappa\partial_x\log(\eta) = u - v$.

The one-dimensional DSW system in (28.4.1) can be rewritten as

$$\partial_t u + u u_x + g(1 - \alpha^2 \partial_x^2)\eta_x + \kappa u_{xx} = 0,$$
$$\partial_t \eta + \partial_x(\eta u) - \kappa \eta_{xx} = 0.$$
(28.4.2)

After transforming the one-dimensional dynamical variables $(m, \eta) \to (u, \eta)$ for the Hamiltonian in (28.3.1) and transforming the Lie–Poisson operator in (28.3.4) into the new variables, the system in (28.4.9) takes the following Hamiltonian form:

$$\partial_t \begin{pmatrix} u \\ \eta \end{pmatrix} = - \begin{bmatrix} 0 & \partial_x \\ \partial_x & 0 \end{bmatrix} \begin{pmatrix} \delta h/\delta u = \eta u - \kappa \eta_x \\ \delta h/\delta \eta = u^2/2 + g(1 - \alpha^2 \partial_x^2)\eta + \kappa u_x \end{pmatrix}.$$
(28.4.3)

For $\alpha^2 = 0$, the system (28.4.2) recovers the Broer–Kaup–Boussinesq–Kupershmidt (BKBK) system:

$$\partial_t u + u u_x + g \eta_x + \kappa u_{xx} = 0,$$
$$\partial_t \eta + \partial_x(\eta u) - \kappa \eta_{xx} = 0,$$
(28.4.4)

and the equation (28.4.3) recovers one of the Hamiltonian formulations of the BKBK system.[2]

Remark 28.4.1 Expressions (28.4.4) and the Hamiltonian form (28.4.3) with $\alpha^2 = 0$ have recovered the celebrated BKBK system, which, for $\kappa = -1/2$, was shown in Ref. [Ku1985] to be a completely integrable nonlinear wave system with three Hamiltonian structures in terms of variables (u, η). In fact, in Ref. [Ku1985], Kupershmidt even called the system in (28.4.4) "the richest integrable system known to date." □

Exercise. Prove that the transformation of variables $(m, \eta) \to (u = m/\eta, \eta)$ takes the 1D version of the

---

[2]See [ChZh2024] for a compendium of nonlinear shallow water waves in one spatial dimension, including the BKBK equation system (28.4.4).

Lie–Poisson structure in (28.3.4) into the constant-coefficient Poisson operator in equation (28.4.3).   ★

Answer. The transformation of the Poisson operator $(m, \eta) \to (u = m/\eta, \eta)$ is done by multiplying the Lie–Poisson operator from the left by the Jacobian matrix and from the right by its transpose. Explicitly, one computes

$$\begin{bmatrix} 1/\eta & -m/\eta^2 \\ 0 & 1 \end{bmatrix} \begin{bmatrix} \partial m + m\partial & \eta\partial \\ \partial\eta & 0 \end{bmatrix} \begin{bmatrix} 1/\eta & 0 \\ -m/\eta^2 & 1 \end{bmatrix} = \begin{bmatrix} 0 & \partial \\ \partial & 0 \end{bmatrix}.$$

▲

Exercise. Linearise the system (28.4.2) around the equilibrium solution $(u, \eta) = (0, \eta_0)$, where η_0 is the constant mean depth. Calculate the dispersion relation $\omega(k)$ and the phase velocity $c_p(k) := \omega(k)/k$ for travelling waves depending on space and time in the vicinity of this equilibrium as $\exp(i(kx - \omega t))$ for wave number k and frequency ω.

In this context, what can go wrong?   ★

Answer. For travelling waves in the vicinity of the constant equilibrium solution $(u, \eta) = (0, \eta_0)$, equation (28.4.3) yields

$$\begin{pmatrix} \omega/k & 0 \\ 0 & \omega/k \end{pmatrix} \begin{pmatrix} \tilde{u} \\ \tilde{\eta} \end{pmatrix} = \begin{pmatrix} ik\kappa & g(1 + \alpha^2 k^2) \\ \eta_0 & -ik\kappa \end{pmatrix} \begin{pmatrix} \tilde{u} \\ \tilde{\eta} \end{pmatrix},$$

(28.4.5)

for which a solution exists, provided

$$\det \begin{pmatrix} k^{-1}\omega - ik\kappa & g(1+\alpha^2 k^2) \\ \eta_0 & k^{-1}\omega + ik\kappa \end{pmatrix} = 0. \qquad (28.4.6)$$

Hence, the dispersion relation is

$$\omega^2(k^2) = k^2(g\eta_0(1+\alpha^2 k^2) - \kappa^2 k^2), \qquad (28.4.7)$$

and the travelling wave solution is linearly well posed (ω^2 remains positive), provided $g\eta_0(1+\alpha^2) > \kappa^2$. ▲

Exercise. In the context of the dispersion relation in equation (28.4.7) in the previous exercise, what can go wrong if $\alpha^2 = 0$? Is this really a problem? ★

Answer. If $\alpha^2 = 0$, then the dispersion relation in equation (28.4.7) becomes

$$\omega^2(k^2) = k^2(g\eta_0 - \kappa^2 k^2), \qquad (28.4.8)$$

and the BKBK system (28.4.4) is *linearly ill posed* for higher wave numbers, $k^2 > g\eta_0/\kappa^2$, independently of the sign of the parameter κ. Regardless of this linear ill-posedness, though, the solutions of the BKBK system in (28.4.4) are completely integrable as a Hamiltonian system for $\kappa = -1/2$, as was shown in Ref. [Ku1985]. ▲

As mentioned above, for $\alpha^2 = 0$, the system (28.4.2) recovers the celebrated BKBK system (28.4.4), rewritten here as

$$\partial_t u + u u_x = -g\eta_x - \kappa u_{xx},$$
$$\partial_t \eta + u\eta_x = -\eta u_x + \kappa \eta_{xx}. \qquad (28.4.9)$$

The linearised dispersion relation is $\omega^2(k^2) = k^2(-\kappa^2 k^2 + g\eta_0)$, which appears to be problematic for the linear travelling waves in this system, as indicated in the previous exercise. However, the non-linear system is in fact completely integrable for $\kappa = -1/2$ [Ku1985]. This type of subtle analytical issue may not be uncommon with Boussinesq long wave equations.

For a recent review of Boussinesq long waves and their "good" and "bad" analytical issues of the Boussinesq equations, see, e.g., Ref. [KlSa2021]. For further discussions of the "good" and "bad" aspects of the BKBK system, see Ref. [KlSa2024].

Remark 28.4.2 Controlling the wave steepness with the α^2 term introduced in the Hamilton–Pontryagin variational principle in (28.1.1) regularises both the "good" and "bad" DSW equations in 1D. They could have also both been regularised by surface tension (introduced by adding $-\tau \eta_{xxxx}$ for the positive parameter τ to the right-hand side of the η-equation in (28.4.9)). However, if this were done, then the system would have no longer been Hamiltonian.   □

Remark 28.4.3 The primary lesson of this lecture is that the dynamical shift in the transport velocity in equation (28.2.2) for the shallow water dynamics in two dimensions introduces additive second-derivative terms in the BKBK system (28.4.9), which produce proportional "bad" Boussinesq wave dispersion in one dimension. For further discussion of the BKBK system, particularly from the viewpoint of generalised two-cocycles, see Refs. [Ku1992, Ku2006]. For recent discussions of "good" versus "bad" Boussinesq equations from the viewpoint of analysis of integrable shallow water equations in one dimension, see, e.g., Refs. [ChLeWa2023, ChZh2024].

□

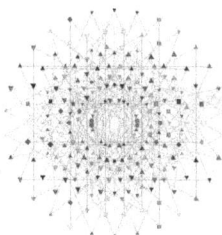

29

ROTATING SHALLOW MAGNETISED WATER (RSW-MHD)

Contents

What is this lecture about? This lecture introduces magnetohydrodynamics (MHD) into rotating shallow water (RSW) dynamics via the Euler–Poincaré variational principle and derives their Hamiltonian formulation.

29.1 RSW-MHD

Following [Gi2000, De2002, De2003], the RSW-MHD motion equation in a thin domain with bathymetry $\hbar(\boldsymbol{x})$ is written as

$$\partial_t \boldsymbol{u} + \boldsymbol{u} \cdot \nabla \boldsymbol{u} + 2\boldsymbol{\Omega} \times \boldsymbol{u} = -g\nabla(\eta - \hbar(\boldsymbol{x})) + \boldsymbol{B} \cdot \nabla \boldsymbol{B}. \qquad (29.1.1)$$

The dynamical RSW-MHD variables in (29.1.1) denote the horizontal velocity $\boldsymbol{u}$, horizontal magnetic field $\boldsymbol{B}$ and local depth $\eta - \hbar(\boldsymbol{x})$.

The parameters for Coriolis force and gravitational acceleration are, respectively, 2Ω and g.

The following two advection relations hold as auxiliary equations for the RSW-MHD motion (29.1.1):

$$\partial_t \eta + \nabla \cdot (\eta \boldsymbol{u}) = 0,$$
$$\partial_t \boldsymbol{B} + \boldsymbol{u} \cdot \nabla \boldsymbol{B} - \boldsymbol{B} \cdot \nabla \boldsymbol{u} = 0. \tag{29.1.2}$$

Together, these auxiliary equations imply preservation of

$$\nabla \cdot (\eta \boldsymbol{B}) = 0, \tag{29.1.3}$$

which can therefore be regarded as a non-dynamical constraint on the initial values.

Exercise. Prove the previous statement.

Hint: See Remark 29.1.1.   ★

To put ourselves into the geometric mechanics framework, we shall begin by showing that the RSW-MHD equations in (29.1.1) and (29.1.2) follow as Euler–Poincaré equations for Hamilton's principle with the action integral

$$S = \int_0^T \ell(\boldsymbol{u}, \eta, \boldsymbol{B}) dt$$
$$= \int_0^T \int_{CS} \left(\frac{1}{2} |\boldsymbol{u}|^2 + \frac{1}{\text{Ro}} \boldsymbol{u} \cdot \boldsymbol{R}(\boldsymbol{x}) - \frac{1}{2} |\boldsymbol{B}|^2 - \frac{1}{2\text{Fr}^2} (\eta - 2\hbar(\boldsymbol{x})) \right)$$
$$\times \eta \, d^2 x dt, \tag{29.1.4}$$

where $CS \in \mathbb{R}^2$ denotes the horizontal cross-section. The RSW-MHD equations in (29.1.1) and (29.1.2) will be derived by first evaluating the variational derivatives for the Lagrangian in the action integral (29.1.4) as[1]

$$\frac{1}{\eta}\frac{\delta l}{\delta u} = \left(u + \frac{1}{\text{Ro}}R(x)\right) \cdot dx =: V(x,t) \cdot dx =: V^\flat$$

$$\frac{\delta l}{\delta \eta} = \left(\frac{1}{2}|u|^2 + \frac{1}{\text{Ro}}u \cdot R(x) - \frac{1}{2}|B|^2 - \frac{1}{\text{Fr}^2}(\eta - \hbar(x))\right) =: \beta(x,t)$$

$$\frac{\delta l}{\delta B} = B \cdot dx \otimes \eta d^2 x =: B^\flat \otimes \eta d^2 x. \tag{29.1.5}$$

The required Euler–Poincaré variations are then given by their transformation properties:

$$\delta u = \partial_t \xi - \text{ad}_u \xi, \quad \delta \eta = -\mathcal{L}_\xi \eta = -\text{div}(\eta \boldsymbol{\xi})\, d^2 x,$$
$$\delta B = -\mathcal{L}_\xi B = \left(-\boldsymbol{\xi} \cdot \nabla B + B \cdot \nabla \boldsymbol{\xi}\right) \cdot \nabla. \tag{29.1.6}$$

The corresponding auxiliary equations are

$$\partial_t \eta = -\mathcal{L}_u \eta = -\text{div}(\eta u)\, d^2 x,$$
$$\partial_t B = -\mathcal{L}_u B = \text{ad}_u B = -[u, B], \tag{29.1.7}$$
$$\partial_t B \cdot \nabla = \left(-u \cdot \nabla B + B \cdot \nabla u\right) \cdot \nabla.$$

Finally, the Euler–Poincaré equations follow by direct calculation as

$$0 = \delta S = \int_0^T \left\langle \frac{\delta \ell}{\delta u}, \partial_t \xi - \text{ad}_u \xi \right\rangle + \left\langle \frac{\delta \ell}{\delta \eta}, -\mathcal{L}_\xi \eta \right\rangle$$

$$+ \left\langle \frac{\delta \ell}{\delta B}, -\text{ad}_\xi B \right\rangle dt$$

$$= \int_0^T \left\langle -(\partial_t + \text{ad}_u^*)\frac{\delta \ell}{\delta u} + \frac{\delta \ell}{\delta \eta} \diamond \eta + \frac{\delta \ell}{\delta B} \diamond B, \xi \right\rangle dt$$

$$= \int_0^T \left\langle -(\partial_t + \text{ad}_u^*)\frac{\delta \ell}{\delta u} + \frac{\delta \ell}{\delta \eta} \diamond \eta + \frac{\delta \ell}{\delta B} \diamond B, \xi \right\rangle$$

---

[1]In Euclidean coordinates, a vector field $u = \mathbf{u} \cdot \nabla$ has an associated 1-form $u^\flat = \mathbf{u} \cdot d\mathbf{x}$, where $d\mathbf{x}$ is the dual basis to ∇ under contraction, $\nabla \lrcorner\, d\mathbf{x} = Id$. With this notation, the terms in the action integral (29.1.4) involving an inner product may be written in coordinate-free notation as $|\mathbf{u}|^2 = u \lrcorner u^\flat$, $|B|^2 = B \lrcorner B^\flat$, and $\mathbf{u} \cdot \mathbf{R} = u \lrcorner R$.

$$
= \int_0^T \Big\langle -(\partial_t + \mathcal{L}_u)(\eta d^2 x \otimes V^\flat) + \eta d^2 x \otimes \mathbf{d}\beta
$$

$$
- \mathcal{L}_B(\eta d^2 x \otimes B^\flat), \xi \Big\rangle
$$

$$
= \int_0^T \Big\langle \eta d^2 x \otimes \big(-(\partial_t + \mathcal{L}_u)V^\flat + \mathbf{d}\beta - \mathcal{L}_B B^\flat \big) \big)
$$

$$
- \mathcal{L}_B(\eta d^2 x) \otimes B^\flat), \xi \Big\rangle
$$

$$
= \int_0^T -\Big\langle \eta d^2 x \otimes \big(\partial_t u + u \cdot \nabla u + 2\Omega \times u
$$

$$
+ g\nabla(\eta - \hbar(\boldsymbol{x})) - \boldsymbol{B} \cdot \nabla \boldsymbol{B} \big) - \mathrm{div}(\eta \boldsymbol{B}) d^2 x \otimes B^\flat, \, \xi \Big\rangle dt.
\tag{29.1.8}
$$

Remark 29.1.1 By virtue of the auxiliary equations in (29.1.7) in geometric form,

$$
(\partial_t + \mathcal{L}_u)(\eta\, d^2 x) = 0 \quad \text{and} \quad (\partial_t + \mathcal{L}_u)B = 0,
$$

one proves the advection of the divergence $\mathrm{div}(\eta \boldsymbol{B})$,

$$
(\partial_t + \mathcal{L}_u)d\big(B \lrcorner (\eta\, d^2 x)\big) = (\partial_t + \mathcal{L}_u)(\mathrm{div}(\eta \boldsymbol{B})d^2 x) = 0. \tag{29.1.9}
$$

Therefore, if the quantity $\mathrm{div}(\eta \boldsymbol{B})$ in (29.1.8) vanishes initially, then it will remain so. Consequently, one may ignore the last term in (29.1.8) by assuming that it has vanished initially and will remain so. This calculation concludes the Euler–Poincaré variational derivation of the motion equation (29.1.1) for RSW-MHD. $\qquad\square$

Remark 29.1.2 The Kelvin–Noether theorem for RSW-MHD is given by

$$
\frac{d}{dt} \oint_{c(u)} V^\flat = \oint_{c(u)} (\partial_t + \mathcal{L}_u)V^\flat = \oint_{c(u)} (\mathbf{d}\beta - \mathcal{L}_B B^\flat) = -\oint_{c(u)} \mathcal{L}_B B^\flat
$$

$$
= -\oint_{c(u)} B \lrcorner \, \mathrm{d}B^\flat = -\oint_{c(u)} \boldsymbol{B} \times \mathrm{curl} \boldsymbol{B} \cdot d\boldsymbol{x}
$$

$$
= \oint_{c(u)} \boldsymbol{J} \times \boldsymbol{B} \cdot d\boldsymbol{x}, \tag{29.1.10}
$$

where V^b is defined in equation (29.1.5). Thus, the circulation of an RSW-MHD flow in terms of V^b is not conserved. □

Exercise. The thermal rotating shallow water MHD (TRSW-MHD) model is an extension of the RSW-MHD model to include horizontally varying buoyancy and an inert lower layer. The TRSW-MHD equations modify the RSW-MHD equations to include the variable (non-negative) buoyancy $\gamma^2(\mathbf{x}, t) = (\bar{\rho} - \rho(\mathbf{x}, t))/\bar{\rho}$, where ρ is the (time- and space-dependent) mass density of the active upper layer and $\bar{\rho}$ is the uniform mass density of the inert lower layer whose boundary is represented by bathymetry, $\hbar(\mathbf{x})$.

Hamilton's principle for the TRSW-MHD equations is the same as for the RSW-MHD equations in (29.1.4) except that in both the Lagrangian in (29.1.4) and the Hamiltonian in (29.2.1), the factor $1/\mathrm{Fr}^2$ is replaced by $\gamma^2(\mathbf{x}, t)/\mathrm{Fr}^2$, where the scalar buoyancy function $\gamma^2(\mathbf{x}, t)$ is advected dynamically by the TRSW-MHD flow velocity.

Calculate the Euler–Poincaré equations and Lie–Poisson Hamiltonian equations for the TRSW-MHD model. ★

29.2   Hamiltonian structure of RSW-MHD equations

Next, we derive the Hamiltonian and Lie–Poisson structures of the RSW-MHD equations inherited from being Euler–Poincaré equations. The Lie–Poisson properties of these equations are interesting (and helpful) in studying a variety of astrophysical phenomena, including the dynamics of gravity waves and Alfvén waves on the solar tachocline [Gi2000, De2002].

The Legendre transformation of the Lagrangian in Hamilton's principle (29.1.4) yields the Hamiltonian in terms of $m = \boldsymbol{m} \cdot d\boldsymbol{x} \otimes d^2x$, $B = \boldsymbol{B} \cdot \nabla$ and ηd^2x:

$$h(m, \eta, B)$$
$$= \int \left(\frac{1}{2\eta^2} |\boldsymbol{m} - \eta \boldsymbol{R}/Ro|^2 + \frac{1}{2}|\boldsymbol{B}|^2 + \frac{1}{2\mathrm{Fr}^2}(\eta - 2\hbar(\boldsymbol{x})) \right) \eta \, d^2x.$$

$$(29.2.1)$$

Variations are obtained as

$$\delta h = \int u \lrcorner \delta m + \eta \delta B \lrcorner B^\flat + \phi \, \delta \eta \, d^2x, \qquad (29.2.2)$$

with the Bernoulli function ϕ defined for this case as

$$\phi := -\frac{|\boldsymbol{u}|^2}{2} + \frac{1}{2}|\boldsymbol{B}|^2 + \frac{1}{\mathrm{Fr}^2}(\eta - \hbar(\boldsymbol{x})). \qquad (29.2.3)$$

The corresponding semidirect-product Lie–Poisson Hamiltonian equations are given by

$$\partial_t \begin{pmatrix} m \\ B \\ \eta \end{pmatrix} = - \begin{bmatrix} \mathrm{ad}^*_\square \, m & -\mathrm{ad}^*_B \, \square & \square \diamond \eta \\ -\mathrm{ad}_\square \, B & 0 & 0 \\ \mathcal{L}_\square \eta & 0 & 0 \end{bmatrix} \begin{pmatrix} u \\ \eta d^2x \otimes B^\flat \\ \phi \end{pmatrix} \quad (29.2.4)$$

$$= - \begin{pmatrix} \mathcal{L}_u(\eta d^2x \otimes u^\flat) + \eta d^2x \otimes \mathcal{L}_B B^\flat + \eta d^2x \otimes d\phi \\ -\mathrm{ad}_u \, B \\ \mathcal{L}_u \eta \end{pmatrix},$$

$$(29.2.5)$$

where we have used the auxiliary equations and $\mathrm{div}(\eta \boldsymbol{B})$ $d^2x \otimes B^\flat = 0$ to write

$$\mathcal{L}_u m = \mathcal{L}_u V^\flat \otimes \eta d^2x + V^\flat \otimes \mathcal{L}_u(\eta d^2x)$$
$$\mathcal{L}_B(\eta d^2x \otimes B^\flat) = \mathrm{div}(\eta \boldsymbol{B}) d^2x \otimes B^\flat + \mathcal{L}_B B^\flat \otimes \eta d^2x$$
$$= \mathcal{L}_B B^\flat \otimes \eta d^2x$$
$$= \left(\boldsymbol{B} \cdot \nabla \boldsymbol{B} + \tfrac{1}{2}\nabla|\boldsymbol{B}|^2 \right) \cdot d\boldsymbol{x} \otimes \eta d^2x.$$

$$(29.2.6)$$

Hence, we obtain the RSW-MHD equations in (29.1.1) and (29.1.2) as

$$
\partial_t \begin{pmatrix} V^\flat \\ B \\ \eta \end{pmatrix} = - \begin{bmatrix} \mathcal{L}_u V^\flat + \mathcal{L}_B B^\flat + \mathrm{d}\phi \\ -\,\mathrm{ad}_u B \\ \mathcal{L}_u \eta \end{bmatrix}. \tag{29.2.7}
$$

This lecture has used the Euler–Poincaré version of Hamilton's principle to derive the equations of RSW-MHD introduced in Ref. [Gi2000]. The Lie–Poisson brackets and Casimirs for the symmetry-reduced Hamiltonian formulation of RSW-MHD have also been considered in Ref. [De2002]. For a review of potential applications of RSW-MHD in solar plasma physics, see Ref. [Mi2005].

Exercise. Because the initial condition $\mathrm{div}(\eta B) = 0$ is advected by the RSW-MHD fluid flow, one may define a stream function for $\eta B = \hat{z} \times \nabla \psi$ and determine its advection relation. Transform variables to eliminate B for $\eta^{-1}\hat{z} \times \nabla \psi$ and recalculate the Hamiltonian structure for RSW-MHD in terms of $(m, \eta.\psi)$. Having calculated the SW-MHD equations, explain the physics of its Kelvin circulation theorem.   ★

Answer. The Hamiltonian in terms of $m = \boldsymbol{m} \cdot d\boldsymbol{x} \otimes d^2x$, $B = \boldsymbol{B} \cdot \nabla$, and $\eta d^2 x$ in equation (29.2.1) transforms into

$$
h(m, \eta, B) = \int \Big(\frac{1}{2\eta^2} |\boldsymbol{m} - \eta \boldsymbol{R}/Ro|^2 + \frac{1}{2\eta^2} |\nabla\psi|^2
$$

$$
+ \frac{1}{2\mathrm{Fr}^2}(\eta - 2\hbar(\boldsymbol{x})) \Big) \eta \, d^2x. \tag{29.2.8}
$$

The variations of this Hamiltonian are obtained as

$$
\delta h = \int \boldsymbol{u} \cdot \delta \boldsymbol{m} - J \, \delta\psi + \tilde{\phi} \, \delta\eta \, d^2x, \tag{29.2.9}
$$

where we have substituted

$$-\text{div}(\eta^{-1}\nabla\psi) = \text{div}(\hat{z} \times B) = -\hat{z} \cdot \text{curl} B = -J. \tag{29.2.10}$$

The modified Bernoulli function $\tilde{\phi}$ in (29.2.9) is defined for this case as

$$\tilde{\phi} := -\frac{|u|^2}{2} - \frac{1}{2}|B|^2 + \frac{1}{\text{Fr}^2}(\eta - \hbar(x)). \tag{29.2.11}$$

The corresponding semidirect-product Lie–Poisson Hamiltonian equations are given by

$$\partial_t \begin{pmatrix} m \\ \psi \\ \eta \end{pmatrix} = -\begin{bmatrix} \text{ad}^*_\Box m & \Box \diamond \psi & \Box \diamond \eta \\ \mathcal{L}_\Box \psi & 0 & 0 \\ \mathcal{L}_\Box \eta & 0 & 0 \end{bmatrix} \begin{pmatrix} u \\ -J \\ \phi \end{pmatrix}$$

$$\partial_t \begin{pmatrix} m_i \\ \psi \\ \eta \end{pmatrix} = -\begin{bmatrix} \partial_j m_i + m_j \partial_i & -\psi_{,i} & \eta \partial_i \\ \psi_{,j} & 0 & 0 \\ \partial_j \eta & 0 & 0 \end{bmatrix} \begin{pmatrix} u^j \\ -J \\ \tilde{\phi} \end{pmatrix}$$

$$\partial_t \begin{pmatrix} u \\ \psi \\ \eta \end{pmatrix} = -\begin{pmatrix} -(u \cdot \nabla)u + 2\Omega \times u + \eta^{-1}J\nabla\psi \\ +\nabla\left(-\frac{1}{2}|B|^2 + \frac{1}{\text{Fr}^2}(\eta - \hbar(x))\right) \\ (u \cdot \nabla)\psi \\ \text{div}(\eta u) \end{pmatrix}.$$

The Kelvin circulation theorem for RSW-MHD takes the following form, cf. equation (29.1.10):

$$\frac{d}{dt}\oint_{c(u)} V \cdot dx = -\oint_{c(u)} \eta^{-1}J\nabla\psi \cdot dx$$

$$= \oint_{c(u)} (J\hat{z} \times B) \cdot dx. \tag{29.2.12}$$

Thus, perhaps not surprisingly, the magnetic force $J\hat{z} \times B = B \cdot \nabla B - \frac{1}{2}\nabla|B|^2$ generates circulation also in the SW-MHD equations. In particular, this means that the RSW-MHD equations do not advect a potential vorticity.

▲

30

INCOMPRESSIBLE 2D MHD ALFVÉN WAVE TURBULENCE

Contents

> **What is this lecture about?** This lecture introduces the system of planar incompressible Hall magnetohydrodynamic (HMHD) nonlinear Alfvén wave dynamics. It then discusses its subsystems and derives their Hamiltonian formulations.

30.1   MHD systems in planar coordinates (x, y)

Alfvén waves [Al1942] sustainable in Hall magnetohydrodynamics (HMHD) are the most common electromagnetic phenomena in magnetised plasmas. Nonlinear Alfvén waves play important roles in dispersing energy in a variety of plasma regimes, ranging from laboratory plasmas to space plasmas [Bi2003]. In particular, the Hall effect in magnetised plasma turbulence plays an important role in regulating the transport of energy in space and astrophysical plasmas, as discussed in, e.g., Ref. [MiAlPo2007].

Planar Alfvén wave turbulence in quasi-neutral plasmas has been modelled by differential equations of the following form [Ha1983, HaHoMo1985, HaMe1985, Ho1985, ShYuRaSp1984]:

$$\partial_t \omega + \{\omega, \phi\} + \{A, J\} = 0,$$

$$\partial_t A + \{A, \phi\} - \alpha\{A, \chi\} = 0, \qquad (30.1.1)$$

$$\partial_t \chi + \{\chi, \phi\} + \{A, J\} = 0.$$

where ω denotes 2D vorticity, A is 2D magnetic "vector" potential, χ is the "stream function" for the divergence-free magnetic field $\boldsymbol{B} = \nabla^\perp A$ and $J = \Delta A$ is the current density $\mathrm{curl}\boldsymbol{B} = \Delta A\, \hat{\boldsymbol{z}}$.

The bracket operation $\{\,\cdot\,,\,\cdot\,\}$ in these equations is the canonical Poisson bracket for the functions a and b in the (x, y) plane:

$$\{a, b\} := a_x b_y - a_y b_x.$$

Incompressible ideal fluid flow in the plane preserves the area element; therefore, it is no surprise that it would be Hamiltonian. Under the L^2 pairing in $\mathbb{R}^2$ for functions (a, b, c) vanishing at infinity, we have the permutation identity

$$\int_{\mathbb{R}^2} c\,\{a, b\}\, dx \wedge dy =: \langle c,\, \{a, b\}\rangle = \langle a,\, \{b, c\}\rangle = \langle b,\, \{c, a\}\rangle.$$

$$(30.1.2)$$

This identity follows easily from integration by parts when one recalls the relation of the canonical Poisson bracket in the plane to the Jacobian $J(a, b) = \{a, b\}$ of an area-preserving transformation,

$$\{a, b\}\, dx \wedge dy = J(a, b)\, dx \wedge dy = da \wedge db = -\,db \wedge da,$$

so that integration by parts and the use of $d^2 b = 0$ yields

$$\int_{\mathbb{R}^2} c\,\{a, b\}\, dx \wedge dy = \int_{\mathbb{R}^2} c\, da \wedge db = \int_{\mathbb{R}^2} c\, d(a\, db)$$

$$= -\int_{\mathbb{R}^2} dc \wedge (a\, db) = \int_{\mathbb{R}^2} a\, db \wedge dc$$

$$= \int_{\mathbb{R}^2} a\,\{b, c\}\, dx \wedge dy.$$

Physically, in 2D Alfvén wave turbulence, the field $\phi(x, y, t)$ in system (30.1.1) represents the electrostatic potential, which acts as a hydrodynamic stream function for Hall drift waves in the plane, whose divergence-free drift velocity v with vorticity $\omega = \hat{z} \cdot \text{curl} v$ is given, respectively, by

$$v = \nabla^{\perp} \phi = (-\phi_y, \phi_x)^T \quad \text{and} \quad \omega = \Delta \phi.$$

The field $A(x, y, t)$ in the system denotes the normalised magnetic flux potential, so that the magnetic field in the plane $B(x, y, t)$ and the current density $J = \hat{z} \cdot \text{curl} B$ are given, respectively, by

$$B = \nabla^{\perp} A = (-A_y, A_x)^T \quad \text{and} \quad J = \Delta A.$$

Finally, the field $\chi(x, y, t)$ in system (30.1.1) denotes the normalised deviation of the charged particle density from its constant equilibrium value.

The details of the physical approximations which lead to the equations in (30.1.1) are discussed in Refs. [HaHoMo1985, HaMe1985].

30.2 Subsystems of the nonlinear planar Alfvén wave equations

A simplification of the nonlinear Alfvén wave equations (30.1.1) applies to plasma physics in the low-beta limit (weak magnetic fields). Subsets of equations (30.1.1) include both ideal reduced magnetohydrodynamics (RMHD; see, e.g., Refs. [KaPo1973, St1976, St1977]) and the Hasegawa–Mima (HM) equation (see, e.g., Refs. [HaMi1978, HaMaKo1979]).

Low-beta RMHD. The low-beta RMHD model[1] results upon neglecting the constant parameter α appearing in the second equation of system (30.1.1) to find

$$\partial_t \omega + \{\omega, \phi\} + \{A, J\} = 0,$$
$$\partial_t A + \{A, \phi\} = 0. \tag{30.2.1}$$

---

[1] Low-beta RMHD means MHD reduced to 2D, in which $\beta = |B|^2/p \ll 1$, where β is the ratio of magnetic field intensity to hydrodynamic pressure.

This limit decouples the field χ in (30.1.1) from the other fields, ω and A. The evolution of ω and A then constitutes the RMHD system. The RMHD system has been used to simulate nonlinear shear Alfvén dynamics in tokamaks; see, e.g., Ref. [HaMe1985].

The RMHD system for ω and A with $\alpha = 0$ in system (30.1.1) has the following associated Lie–Poisson bracket:

$$\partial_t \begin{pmatrix} \omega \\ A \end{pmatrix} = - \begin{bmatrix} \{\Box, \omega\} & \{\Box, A\} \\ \{\Box, A\} & 0 \end{bmatrix} \begin{pmatrix} \delta H/\delta \omega = -\phi \\ \delta H/\delta A = -J \end{pmatrix}. \tag{30.2.2}$$

This Lie–Poisson operator has the same structure as for a 2D Euler flow with Hodge decomposition in (27.4.7).

The HM equation is recovered from system (30.1.1) by assuming the linear relation

$$\phi - \alpha \chi = 0.$$

The HM equation arises physically upon linearising the adiabatic Maxwell–Boltzman limit for the electrons; see, e.g., Refs. [Ha1983, ShYuRaSp1984]. After assuming this linear relation, system (30.1.1) implies that $\partial_t \chi + [A, J] = 0$ and $\partial_t A = 0$. Hence, the field A decouples from the RMHD system in (30.2.1). The difference between the first and third equations in (30.1.1) then becomes

$$\partial_t q + \{\phi, q\} = 0 \quad \text{with } q := \omega - \chi \quad \text{and} \quad \chi = \phi/\alpha, \tag{30.2.3}$$

which is the HM equation for the electrostatic field. The HM equation describes ideal drift wave turbulence in a low-beta plasma; see, e.g., Refs. [HaMi1978, HaMaKo1979].

Remark 30.2.1 The HM equation also has a hydrodynamic interpretation for geostrophic fluid dynamics, for example in oceanic and atmospheric physics, where the quantity $q := \omega - \alpha^{-1}\phi$ is known as *potential vorticity* (PV); see, e.g., Refs. [St1975, LaRe1976, FlLaMcRe1980]. □

30.3 Hamiltonian structure of nonlinear planar Alfvén wave dynamics

The nonlinear Alfvén wave equations (30.1.1) comprises a Hamiltonian system with the quadratic Hamiltonian

$$
H(\omega, A, \chi) = \frac{1}{2} \int_{\mathbb{R}^2} |v|^2 + |B|^2 + \alpha \chi^2 \, dx dy
$$
$$
= \frac{1}{2} \int_{\mathbb{R}^2} -\omega \Delta^{-1} \omega + |\nabla^{\perp} A|^2 + \alpha \chi^2 \, dx dy. \tag{30.3.1}
$$

Rearranging the equations into Hamiltonian form yields

$$
\partial_t \begin{pmatrix} \omega \\ \chi \\ A \end{pmatrix} = - \begin{bmatrix} \{\Box, \omega\} & \{\Box, \chi\} & \{\Box, A\} \\ \{\Box, \chi\} & \{\Box, \chi\} & \{\Box, A\} \\ \{\Box, A\} & \{\Box, A\} & 0 \end{bmatrix} \begin{pmatrix} \frac{\delta H}{\delta \omega} = -\Delta^{-1}\omega = -\phi \\ \frac{\delta H}{\delta \chi} = \alpha \chi \\ \frac{\delta H}{\delta A} = -\Delta A = -J \end{pmatrix}
$$
$$
\tag{30.3.2}
$$
$$
= - \begin{pmatrix} \{\omega, \phi\} + \{\alpha\chi, \chi\} + \{A, J\} \\ \{\chi, \phi - \alpha\chi\} + \{A, J\} \\ \{A, \phi - \alpha\chi\} \end{pmatrix}. \tag{30.3.3}
$$

Remark 30.3.1 The matrix operator in square brackets in (30.3.2) defines a Lie–Poisson bracket $\{F, H\}(\omega, \chi, A)$ on the dual of the following *nested* semidirect-product Lie algebra:

$$
\mathfrak{g}_1 \, \text{Ⓢ} \, (\mathfrak{g}_2 \oplus V) \oplus (\mathfrak{g}_2 \, \text{Ⓢ} \, V), \tag{30.3.4}
$$

with the dual coordinates $\omega \in \mathfrak{g}_1^*$ and $\chi \in \mathfrak{g}_2^*$ and $A \in V^*$. Equation (30.3.4) represents the nested direct sum of two semidirect-product Lie algebras. □

Remark 30.3.2 Exercise 10.4 provides a finite-dimensional example of such a Lie–Poisson bracket defined on the dual of a nested semidirect-product Lie algebra. □

Here, the semidirect-product actions are defined as follows. Let an element of $\mathfrak{g}_1 \circledS (\mathfrak{g}_2 \oplus V)$ be written as $(X_1; (X_2; a_2))$. For another element $(\bar{X}_1; (\bar{X}_2; \bar{a}_2))$ with $(a_2, \bar{a}_2) \in V$, the first semidirect-product action in equation (30.3.4) is defined by

$$
\begin{aligned}
\Big[&(X_1; (X_2; a_2)), (\bar{X}_1; (\bar{X}_2; \bar{a}_2)) \Big] \\
&:= \Big([X_1, \bar{X}_1]; \big([X_1, \bar{X}_2] - [\bar{X}_1, X_2] \\
&\quad + [X_2, \bar{X}_2]; X_1(\bar{X}_2 \bar{a}_2) - \bar{X}_1(X_2 a_2) \big) \Big),
\end{aligned}
\tag{30.3.5}
$$

where the action of the elements of $\mathfrak{g}_1$ and $\mathfrak{g}_2$ on the elements of V is referred to as concatenation.

With these definitions of the semidirect-product actions in equation (30.3.5), the dual coordinates of (30.3.4) are identified as follows: $\omega \in \mathfrak{g}_1^*$ is dual to $\mathfrak{g}_1$; $\chi \in \mathfrak{g}_2^*$ is dual to $\mathfrak{g}_2$; and $A \in V^*$ is dual to V. For further discussion of Poisson brackets in continuum physics which are associated with the duals of semidirect-product Lie algebras, see, e.g., Ref. [HoMaRa1998a].

The conservation of energy $H(\omega, A, \chi)$ in (30.3.1) under the dynamics of system (30.3.2) is now an immediate consequence of the Hamiltonian formulation and skew symmetry of the Lie–Poisson bracket in (30.3.2).

Casimir functionals for $\{F, H\}(\omega, \chi, A)$. As for the well-known conservation of integrals of the PV in (30.2.3), system (30.1.1) conserves Casimir functionals of the form

$$
C_{FGK} = \int_{\mathbb{R}^2} \big[F(A) + \chi G(A) + K(\omega - \chi) \big] \, dx \, dy,
\tag{30.3.6}
$$

for arbitrary differentiable functions F, G and K. For example, taking $F = 0, G = 0$ and $K(\xi) = \xi^2$ yields the conserved quantity

$$
\int_{\mathbb{R}^2} (\omega - \chi)^2 \, dx \, dy,
$$

which is analogous to the enstrophy invariant in geophysical fluid dynamics for the PV q-equation (30.2.3).

Furthermore, being Casimirs of the Poisson bracket (30.3.2), the functionals C_{FGK} in (30.3.6) are conserved in the sense that

$$\{C_{FGK}, K\} = 0, \quad \forall K(\omega, \chi, A).$$

That is, the functionals C_{FGK} are conserved for any Hamiltonian $K(\omega, \chi, A)$, not just for the Hamiltonian H in (30.3.1).

Exercise. Show that the change in field variables in the Lie–Poisson matrix of (30.3.2) from (ω, χ, A) to (μ, χ, A), with $\mu = \omega - \chi$, yields the following Lie–Poisson operator:

$$\partial_t \begin{pmatrix} \mu \\ \chi \\ A \end{pmatrix} = - \begin{bmatrix} \mathrm{ad}^*_{\square}\mu & 0 & 0 \\ 0 & \mathrm{ad}^*_{\square}\chi & \mathrm{ad}^*_{\square}A \\ 0 & \mathrm{ad}^*_{\square}A & 0 \end{bmatrix} \begin{pmatrix} \frac{\delta H}{\delta \mu} \\ \frac{\delta H}{\delta \chi} \\ \frac{\delta H}{\delta A} \end{pmatrix}, \quad (30.3.7)$$

which is dual to the direct sum of the volume-preserving vector fields $\mathfrak{g}_1 \in \mathfrak{X}(\mathbb{R}^2)$ and the semidirect-product Lie algebra of the volume-preserving vector fields $\mathfrak{g}_2 \in \mathfrak{X}(\mathbb{R}^2)$ acting on the vector space of scalar densities, V. Namely,

$$\mathfrak{g}_1 \oplus (\mathfrak{g}_2 \circledS V), \quad (30.3.8)$$

with dual coordinates $\mu \in \mathfrak{g}_1^*$, $\chi \in \mathfrak{g}_2^*$ and $A \in V^*$.   ★

Answer. To prove this statement, note that the Jacobian matrix for the transformation from (ω, χ, A) to (μ, χ, A), with $\mu = \omega - \chi$, is given by

$$J = \begin{bmatrix} 1 & -1 & 0 \\ 0 & 1 & 0 \\ 0 & 0 & 1 \end{bmatrix}.$$

Multiplying the Poisson matrix in (30.3.2) by Jacobian J from the left and by its transpose J^T from the right produces the transformed Poisson matrix in (30.3.7). This linear change of variables preserves the eigenvalues of the Poisson matrix. In particular, such linear transformations preserve the matrix null eigenvectors. Hence, this linear change of variables preserves Casimirs.

Consequently, one may check that the variational derivatives of the following functions C_1, C_2 and C_3 are Casimirs:

$$C_1 = F_1(\mu), \quad C_2 = \chi F_2(A) \quad \text{and} \quad C_3 = F_3(A),$$
$$(30.3.9)$$

where F_1, F_2 and F_3 are arbitrary differentiable functions of their arguments, respectively. That is, their variational derivatives are null eigenvectors of the Lie–Poisson bracket in (30.3.7) as well as the transformed Lie–Poisson bracket in (30.3.7). This transformation mirrors the discussion in Section 30.2 of the physical arguments for reducing the system of Alfvén turbulence equations to simpler subsystems.   ▲

Exercise. Show that Exercise 10.4 also admits a finite-dimensional example of transforming a Lie–Poisson bracket defined on the dual of a nested semidirect-product Lie algebra into another Lie–Poisson bracket defined on the dual of a direct sum as in (30.3.7).   ★

31

LÜST HALL MAGNETO-
HYDRODYNAMICS

Contents

What is this lecture about? This lecture demonstrates the process of constructing step-by-step the series of nested Lie–Poisson brackets for ideal hydrodynamic equations leading successively from Euler's fluid equations to magnetohydrodynamics (MHD), to Hall MHD (HMHD) and then to Lüst Hall MHD (LHMHD). In this series of models, each step breaks the original symmetry of ideal hydrodynamics further to introduce additional physical effects and additional semidirect-product Poisson structures. For simplicity, we study reduced solutions depending only on planar (x, y) coordinates.

31.1　Incompressible LHMHD in two dimensions

The Hall effect in a quasi-neutral plasma is a classical problem in plasma physics [Bi2003, Li1960]. The Hall effect produces an additional advective drift of the magnetic field lines induced by the electron fluid motion. The Hall drift velocity is proportional to the current density, which generates the magnetic force, which in turn generates fluid vorticity, leading to MHD turbulence. The Hall effect is particularly interesting in the physics of the solar wind and space weather because it can create small-scale structures that seed high-wavenumber instabilities as well as energy and magnetic field transfer from smaller to larger scales [MiAlPo2007].

The nonlinear interactions of the slow-fast, large-small, resolved-unresolved decomposition of the mean and the fluctuating or turbulent components of the physical processes in MHD plasmas require a structure-preserving approach. Geometric mechanics models Hall MHD as a two-fluid ion–electron plasma [Ho1987a, HoKu1987].

LHMHD is a further generalisation that accounts for both the impact of Hall drift velocity and the influence of electron inertial effects [Lü1959].

Upon introducing the following incompressible flow *ansatz* for 3D velocity $\mathbf{u}(x, y, t)$ and magnetic field $\mathbf{B}(x, y, t)$, LHMHD solutions which depend only on planar (x, y) coordinates take the following form:

$$\mathbf{B}(x, y, t) = \nabla A(x, y, t) \times \widehat{z} + b(x, y, t)\widehat{z},$$
$$\mathbf{u}(x, y, t) = -\nabla \phi(x, y, t) \times \widehat{z} + v(x, y, t)\widehat{z}. \tag{31.1.1}$$

The full 3D LHMHD equations originally derived in Ref. [Lü1959] may be written for the solutions in (31.1.1), depending only on planar (x, y) coordinates via the canonical Poisson bracket $\{\cdot, \cdot\}$, as [HoHuSt2024]

$$\partial_t \omega + \{\phi, \omega\} + \{\triangle A, A\} - \gamma^2 \{\triangle^2 A, A\} - \gamma^2 \{b, \triangle b\} = 0,$$

$$\partial_t v + \{\phi, v\} + \{A, b\} - \gamma^2 \{A, \triangle b\} = 0,$$

$$\partial_t A + \{\phi, A\} - \frac{R}{a} \{A, b\} + \frac{\gamma^2 R}{a} \{A, \triangle b\} = 0,$$

$$\partial_t b + \{\phi, b\} + \{A, v\} - \frac{R}{a} \{\triangle A, A\} + \frac{\gamma^2 R}{a} \{\triangle^2 A, A\}$$

$$- \frac{\gamma^2 R}{a} \{\triangle b, b\} = 0, \tag{31.1.2}$$

in which the Hall parameter R/a and the inertial parameter γ^2 are prescribed constants. In the reduced dynamical variables in (31.1.1), the conserved energy Hamiltonian for ideal 3D LHMHD in the literature, such as in Ref. [HoHuSt2024], may be expressed as

$$H = \frac{1}{2} \int_D -\phi\omega + v^2 + \left(-A\triangle A + b^2\right) + \gamma^2 \left(-b\triangle b + (\triangle A)^2\right) \, d^2x. \tag{31.1.3}$$

The variational derivatives of this Hamiltonian are given by

$$\frac{\delta H}{\delta \omega} = -\phi, \quad \frac{\delta H}{\delta v} = v, \quad \frac{\delta H}{\delta b} = b - \gamma^2 \triangle b, \quad \frac{\delta H}{\delta A} = -\triangle A + \gamma^2 \triangle^2 A. \tag{31.1.4}$$

In terms of these Hamiltonian variations, the reduced LHMHD equations take the following Lie–Poisson form:

$$\partial_t \omega + \{\phi, \omega\} - \left\{ \frac{\delta H}{\delta A}, A \right\} - \left\{ \frac{\delta H}{\delta b}, b \right\} = 0,$$

$$\partial_t v + \{\phi, v\} - \left\{ \frac{\delta H}{\delta b}, A \right\} = 0,$$

$$\partial_t b + \{\phi, b\} + \{A, v\} + \frac{R}{a} \left\{ \frac{\delta H}{\delta A}, A \right\} + \frac{R}{a} \left\{ \frac{\delta H}{\delta b}, b \right\} = 0, \tag{31.1.5}$$

$$\partial_t A + \{\phi, A\} + \frac{R}{a} \left\{ \frac{\delta H}{\delta b}, A \right\} = 0.$$

One may write these equations in terms of a Lie–Poisson operator as

$$
\partial_t \begin{pmatrix} \omega \\ v \\ b \\ A \end{pmatrix} = \begin{bmatrix} \{\Box, \omega\} & \{\Box, v\} & \{\Box, b\} & \{\Box, A\} \\ \{\Box, v\} & 0 & \{\Box, A\} & 0 \\ \{\Box, b\} & \{\Box, A\} & -Ra^{-1}\{\Box, b\} & -Ra^{-1}\{\Box, A\} \\ \{\Box, A\} & 0 & -Ra^{-1}\{\Box, A\} & 0 \end{bmatrix}
$$

$$
\times \begin{pmatrix} -\phi \\ \delta H/\delta v \\ \delta H/\delta b \\ \delta H/\delta A \end{pmatrix}. \tag{31.1.6}
$$

The Lie–Poisson bracket implied by this Lie–Poisson matrix is a dual of the following nested semidirect product of two semidirect-product Lie algebras:

$$
\mathfrak{s}_1 \text{Ⓢ} \mathfrak{s}_2, \quad \text{where } \mathfrak{s}_i = \mathfrak{g}_i \text{Ⓢ} V_i \quad i = 1, 2. \tag{31.1.7}
$$

The skew symmetry of the Lie–Poisson operator (31.1.6) under L^2 pairing guarantees that the Hamiltonian $H(\omega, A, b, v)$ in (31.1.3) will be preserved by the dynamics it generates through the Lie–Poisson bracket associated with the Lie–Poisson matrix operator in (31.1.6).

Remark 31.1.1 (Casimir integrals for reduced LHMHD) Casimir integrals are conserved by any Hamiltonian because their variational derivatives are null eigenvectors of the corresponding Lie–Poisson operator. In particular, the Casimir integrals of the Lie–Poisson bracket arising from the Lie–Poisson matrix in (31.1.6) comprise a set of constants of motion for the system of 2D reduced LHMHD equations appearing in (31.1.2):

$$
C_\Phi(\omega, v) = \int_{\mathcal{D}} \omega \Phi_1(v) + \Phi_2(v)\, d^2x \quad \text{and}
$$

$$
C_\Theta(b, A) = \int_{\mathcal{D}} b\Theta_1(A) + \Theta_2(A)\, d^2x,
$$

where $\Phi_1(v)$, $\Phi_2(v)$, $\Theta_1(A)$ and $\Theta_2(A)$ are differentiable functions of their respective arguments. $\qquad\qquad\qquad\qquad\qquad\qquad\qquad\qquad\square$

31.2 Known subsystems of the reduced LHMHD equations

- The reduced Hall MHD equations emerge in the limit $\gamma^2 \to 0$ in the reduced LHMHD Hamiltonian in (31.1.3). These equations may also be obtained by modifying the Poisson structure in (31.1.6) into block diagonal semidirect-product form as follows:

$$
\partial_t \begin{pmatrix} \omega \\ v \\ b \\ A \end{pmatrix}
$$

$$
= \begin{bmatrix} \{\square, \omega\} & \{\square, v\} & 0 & 0 \\ \{\square, v\} & 0 & 0 & 0 \\ 0 & 0 & -Ra^{-1}\{\square, b\} & -Ra^{-1}\{\square, A\} \\ 0 & 0 & -Ra^{-1}\{\square, A\} & 0 \end{bmatrix}
$$

$$
\times \begin{pmatrix} -\phi \\ \delta H/\delta v \\ \delta H/\delta b \\ \delta H/\delta A \end{pmatrix}. \qquad\qquad (31.2.1)
$$

- The reduced inertial MHD equations emerge in the limit $R/a \to 0$ in the Poisson operator in (31.1.6) as

$$
\partial_t \begin{pmatrix} \omega \\ v \\ b \\ A \end{pmatrix} = \begin{bmatrix} \{\square, \omega\} & \{\square, v\} & \{\square, b\} & \{\square, A\} \\ \{\square, v\} & 0 & \{\square, A\} & 0 \\ \{\square, b\} & \{\square, A\} & 0 & 0 \\ \{\square, A\} & 0 & 0 & 0 \end{bmatrix} \begin{pmatrix} -\phi \\ \delta H/\delta v \\ \delta H/\delta b \\ \delta H/\delta A \end{pmatrix}.
$$

$$
\qquad\qquad\qquad\qquad\qquad\qquad\qquad\qquad\qquad (31.2.2)
$$

- In the absence of the variable v, the Lie–Poisson matrix operator in (31.1.6) reduces to the Lie–Poisson matrix operator in (30.3.2), derived in Refs. [HaMe1985, HaHoMo1985] for modelling Alfvén waves in plasmas:

$$
\partial_t \begin{pmatrix} \omega \\ b \\ A \end{pmatrix} = \begin{bmatrix} \{\Box, \omega\} & \{\Box, b\} & \{\Box, A\} \\ \{\Box, b\} & -Ra^{-1}\{\Box, b\} & -Ra^{-1}\{\Box, A\} \\ \{\Box, A\} & -Ra^{-1}\{\Box, A\} & 0 \end{bmatrix}
$$

$$
\times \begin{pmatrix} -\phi \\ \delta H/\delta b \\ \delta H/\delta A \end{pmatrix}.
\qquad (31.2.3)
$$

- The reduced MHD equations [KaPo1973, St1976, St1977] arise in this notation when b is absent in (31.2.3) and with a relative minus sign convention:

$$
\partial_t \begin{pmatrix} \omega \\ A \end{pmatrix} = - \begin{bmatrix} \{\Box, \omega\} & \{\Box, A\} \\ \{\Box, A\} & 0 \end{bmatrix} \begin{pmatrix} \delta H/\delta \omega = -\phi \\ \delta H/\delta A = -J \end{pmatrix}.
\qquad (31.2.4)
$$

- Finally, when all magnetic effects are suppressed, one returns to the Poisson structure for the vorticity dynamics of the planar Euler fluid equations.

Exercise. Show that when v is absent in the reduced LHMHD Lie–Poisson operator in (31.1.6), then it reduces to the Alfvén Lie–Poisson operator in (30.3.2).

★

Exercise. Show that Exercise 10.5 in paragraph 10.5 provides a finite-dimensional example of a Lie–Poisson

> operator dual to the nested semidirect product of three
> semidirect-product Lie algebras, as opposed to equation
> (31.1.7) for LHMHD, which involves the nested semidi-
> rect product of two semidirect-product Lie algebras. ★

31.3 Summary: Whorls within whorls, within whorls

"Big whorls have little whorls,..., and so on to viscosity." The
whorls in this rhyme verse by L. F. Richardson [Ri1922] refer to
coherent circulations governed by Euler's fluid equation (1765)
for ideal volume-preserving fluid flow. Euler's equation implies
Kelvin's theorem [Ke1869], which states that time-dependent Euler
fluid flows ϕ_t preserve functionals of the fluid velocity 1-form $u^\flat :=$
$\boldsymbol{u} \cdot d\boldsymbol{x}$, defined on a space of loops $c_t = \phi_{t*}c_0$ pushed forward by ϕ_t:

$$\frac{d}{dt} \oint_{\phi_{t*}c_0} u^\flat = \oint_{\phi_{t*}c_0} (\partial_t + \mathcal{L}_u)u^\flat = 0.$$

Here, $\phi_{t*}c_0$ with $\phi_0 = Id$ is the push-forward of the initial material
loop c_0 by the smooth invertible flow map ϕ_t depending on time t.[1]
The proof of Kelvin's theorem follows from the *Lie chain rule* for the
push-forward, recalled here from equation (18.4.2):

$$\frac{d}{dt}\phi_{t*}u^\flat = -\phi_{t*}(\mathcal{L}_u u^\flat).$$

The Lie chain rule defines the coordinate-free *Lie derivative* $\mathcal{L}_u$ as
(minus) the tangent to the push-forward along the vector field $u :=$
$\boldsymbol{u} \cdot \nabla$, which generates the flow ϕ_t.[2]

---

[1]The semidirect-product action $G \circledS V$ of smooth invertible maps G on vector
spaces V comprises the configuration manifold of a Euler flow. Eulerian fluid
dynamics is defined on $T(G \circledS V)/G = \mathfrak{g} \circledS V$ and $T^*(G \circledS V)/G = \mathfrak{g}^* \circledS V$.

[2]In general, one uses the Lie derivative and differential notation to make one's
calculations coordinate-free. This notation is appropriate here since solutions of
Euler's equations for homogeneous fluids follow geodesic paths on the manifold
of smooth volume-preserving invertible maps (diffeomorphisms) with respect to
the metric given by the fluid kinetic energy [Ar2014].

Vector-space valued advected fluid quantities evolve by the push-forward $a_t = \phi_{t*}a_0 \in V$ and satisfy $(\partial_t + \mathcal{L}_u)a_t = 0$. Their introduction reduces the Lie group symmetry of the Lagrangian $\ell(u, a_t)$ in Hamilton's principle $\delta S = \delta \int_0^T \ell(u, a_t)dt$ for a Euler flow to the isotropy subgroup of the initial condition a_0. The remaining symmetry leads to the Kelvin–Noether theorem [HoMaRa1998a]:

$$\frac{d}{dt} \oint_{\phi_{t*}c_0} u^\flat = \oint_{\phi_{t*}c_0} (\partial_t + \mathcal{L}_u)u^\flat = \oint_{\phi_{t*}c_0} \frac{\delta\ell}{\delta a_t} \diamond a_t,$$

in which the diamond operation ($\diamond$) is defined in terms of two pairings: $\langle b \diamond a, \xi \rangle_{\mathfrak{g}^* \times \mathfrak{g}} := \langle b, -\mathcal{L}_\xi a_t \rangle_{V^* \times V}$ for $\xi \in \mathfrak{X}(\mathcal{D})$, a vector field defined on the flow domain, $\mathcal{D}$.

The Lie symmetry-reduced Legendre transformation defines the corresponding Hamiltonian,

$$h(m, a) := \langle m, u \rangle_{\mathfrak{g}^* \times \mathfrak{g}} - \ell(u, a) \quad \& \quad m := \delta\ell/\delta u,$$

on the Poisson manifold $(m, a) \in P = \mathfrak{g}^* \circledS V$, where $\circledS$ denotes semidirect action of the Lie algebra of vector fields $\mathfrak{g} = \mathfrak{X}(\mathcal{D})$ on the vector space of advected quantities $a \in V(\mathcal{D})$ with the Lie–Poisson bracket in L^2 pairing

$$\{f, h\}_{LP} = - \langle \mu, [\delta f/\delta(m, a), \delta h/\delta(m, a)] \rangle_{L^2}.$$

Its semidirect-product dynamics is denoted as

$$\frac{df}{dt} = \{f, h\}_{LP} = - \begin{pmatrix} \delta f/\delta m \\ \delta f/\delta a \end{pmatrix}^T \begin{bmatrix} \mathcal{L}_\square m & \square \diamond a \\ \mathcal{L}_\square a & 0 \end{bmatrix} \begin{pmatrix} \delta h/\delta m \\ \delta h/\delta a \end{pmatrix}.$$

This is the Lie–Poisson Hamiltonian formulation of ideal fluid dynamics on Poisson manifolds obtained via the Lie-symmetry reduction of Hamilton's principle in Ref. [HoMaRa1998a].

Nested semidirect-product action for MHD plasma dynamics. In plasma dynamics approximated by the MHD equations, the magnetic field breaks Euler symmetry and is advected by the fluid motion map ϕ_t. Hence, the Lie–Poisson Hamiltonian formulation derived in Ref. [HoMaRa1998a] applies to MHD [Ho1987a].

Allowing for finite electron inertia breaks the symmetry further, as the electrons in Hall MHD are regarded as a second fluid whose charge density is carried by the B-field. This is the classical Hall effect. The Hall effect shifts momentum to account for the composition of flow maps as the fluid velocity carries the B-field lines, and the B-field carries the electron charge density. Thus, Hall MHD produces whorls within whorls, within whorls [HoHuSt2023].

However, at the level of Hall MHD, the Poisson bracket is block diagonal. When Hall MHD is extended further to Lüst Hall MHD to account for finite electron inertia effects, the remaining off-diagonal parts of the Lie–Poisson bracket get involved, as the nested semidirect-product bracket (31.1.7) becomes fully entwined by the additional physics of the finite electron inertia.

In summary, this lecture has demonstrated the process of constructing step-by-step the series of nested Lie–Poisson brackets for ideal hydrodynamics equations, leading successively from Euler's fluid equations to MHD, to Hall MHD and then to Lüst Hall MHD, after having reduced the full versions of the theory from 3D on $(x, y, z) \in \mathbb{R}^3$ to 2D on $(x, y) \in \mathbb{R}^2$ by using the Hodge decomposition in equation (27.4.1). For a review of potential applications of LHMHD in plasma physics, see Ref. [Mi2005]. For derivations using Hamilton's principle of the Hamiltonian dynamics of a charged fluid, including electro- and magnetohydrodynamics, see Refs. [Ho1986, Ho1987b]. For recent developments aimed at making the transport operator in Hall MHD stochastic, see Ref. [HoHuSt2024].

Appendix A

GEOMETRIC MECHANICS: DEFINITIONS AND TOPICS

Contents

A.1   A sketch of the lecture topics

- Spaces – Smooth manifolds: locally, $\mathbb{R}^n$ spaces on which the rules of calculus apply.

- Motion – Flows $\phi_t \circ \phi_s = \phi_{t+s}$ of Lie groups acting on smooth manifolds.

- Laws of motion and discussion of solutions.

- Newton's laws:

    - Newton: $d\mathbf{p}/dt = \mathbf{F}$, for momentum $\mathbf{p}$ and prescribed force $\mathbf{F}$ (on $\mathbb{R}^n$ historically).
    - Lie group invariant variational principles:
        * Euler–Lagrange equations: optimal "action" (Hamilton principle);
        * Geodesic motion – optimal dynamics with respect to kinetic energy metric.

- Lagrangian and Hamiltonian formalism:

    - Newton's law of motion;
    - Euler–Lagrange theorem;
    - Noether's theorem (Lie symmetries imply conservation laws);
    - Euler–Poincaré theorem;
    - Kelvin–Noether theorem.

- Applications and examples:

    - Geodesic motion on a Riemannian manifold[1];
    - Rigid body – geodesic motion on the Lie group $SO(3)$;
    - Other geodesic motion, for example, Riemann ellipsoids on Lie group $GL(3, R)$;
    - Heavy top.

- Lagrangian mechanics on Lie groups and Euler–Poincaré (EP) equations:

    - EP(G), EP equations for geodesics on a Lie group, G;
    - EPDiff$(\mathbb{R})$ for geodesics on Diff$(\mathbb{R})$;

---

[1] A Riemannian manifold is a smooth manifold Q endowed with a symmetric non-degenerate covariant tensor g, which is positive definite.

- Pulsons, the singular solutions of EPDiff($\mathbb{R}$)) with respect to any norm;

- Peakons, the singular solitons for EPDiff($\mathbb{R}, H^1$), with respect to the H^1 norm;

- EPDiff($\mathbb{R}^n$) and singular geodesics;

- Diffeons and momentum maps for EPDiff($\mathbb{R}^n$).

• Euler–Poincaré (EP) equations for continua:

- EP semidirect-product reduction theorem;

- Kelvin–Noether circulation theorem;

- EP equations with advected parameters for geophysical fluid dynamics.

A.2 Hamilton's principle of stationary action

Lagrangians on $T\mathbb{R}^{3N}$	G-invariant Lagrangians on TG
Euler–Lagrange equations	Euler–Poincaré equations
Noether's theorem	Kelvin–Noether theorem
Symmetry $\implies$ cons. laws	Conservation laws emerge
Legendre transformation	Legendre transformation
Hamilton's canonical equations	Lie–Poisson equations
Poisson brackets	Lie–Poisson brackets
Symplectic manifold	Poisson manifold
Symplectic momentum map	Cotangent-lift momentum map
Reduction by symmetry	Reduction to coadjoint orbits

A.3 Motivation for the geometric approach

We begin with a series of outline sketches to motivate the geometric approach taken in the course and explain more about its content.

A.3.1   Why is the geometric approach interesting to know?

- Defines problems on manifolds

    - coordinate-free

        * re-do of calculations not required when changing coordinates,
        * more compact,
        * unified framework for expressing ideas and using symmetry.

- "First principles" approach

    - variational principles — systematic — unified approach; for example, similarity between tops and fluid dynamics (semi-direct product), and magnetohydrodynamics (MHD) and ...

    - POWER
      Geometric constructions can give useful answers without us having to find and work with complicated explicit solutions. For example, classifying rigid body equilibria and fluid equilibria, for example, as critical points of the sum over their constants of motion, and then taking second variations to derive conditions for their stability.

A.4   Potential master's/PhD research topics in geometric mechanics

Rigid body

- Euler–Lagrange and Euler–Poincaré equations

- Kelvin–Noether theorem

- Lie–Poisson bracket, Casimirs and coadjoint orbits

- Reconstruction and momentum maps

- The symmetric form of the rigid body equations ($\dot{Q} = Q\Omega$, $\dot{P} = P\Omega$)

- $\mathbb{R}^3$ bracket and intersecting level surfaces

$$\dot{\mathbf{x}} = \nabla C \times \nabla H = \nabla(\alpha C + \beta H) \times \nabla(\gamma C + \epsilon H), \quad \text{for } \alpha\epsilon - \beta\gamma = 1$$

Examples:

(1) Rigid body motion: rotations and translations

(2) Fermat's principle and ray optics

(3) Multicomponent dynamics

(4) Composition of maps

- $SU(2)$ rigid body, Cayley–Klein parameters and Hopf fibration

- Higher-dimensional rigid bodies

    - Manakov integrable top on $O(n)$ and its spectral problem

Heavy top — symmetry breaking

- Euler–Poincaré variational principle for the heavy top
- Kaluza–Klein formulation of the heavy top

Geometric formulations of physical models

- Kirchhoff elastica, underwater vehicles, liquid crystals, stratified flows, polarization dynamics of telecom optical pulses, ideal fluid flows and geophysical fluid dynamics

General theory

- Euler–Poincaré semidirect-product reduction theorem
- Semidirect-product Lie–Poisson formulation

Shallow water waves and ideal fluid flows

- KdV and CH equation — solitons and peakons (geodesics on the Bott–Virasoro group)

- EPDiff equation (geodesics on the manifold of smooth invertible maps)

Fluid dynamics

- Euler–Poincaré variational principle for incompressible and compressible ideal fluids

- Preparation for stochastic geometric mechanics (discussed elsewhere, see Ref. [HoHuSt2024])

A.5   Course outline

- Geometric Structures in Classical Mechanics

  - Vocabulary of counterpoints between mathematics and physics
  - Smooth manifolds
    * locally $\mathbb{R}^n$ spaces which admit calculus, tangent vectors, action principles
  - Lie groups: Groups of transformations that depend smoothly on a set of parameters.
    Lie groups are also manifolds
    * uniqueness of Lie group product implies the flow property $\phi_{t+s} = \phi_t \circ \phi_s$
    * Lie symmetries encode conservation laws into geometry
  - Variational principles with Lie symmetries
    * Euler–Lagrange equations $\rightarrow$ Euler–Poincaré equations (compact)

∗ Two formulations (Lagrangian and Hamiltonian) which are mutually beneficial

Lagrangian side	Hamiltonian side
Hamilton's principle	Lie–Poisson brackets
Noether's theorem	Lie symmetries $\implies$ cons. laws
Lie symmetries $\implies$ cons. laws	Momentum maps
Composed of momentum maps	Jacobi identity

- Geometric Formulations of Physical Applications and Modelling

  - Newtonian mechanics
  - Rigid bodies
  - Heavy tops — integrable cases
  - Geometric ray optics
  - Fluids — Kelvin–Noether theorem
  - Waves $\begin{cases} \text{shallow water waves} \\ \text{solitons} \end{cases}$
  - Dispersive shallow water waves
  - MHD
  - Magnetohydrodynamic rotating shallow water
  - 2D Hall MHD and Lüst Hall MHD

Bibliography

[ArDeHo2018] Arnaudon, A., De Castro, A. L. and Holm, D. D. [2018] Noise and dissipation on coadjoint orbits. *J. Nonlinear Sci.* **28**, 91–145.

[AlHo1996] Allen, J. S. and Holm, D. D. [1996] Extended-geostrophic Hamiltonian models for rotating shallow water motion. *Phys. D.* **98**(2–4), 229–248.

[ArKh1998] Arnold, V. I. and Khesin, B. A. [1998] *Topological Methods in Hydrodynamics.* Applied Mathematical Sciences, Vol. 125. New York: Springer.

[Al1942] Alfvén, H. [1942] Existence of electromagnetic-hydro-dynamic waves. *Nature* **150**(3805), 405–406.

[AbMa1978] Abraham, R. and Marsden, J. E. [1978] *Foundations of Mechanics.* Reading, MA: Benjamin/Cummings Publishing Co. Advanced Book Program. 2nd Ed., revised and enlarged, with the assistance of Tudor Raţiu and Richard Cushman.

[An1972] Anderson, P. W. [1972] More is different: Broken symmetry and the nature of the hierarchical structure of science. *Science* **177**(4047), 393–396.

[Ar1966] Arnold, V. I. [1966] Sur la géométrie différentielle des groupes de Lie de dimension infinie et ses applications à l'hydrodynamique des fluides parfaits. *Ann. Inst. Fourier (Grenoble)* **16**(fasc. 1), 319–361.

[Ar1989] Arnold, V. I. [1989]   *Mathematical Methods of Classical Mechanics*, 2nd Ed. Graduate Texts in Mathematics, Vol. 60. New York: Springer. Translated from the Russian by K. Vogtmann and A. Weinstein.

[Ar2013] Arnold, V. I. [2013]   *Mathematical Methods of Classical Mechanics*, Vol. 60.   Springer Science & Business Media: New York.

[Ar2014] Arnold, V. I. [2014]   On the differential geometry of infinite-dimensional lie groups and its application to the hydrodynamics of perfect fluids. *Vladimir I. Arnold-Collected Works: Hydrodynamics, Bifurcation Theory, and Algebraic Geometry 1965-1972*, pp. 33–69.

[Ba2020] Baez, J. C. [2020] Getting to the bottom of Noether's theorem. *arXiv preprint* arXiv:2006.14741.

[BlBrCr1997] Bloch, A. M., Brockett, R. W. and Crouch, P. E. [1997] Double bracket equations and geodesic flows on symmetric spaces. *Commun. Math. Phys.* **187**(2), 357–373.

[BlCr1996] Bloch, A. M. and Crouch, P. E. [1996] Optimal control and geodesic flows. *Syst. Control Lett.* **28**(2), 65–72.

[BaGa2020] Barbaresco, F. and Gay-Balmaz, F. [2020] Lie group cohomology and (multi) symplectic integrators: New geometric tools for lie group machine learning based on souriau geometric statistical mechanics. *Entropy* **22**(5), 498.

[BrGaHoRa2011] Bruveris, M., Gay-Balmaz, F., Holm, D. D. and Raţiu, T. S. [2011] The momentum map representation of images. *J. Nonlinear Sci.* **21**, 115–150.

[Bi2003] Biskamp, D. [2003] *Magnetohydrodynamic Turbulence*. Cambridge, UK: Cambridge University Press.

[Bl2003] Bloch, A. M. [2003] *Nonholonomic Mechanics and Control*. Interdisciplinary Applied Mathematics, Vol. 24. New York: Springer. With the collaboration of J. Baillieul, P. Crouch and

J. Marsden. With scientific input from P. S. Krishnaprasad, R. M. Murray and D. Zenkov.

[BoWo2013] Born, M. and Wolf, E. [2013] *Principles of Optics: Electromagnetic Theory of Propagation, Interference and Diffraction of Light*. Elsevier: Amsterdam.

[Ca1995] Calogero, F. [1995] An integrable Hamiltonian system. *Phys. Lett. A*. **201**(4), 306–310.

[CaFr1996] Calogero, F. and Françoise, J.-P. [1996] A completely integrable Hamiltonian system. *J. Math. Phys.* **37**(6), 2863–2871.

[CaHo1993] Camassa, R. and Holm, D. D. [1993] An integrable shallow water equation with peaked solitons. *Phys. Rev. Lett.* **71**(11), 1661–1664.

[CoLe1984] Coddington, E. A. and Levinson, N. [1984] *Theory of Ordinary Differential Equations*. Robert E. Krieger, Malabar.

[ChLeWa2023] Charlier, C., Lenells, J. and Wang, D.-S. [2023] The "good" Boussinesq equation: Long-time asymptotics. *Anal. & PDE* **16**(6), 1351–1388.

[ChZh2024] Cheviakov, A. and Zhao, P. [2024] *Analytical Properties of Nonlinear Partial Differential Equations: With Applications to Shallow Water Models*, Vol. 10. Springer Nature, New York.

[De2002] Dellar, P. J. [2002] Hamiltonian and symmetric hyperbolic structures of shallow water magnetohydrodynamics. *Phys. Plasmas* **9**(4), 1130–1136.

[De2003] Dellar, P. J. [2003] Common Hamiltonian structure of the shallow water equations with horizontal temperature gradients and magnetic fields. *Phys. Fluids* **15**(2), 292–297.

[DuGoHo2001] Dullin, H., Gottwald, G. and Holm, D. D. [2001] An integrable shallow water equation with linear and nonlinear dispersion. *Phys. Rev. Lett.* **87**, 194501–04.

[DuGoHo2003] Dullin, H. R., Gottwald, G. A. and Holm, D. D. [2003] Camassa–Holm, Korteweg-de Vries-5 and other asymptotically equivalent equations for shallow water waves. *Fluid Dyn. Res.* **33**(1–2), 73–95. In Memoriam Prof. Philip Gerald Drazin 1934–2002.

[DuGoHo2004] Dullin, H. R., Gottwald, G. A. and Holm, D. D. [2004] On asymptotically equivalent shallow water wave equations. *Phys. D.* **190**(1–2), 1–14.

[EbMa1970] Ebin, D. G. and Marsden, J. [1970] Groups of diffeomorphisms and the motion of an incompressible fluid. *Ann. Math.* (2) **92**, 102–163.

[FrHo2001] Fringer, O. B. and Holm, D. D. [2001] Integrable vs. nonintegrable geodesic soliton behavior. *Phys. D.* **150**(3–4), 237–263.

[Fl2015] Flaschka, H. [2015] Henry P. McKean Jr. and Integrable Systems. In *Henry P. McKean Jr. Selecta*, F. Alberto Grünbaum, Pierre van Moerbeke, Victor H. Moll (eds.), New York: Springer, pp. 15–30.

[FlLaMcRe1980] Flierl, G. R., Larichev, V. D., McWilliams, J. C. and Reznik, G. M. [1980] The dynamics of baroclinic and barotropic solitary eddies. *Dyn. Atmos. Oceans* **5**(1), 1–41.

[GaMaRa2012] Gay-Balmaz, F., Marsden, J. E. and Raţiu, T. S. [2012] Reduced variational formulations in free boundary continuum mechanics. *J. Nonlinear Sci.* **22**, 463–497.

[GaVi2012] Gay-Balmaz, F. and Vizman, C. [2012] Dual pairs in fluid dynamics. *Ann. Glob. Anal. Geom.* **41**, 1–24.

[GiHoKu1982] Gibbons, J., Holm, D. D. and Kupershmidt, B. [1982] Gauge-invariant Poisson brackets for chromohydrodynamics. *Phys. Lett. A.* **90**(6), 281–283.

[Gi2000] Gilman, P. A. [2000] Magnetohydrodynamic 'shallow water' equations for the solar tachocline. *Astrophys. J.* **544**(1):L79.

[GuSt1984] Guillemin, V. and Sternberg, S. [1984] *Symplectic Techniques in Physics*. Cambridge: Cambridge University Press.

[Ha1984] Haine, L. [1984] The algebraic complete integrability of geodesic flow on $SO(n)$. *Commun. Math. Phys.* **94**(2), 271–287.

[Ha1983] Hazeltine, R. D. [1983] Reduced magnetohydrodynamics and the Hasegawa–Mima equation. *Phys. Fluids* **26**(11), 3242.

[HaHoMo1985] Hazeltine, R. D., Holm, D. D. and Morrison, P. J. [1985] Electromagnetic solitary waves in magnetized plasmas. *J. Plasma Phys.* **34**(1), 103–114.

[HoHuSt2023] Holm, D. D., Hu, R. and Street, O. D. [2023] Lagrangian reduction and wave mean flow interaction. *Phys. D.* **454**, 133847.

[HoHuSt2024] Holm, D. D., Hu, R. and Street, O. D. [2024] Deterministic and stochastic geometric mechanics for Hall MHD. *arXiv preprint* arXiv:2404.06528.

[Hi1935] Hilbert, D. [1935] Naturerkennen und Logik. *Naturwissenschaften*. New York: Springer, pp. 959–963.

[HoKu1982] Holm, D. D. and Kupershmidt, B. A. [1982] Poisson structures of superfluids. *Phys. Lett. A.* **91**(9), 425–430.

[HoKu1983] Holm, D. D. and Kupershmidt, B. A. [1983] Poisson brackets and Clebsch representations for magnetohydrodynamics, multifluid plasmas, and elasticity. *Physics D* **6**(3), 347–363.

[HoKu1987] Holm, D. D. and Kupershmidt, B. A. [1983] Superfluid plasmas: Multivelocity nonlinear hydrodynamics of superfluid solutions with charged condensates coupled electromagnetically. *Phys. Rev. A.* **36**(8), 3947.

[HoKu1988] Holm, D. D. and Kupershmidt, B. A. [1988] The analogy between spin glasses and Yang–Mills fluids. *J. Math. Phys.* **29**(1), 21–30.

[HaMi1978] Hasegawa, A. and Mima, K. [1978] Pseudo-three-dimensional turbulence in magnetized nonuniform plasma. *Phys. Fluids* **21**(1), 87–92.

[HaMaKo1979] Hasegawa, A., Maclennan, C. G. and Kodama, Y. [1979] Nonlinear behavior and turbulence spectra of drift waves and rossby waves. *Phys. Fluids* **22**(11), 2122–2129.

[HaMe1985] Hazeltine, R. D. and Meiss, J. D. [1985] Shear-Alfvén dynamics of toroidally confined plasmas. *Phys. Rep.* **121**(1–2), 1–164.

[HoMa2005] Holm, D. D. and Marsden, J. E. [2005] Momentum maps and measure-valued solutions (peakons, filaments, and sheets) for the EPDiff equation. In *The Breadth of Symplectic and Poisson Geometry*, Jerrold E. Marsden, Tudor S. Ratiu (eds.), Progress in Mathematics, Vol. 232. Boston: Birkhäuser, pp. 203–235.

[HoMaRa1998a] Holm, D. D., Marsden, J. E. and Raţiu, T. S. [1998] The Euler–Poincaré equations and semidirect products with applications to continuum theories. *Adv. Math.* **137**(1), 1–81.

[HoMaRa1998b] Holm, D. D., Marsden, J. E. and Raţiu, T. S. [1998] Euler–Poincaré models of ideal fluids with nonlinear dispersion. *Phys. Rev. Lett.* **80**, 4173–4177.

[HoMaRaWe1985] Holm, D. D., Marsden, J. E., Raţiu, T. S. and Weinstein, A. [1985] Nonlinear stability of fluid and plasma equilibria. *Phys. Rep.* **123**(1–2), 1–116.

[Ho1985] Holm, D. D. [1985] Hamiltonian structure for Alfvén wave turbulence equations. *Phys. Lett. A.* **108**(9), 445–447.

[Ho1986] Holm, D. D. [1986] Hamiltonian dynamics of a charged fluid, including electro-and magnetohydrodynamics. *Phys. Lett. A.* **114**(3), 137–141.

[Ho1987a] Holm, D. D. [1987] Hall magnetohydrodynamics: conservation laws and Lyapunov stability. *Phys. Fluids* **30**(5), 1310–1322.

[Ho1987b] Holm, D. D. [1987] Hamiltonian dynamics and stability analysis of neutral electromagnetic fluids with induction. *Phys. D* **25**(1–3), 261–287.

[Ho2002a] Holm, D. D. [2002] Euler–Poincaré dynamics of perfect complex fluids. In *Geometry, Mechanics, and Dynamics*, Paul Newton, Philip Holmes, Alan (eds.), New York: Springer, pp. 113–167.

[Ho2002b] Holm, D. D. [2002] Euler-Poincaré dynamics of perfect complex fluids. In *Geometry, Mechanics, and Dynamics*. New York: Springer, pp. 169–180.

[Ho2008] Holm, D. D. [2008] *Geometric Mechanics, Part II: Rotating, Translating and Rolling*, 1st Ed. Imperial College Press, Singapore.

[Ho2011a] Holm, D. D. [2011] Applications of Poisson geometry to physical problems. *Geom. Topol. Monogr.* **17**, 221–384.

[Ho2011b] Holm, D. D. [2011] *Geometric Mechanics-Part I: Dynamics and Symmetry*, 2nd Ed. World Scientific Publishing Company, Singapore.

[Ho2011c] Holm, D. D. [2011] *Geometric Mechanics-Part II: Rotating, Translating and Rolling*, 2nd Ed. World Scientific, Singapore.

[Ho2015] Holm, D. D. [2015] Variational principles for stochastic fluid dynamics. *Proc. R. Soc. A: Math. Phys. Eng. Sci.* **471**(2176), 20140963.

[HoScSt2009] Holm, D. D., Schmah, T. and Stoica, C. [2009] *Geometric Mechanics and Symmetry: From Finite to Infinite Dimensions*. Oxford: Oxford University Press.

[HuZh1994] Hunter, J. K. and Zheng, Y. X. [1994] On a completely integrable nonlinear hyperbolic variational equation. *Phys. D.* **79**(2–4), 361–386.

[JoSa1998] José, J. V. and Saletan, E. J. [1998] *Classical Dynamics: A Contemporary Approach*. Cambridge: Cambridge University Press.

[Ke1869] Kelvin, L. [1869] On vortex motion. *Trans. R. Soc. Edinb.* **25**, 217–260.

[Ko1966] Kostant, B. [1966] Orbits, symplectic structures and representation theory. In *Proceedings of the United States–Japan Seminar in Differential Geometry (Kyoto, 1965)*. Tokyo: Nippon Hyoronsha, p. 71.

[KaPo1973] Kadomtsev, B. B. and Pogutse, O. P. [1973] Nonlinear helical perturbations of a plasma in the tokamak. *Sov. Phys. JETP* **5**, 575–590.

[KlSa2021] Klein, C. and Saut, J.-C. [2021] *Nonlinear Dispersive Equations: Inverse Scattering and PDE Methods*. Applied Mathematical Sciences, Vol. 209. Springer: Cham.

[KlSa2024] Klein, C. and Saut, J.-C. [2024] On the Kaup-Broer-Kupershmidt systems. *arXiv preprint* arXiv:2402.17576.

[Ku1985] Kupershmidt,B. A. [1985] Mathematics of dispersive water waves. *Commun. Math. Phys.* **99**, 51–73.

[Ku1992] Kupershmidt, B. A. [1992] *The Variational Principles of Dynamics*, Vol. 13. World Scientific Publishing Company: Singapore.

[Ku2006] Kupershmidt, B. A. [2006] Extended equations of long waves. *Stud. Appl. Math.* **116**(4), 415–434.

[Le2003] Lee, J. M. [2003] *Introduction to Smooth Manifolds*. Graduate Texts in Mathematics, Vol. 218. New York: Springer.

[Li1890] Lie, S. [1890] *Theorie der Transformationsgruppen. Zweiter Abschnitt*. Leipzig: Teubner.

[Li1960] Lighthill, M. J. [1960] Studies on magneto-hydrodynamic waves and other anisotropic wave motions. *Philos. Trans. R. Soc. Lond. Ser. A, Math. Phys. Sci.* **252**(1014), 397–430.

[LiMa1987] Libermann, P. and Marle, C.-M. [1987] *Symplectic Geometry and Analytical Mechanics*. Mathematics and its Applications, Vol. 35. Dordrecht: D. Reidel Publishing Co. Translated from the French by Bertram Eugene Schwarzbach.

[LaRe1976] Larichev, V. D. and Reznik, G. [1976] Strongly nonlinear two-dimensional isolated rossby waves. *Oceanologia* **16**(550), 1077–1079.

[LeRaSiMa1992] Lewis, D., Raţiu, T. S., Simo, J. C. and Marsden, J. E. [1992] The heavy top: A geometric treatment. *Nonlinearity* **5**(1), 1.

[LuSz2022] Lundmark, H. and Szmigielski, J. [2022] A view of the peakon world through the lens of approximation theory. *Phys. D.* **440**, 133446.

[Lü1959] Lüst, R. [1959] Über die ausbreitung von wellen in einem plasma. *Fortschr. Phys.* **7**(9), 503–558.

[Ma1978] Magri, F. [1978] A simple model of the integrable Hamiltonian equation. *J. Math. Phys.* **19**(5), 1156–1162.

[Ma1976] Manakov, S. V. [1976] A remark on the integration of the Eulerian equations of the dynamics of an n–dimensional rigid body. *Funkc. Anal. Pril.* **10**(4), 93–94.

[MiAlPo2007] Mininni, P. D., Alexakis, A. and Pouquet, A. [2007] Energy transfer in Hall-MHD turbulence: Cascades, backscatter, and dynamo action. *J. Plasma Phys.* **73**(3), 377–401.

[Ma1992] Marsden, J. E. [1992] *Lectures on Mechanics*. London Mathematical Society Lecture Note Series, Vol. 174. Cambridge: Cambridge University Press.

[Ma1997] Marsden, J. E. [1997] *Lectures on Mechanics*, 2nd Ed. Cambridge: Cambridge University Press.

[Ma2013] Marle, C.-M. [2013] On Henri Poincaré's note "sur une forme nouvelle des équations de la mécanique". *J. Geom. Symmetry Phys.* **29**, 1–38.

[MiFo1978] Miščenko, A. S. and Fomenko, A. T. [1978] Euler equation on finite-dimensional Lie groups. *Izv. Akad. Nauk SSSR Ser. Mat.* **42**(2), 396–415, 471.

[MaHu1994] Marsden, J. E. and Hughes, T. J. R. [1994] *Mathematical Foundations of Elasticity*. New York: Dover Publications. Corrected reprint of the 1983 original.

[Mi2005] Miesch, M. S. [2005] Large-scale dynamics of the convection zone and tachocline. *Living Rev. Sol. Phys.* **2**(1), 1.

[MaMoRa1990] Marsden, J. E., Montgomery, R. and Raţiu, T. S. [1990] *Reduction, Symmetry, and Phases in Mechanics*, Vol. 436. American Mathematical Society, Providence, RI.

[Mo1993] Montgomery, R. [1993] Gauge theory of the falling cat. *Fields Inst. Commun.* **1**(10.1090).

[MaRa1994] Marsden, J. E. and Raţiu, T. S. [1994] *Introduction to Mechanics and Symmetry: A Basic Exposition of Classical Mechanical Systems*. Texts in Applied Mathematics, Vol. 17. New York: Springer.

[MaRaWe1984a] Marsden, J. E., Raţiu, T. S. and Weinstein, A. [1984] Reduction and Hamiltonian structures on duals of semidirect product Lie algebras. In *Fluids and Plasmas: Geometry and Dynamics (Boulder, Colo., 1983)*. Contemporary Mathematics, Vol. 28. Providence, RI: American Mathematical Society, pp. 55–100.

[MaRaWe1984b] Marsden, J. E., Raţiu, T. S. and Weinstein, A. [1984] Semidirect products and reduction in mechanics. *Trans. Am. Math. Soc.* **281**(1), 147–177.

[McSa1995] McDuff, D. and Salamon, D. [1995] *Introduction to Symplectic Topology*. Oxford Mathematical Monographs. New York: The Clarendon Press Oxford University Press.

[MaWe1974] Marsden, J. and Weinstein, A. [1974] Reduction of symplectic manifolds with symmetry. *Rep. Math. Phys.* **5**(1), 121–130.

[MaWe1983a] Marsden, J. and Weinstein, A. [1983] Coadjoint orbits, vortices, and Clebsch variables for incompressible fluids. *Phys. D.* 7(1–3), 305–323.

[MaWe1983b] Marsden, J. and Weinstein, A. [1983] Coadjoint orbits, vortices, and Clebsch variables for incompressible fluids. *Phys. D.* 7(1–3), 305–323.

[No1918] Noether, E. [1918] Invariante variationsprobleme, nachrichten von der königlichen. *Gesellschaft der Wissenschaften zu Göttingen.*

[Ol1993] Olver, P. J. [1993] *Applications of Lie Groups to Differential Equations*, 2nd Ed. Graduate Texts in Mathematics, Vol. 107. New York: Springer.

[OrRa2004] Ortega, J.-P. and Raţiu, T. S. [2004] *Momentum Maps and Hamiltonian Reduction.* Progress in Mathematics, Vol. 222. Boston: Birkhäuser.

[Pa1953] Pauli, W. [1953] On the Hamiltonian structure of non-local field theories. *Il Nuovo Cimento (1943-1954)* **10**, 648–667.

[Pe2013] Pedlosky, J. [2013] *Geophysical Fluid Dynamics*. New York: Springer Science & Business Media.

[Po1901] Poincaré, H. [1901] Sur une forme nouvelle des équations de la mécanique. *C. R. Acad. Sci.* **132**, 369–371.

[Ra1980] Raţiu, T. [1980] The motion of the free n-dimensional rigid body. *Indiana Univ. Math. J.* **29**(4), 609–629.

[Ri1922] Richardson, L. F. [1922] *Weather Prediction by Numerical Process.* Cambridge: Cambridge University Press.

[RaTuSoTe2005] Raţiu, T. S., Tudoran, R., Sbano, L., Sousa Dias, E. and Terra, G. [2005] A crash course in geometric mechanics. In *Geometric Mechanics and Symmetry*. London Mathematical Society Lecture Notes Series, Vol. 306. Cambridge: Cambridge University Press, pp. 23–156. Notes of the courses given by Raţiu.

[Sa2023] Sachkov, Y. L. [2023]   Lorentzian distance on the Lobachevsky plane. *arXiv preprint* arXiv:2307.07706.

[SuMu1974] Sudarshan, E. C. G. and Mukunda, N. [1974] *Classical Dynamics: A Modern Perspective*. World Scientific, Singapore.

[Sm1970a] Smale, S. [1970]   Topology and mechanics. I: *Invent. Math.* **10**, 305–331.

[Sm1970b] Smale, S. [1970] Topology and mechanics II: The planar n-body problem. *Invent. Math.* **11**, 45–64.

[So1970] Souriau, J.-M. [1970]   *Structure des systèmes dynamiques*. Maîtrises de mathématiques. Paris: Dunod.

[St1975] Stern, M. E. [1975] Minimal properties of planetary eddies. *J. Mar. Res.* **33**(1), 1.

[St1976] Strauss, H. R. [1976]   Nonlinear, three-dimensional magnetohydrodynamics of noncircular tokamaks. *Phys. Fluids* **19**(1), 134–140.

[St1977] Strauss, H. R. [1977] Dynamics of high β tokamaks. *Phys. Fluids* **20**(8), 1354–1360.

[ShYuRaSp1984] Shukla, P. K., Yu, M. Y., Rahman, H. U. and Spatschek, K. H. [1984]   Nonlinear convective motion in plasmas. *Phys. Rep.* **105**(4–5), 227–328.

[StZe1996] Stegner, A. and Zeitlin, V. [1996]   Asymptotic expansions and monopolar solitary rossby vortices in barotropic and two-layer models. *Geophys. Astrophys. Fluid Dyn.* **83**(3–4), 159–194.

[Tr2010] Tronci, C. [2010] Hamiltonian approach to hybrid plasma models. *J. Phys. A: Math. Theor.*, **43**(37), 375501.

[Vi1981] Virasoro, M. A. [1981]   Variational principle for two-dimensional incompressible hydrodynamics and quasi-geostrophic flows. *Phys. Rev. Lett.* **47**(17), 1181.

[We1983a] Weinstein, A. [1983]   Hamiltonian structure for drift waves and geostrophic flow. *Phys. Fluids* **26**(2), 388–390.

[We1983b] Weinstein, A. [1983] Sophus Lie and symplectic geometry. *Exposition. Math.* **1**(1), 95–96.

[Wi1990] Wigner, E. P. [1990]   The unreasonable effectiveness of mathematics in the natural sciences. In *Mathematics and Science*. Ronald E Mickens (ed.), World Scientific, Singapore, pp. 291–306.

[ZePa1994] Zeitlin, V. and Pasmanter, R. A. [1994] On the differential geometry approach to geophysical flows. *Phys. Lett. A.* **189**(1–2), 59–63.

Index